agr 470
LGc 476
5. Expl.

Ausgeschieden im Jahr 2024

Bei Überschreitung der Leihfrist wird dieses Buch sofort gebührenpflichtig angemahnt (ohne vorhergehendes Erinnerungsschreiben).

# Precision in Crop Farming

Hermann J. Heege
Editor

# Precision in Crop Farming

Site Specific Concepts and Sensing
Methods: Applications and Results

*Editor*
Hermann J. Heege
Agricultural Systems Engineering
University of Kiel
Kiel, Germany

ISBN 978-94-007-6759-1           ISBN 978-94-007-6760-7 (eBook)
DOI 10.1007/978-94-007-6760-7
Springer Dordrecht Heidelberg New York London

Library of Congress Control Number: 2013942468

© Springer Science+Business Media Dordrecht 2013
This work is subject to copyright. All rights are reserved by the Publisher, whether the whole or part of the material is concerned, specifically the rights of translation, reprinting, reuse of illustrations, recitation, broadcasting, reproduction on microfilms or in any other physical way, and transmission or information storage and retrieval, electronic adaptation, computer software, or by similar or dissimilar methodology now known or hereafter developed. Exempted from this legal reservation are brief excerpts in connection with reviews or scholarly analysis or material supplied specifically for the purpose of being entered and executed on a computer system, for exclusive use by the purchaser of the work. Duplication of this publication or parts thereof is permitted only under the provisions of the Copyright Law of the Publisher's location, in its current version, and permission for use must always be obtained from Springer. Permissions for use may be obtained through RightsLink at the Copyright Clearance Center. Violations are liable to prosecution under the respective Copyright Law.

The use of general descriptive names, registered names, trademarks, service marks, etc. in this publication does not imply, even in the absence of a specific statement, that such names are exempt from the relevant protective laws and regulations and therefore free for general use.

While the advice and information in this book are believed to be true and accurate at the date of publication, neither the authors nor the editors nor the publisher can accept any legal responsibility for any errors or omissions that may be made. The publisher makes no warranty, express or implied, with respect to the material contained herein.

Printed on acid-free paper

Springer is part of Springer Science+Business Media (www.springer.com)

"In the landscape of extinction, precision is next to godliness" *Samuel Beckett*

# Contents

1 **Introduction** ............................................................................. 1
   Hermann J. Heege

2 **Heterogeneity in Fields: Basics of Analyses** ......................... 3
   Hermann J. Heege

   2.1 Variation and Resolution ................................................ 3
   2.2 Semivariance and Semivariogram .................................. 6
   2.3 Cell Sizes ....................................................................... 8
   2.4 Processing and Adjusting the Resolution ..................... 10
   References ............................................................................ 13

3 **Sensing by Electromagnetic Radiation** ................................ 15
   Hermann J. Heege

   3.1 Basics in Sensing by Electromagnetic Radiation ......... 15
   3.2 Emitted, Absorbed, Reflected and Transmitted Radiation ............... 17
   3.3 Atmospheric Windows and Clouds .............................. 19
   3.4 Sensing from Satellites, Aerial Platforms
       and Field Machines ....................................................... 20
   3.5 Microwaves or Radar Instead of Visible
       or Infrared Waves .......................................................... 23
   3.6 Using Maps or On-The-Go Control in Real-Time ....... 26
   3.7 Georeferencing by Positioning Systems ....................... 28
   References ............................................................................ 32

4 **Precision in Guidance of Farm Machinery** .......................... 35
   Hermann J. Heege

   4.1 Principles of Guidance ................................................. 35
   4.2 Techniques of GNSS Based Guidance .......................... 38
   4.3 Economics of GNSS Based Guidance .......................... 42

|  |  |  | |
|---|---|---|---|
| | 4.4 | Problems and Solutions on Slopes | 44 |
| | | 4.4.1 Tractors and Self-Propelled Implements | 44 |
| | | 4.4.2 Drawn- and Mounted Implements | 47 |
| | 4.5 | Precision in Section Control of Farm Machines | 49 |
| | References | | 50 |

## 5 Sensing of Natural Soil Properties ... 51
Hermann J. Heege

|  |  |  | |
|---|---|---|---|
| | 5.1 | Sensing of Topography | 52 |
| | 5.2 | Sensing of Soil Properties on a Volume Basis | 55 |
| | | 5.2.1 Methods for Sensing of Electrical Conductivities | 58 |
| | | 5.2.2 Electrical Conductivities, Soil Properties and Yields | 65 |
| | | 5.2.3 Water Sensing Based on Permittivity and Capacitance | 76 |
| | 5.3 | Sensing of Soil Properties on a Surface Basis by Reflectance | 87 |
| | | 5.3.1 Basics of Surface Sensing | 89 |
| | | 5.3.2 Results of Surface Sensing in Laboratories | 91 |
| | | 5.3.3 Concepts and Results for Surface Sensing in Fields | 95 |
| | 5.4 | Summary | 99 |
| | References | | 99 |

## 6 Sensing of Crop Properties ... 103
Hermann J. Heege and Eiko Thiessen

|  |  |  | |
|---|---|---|---|
| | 6.1 | Basics of Sensing by Visible and Infrared Reflectance | 104 |
| | 6.2 | Defining the Reflectance by Indices | 108 |
| | | 6.2.1 Precision in Sensing of Chlorophyll | 112 |
| | 6.3 | Sensing Yield Potential of Crops by Reflectance | 114 |
| | 6.4 | Fluorescence Sensing | 118 |
| | | 6.4.1 Fluorescence Sensing in a Steady State Mode | 120 |
| | | 6.4.2 Fluorescence Sensing in a Non-Steady State Mode | 123 |
| | | 6.4.3 Fluorescence or Reflectance | 124 |
| | 6.5 | Sensing the Water Supply of Crops by Infrared Radiation | 126 |
| | | 6.5.1 Sensing Water by Near- and Shortwave-Infrared Reflectance | 126 |
| | | 6.5.2 Sensing Water by Emitted Thermal Infrared Radiation | 130 |
| | 6.6 | Sensing Properties of Crops by Microwaves | 133 |
| | References | | 138 |

## 7 Site-Specific Soil Cultivation ... 143
Hermann J. Heege

|  |  |  | |
|---|---|---|---|
| | 7.1 | Basic Needs | 143 |
| | 7.2 | Primary Cultivation | 144 |
| | | 7.2.1 Factors for the Depth of Primary Cultivation | 145 |
| | | 7.2.2 Site-Specific Control of the Primary Cultivation Depth | 150 |

|  |  |  |  |
|---|---|---|---|
| | 7.3 | Secondary Cultivation | 152 |
| | | 7.3.1 Sensing Soil Tilth | 153 |
| | | 7.3.2 Precision in the Vertical Direction Within the Seedbed | 160 |
| | 7.4 | Stubble- and Fallow Cultivation | 162 |
| | 7.5 | No-Tillage: Prerequisites, Consequences and Prospects | 165 |
| | References | | 168 |
| **8** | **Site-Specific Sowing** | | **171** |
| | Hermann J. Heege | | |
| | 8.1 | Seed-Rate or Seed-Density | 172 |
| | | 8.1.1 Site-Specific Control of Seed-Density | 173 |
| | 8.2 | Seed Distribution over the Area | 176 |
| | 8.3 | Seeding Depth | 178 |
| | | 8.3.1 Control of the Seeding Depth | 179 |
| | 8.4 | Less Tillage, Crop Residues and Sowing Methods | 184 |
| | | 8.4.1 Vertical Discs, Cleaned Rows or Inter-Row Sowing | 185 |
| | | 8.4.2 Seeding into Cover Crops | 187 |
| | | 8.4.3 Seeding and Loose Residue Sizes | 188 |
| | | 8.4.4 Seeding Underneath Undercutters | 190 |
| | References | | 190 |
| **9** | **Site-Specific Fertilizing** | | **193** |
| | Hermann J. Heege | | |
| | 9.1 | Fertilizing Based on Nutrient Removal by Previous Crops | 194 |
| | 9.2 | Fertilizing Based on Soil Sensing by Ion-Selective Electrodes | 197 |
| | | 9.2.1 Basics | 197 |
| | | 9.2.2 Sensing pH and Nutrients in Naturally Moist Soils | 198 |
| | | 9.2.3 Sensing Nitrate in Slurries of Soil | 202 |
| | | 9.2.4 General Prerequisites and Prospects | 206 |
| | 9.3 | Fertilizing Based on Reflectance of Soils | 209 |
| | | 9.3.1 General Remarks | 209 |
| | | 9.3.2 Sensing the Lime Requirement | 212 |
| | | 9.3.3 Sensing the Phosphorus Requirement | 214 |
| | | 9.3.4 Sensing the Potassium- and the Nitrate Requirement | 218 |
| | 9.4 | Fertilizing Nitrogen Based on In-Season Crop Properties | 220 |
| | | 9.4.1 Fundamentals of Nitrogen Sensing by Reflectance | 223 |
| | | 9.4.2 Sensing by Standard Reflectance Indices and Natural Light | 225 |
| | | 9.4.3 Sensing Nitrogen by Reflectance Based on Artificial Light | 232 |
| | | 9.4.4 Soil or Plants in the Field of View | 239 |
| | | 9.4.5 Sensing Nitrogen by Fluorescence | 240 |
| | | 9.4.6 Sensing Nitrogen Based on Bending Resistance or Height | 244 |

| | | | |
|---|---|---|---|
| | 9.4.7 | Cell Sizes or Resolution | 246 |
| | 9.4.8 | Distance- and Time Lag in Site-Specific Application | 249 |
| | 9.4.9 | Sensed Signals and the Control of Nitrogen Application | 251 |
| | 9.4.10 | Interactions Between Nitrogen and Water | 257 |
| | 9.4.11 | Benefits, Costs and Economics | 260 |
| 9.5 | Summary | | 265 |
| References | | | 266 |

## 10 Site-Specific Weed Control ... 273
Roland Gerhards

| | | |
|---|---|---|
| 10.1 | Introduction | 273 |
| 10.2 | Weed Mapping | 274 |
| | 10.2.1 Spectrometers | 276 |
| | 10.2.2 Fluorescence Sensors | 277 |
| | 10.2.3 Digital Image Analysis Based on Shape Features | 278 |
| 10.3 | Temporal and Spatial Dynamics of Weed Population | 284 |
| 10.4 | Site-Specific Weed Control | 288 |
| 10.5 | Outlook and Perspectives | 290 |
| References | | 292 |

## 11 Site-Specific Sensing for Fungicide Spraying ... 295
Eiko Thiessen and Hermann J. Heege

| | | |
|---|---|---|
| 11.1 | The Situation for Site-Specific Fungicide Applications | 296 |
| 11.2 | The Full-Area Preventive Concept Based on Biomass | 296 |
| 11.3 | The Discrete-Spot Sensing Concept Based on Reflectance | 298 |
| | 11.3.1 The Pinpointing Approach and the Field of View | 298 |
| | 11.3.2 Spectra and Indices of Reflectance | 299 |
| 11.4 | The Discrete-Spot Sensing Concept Based on Fluorescence | 301 |
| | 11.4.1 Indirect Measuring with In-Situ Sensor System | 301 |
| | 11.4.2 Fungi-Plant-Interaction and Physiology of Infected Plants | 303 |
| | 11.4.3 Fluorescence Indices Related to Infection | 304 |
| | 11.4.4 Problems and Discussion | 305 |
| | 11.4.5 Sensors for Practice and Research | 307 |
| 11.5 | Differentiation Between N Deficiency and Effects of Fungi | 307 |
| 11.6 | Summary and Prospects | 309 |
| References | | 310 |

## 12 Site-Specific Recording of Yields ... 313
Markus Demmel

| | | |
|---|---|---|
| 12.1 | Introduction | 314 |
| 12.2 | Principle of Site-Specific Yield Recording | 314 |
| 12.3 | Yield Measurement for Combinable Crops | 315 |
| 12.4 | Yield Measurement for Forage Crops | 319 |

|  |  |  |
|---|---|---|
| 12.5 | Yield Sensing for Root Crops | 322 |
| 12.6 | Yield Measurement for Other Crops | 323 |
| 12.7 | Quality Sensing of Harvested Material | 324 |
| 12.8 | Processing and Mapping of Yield Data | 325 |
| | References | 325 |

## 13 Fusions, Overlays and Management Zones .... 331
Hermann J. Heege

|  |  |  |
|---|---|---|
| 13.1 | Crop Growth Factors, Sensing and Information Use | 331 |
| 13.2 | Sensor Fusion – Solutions and Approaches | 333 |
| 13.3 | From Properties to Treatment – Map Overlay | 337 |
| 13.4 | From Properties to Treatment – Management Zones | 342 |
| | References | 344 |

## 14 Summary and Perspectives .... 345
Hermann J. Heege

**Index** .... 349

# Chapter 1
# Introduction

Hermann J. Heege

Sustainable and thus also economical farming requires precise adaptation to the natural and economic conditions. The irradiation of the sun, the natural water supply, the soil properties and the demand of the market just are not uniform at all.

Even within small regions or within a farm, the soil qualities or the slope of the fields can be different. Farmers have adapted their practices to this for centuries.

However, the differences existing within single fields also deserve attention. Within fields, there certainly are differences in the soil properties, in the slope, in the water supply and consequently in the development of crops in many cases. But as long as farming operations were done manually or by means of small implements, the farmers succeeded rather easily to adjust to these differences.

Yet this situation has changed fundamentally in many areas of the globe, where now machines with a working width of up to 40 m operate in fields that are much larger than those of the past. Under these conditions, the farmer has lost very much in immediate contact with the soil and the crop. The high capacity machinery treats large fields in a uniform way. This method cannot be regarded as being sustainable, since in most cases neither the soil nor the crop are uniform within a field. Permanent and precise adjusting to the varying, site-specific soil and crop conditions in the field can address the environmental needs much better and ensure high yields. This is the rationale for site-specific precision farming.

This rationale needs means to get to its objectives. Human visual observation and subsequent manual adjustment on-the-go hardly is possible when operating with high speeds and wide machinery. Sensors which record and computers that process the signals about the respective site-specific situations of the soils or crops within a field can and will overcome the challenges. When used in combination with actuators – which adjust the machinery correspondingly – they allow for automated

H.J. Heege (✉)
Department of Agricultural Systems Engineering, University of Kiel,
24098 Kiel, Germany
e-mail: hheege@ilv.uni-kiel.de

and on-the-go corrected site-specific farming operations. The results of the sensing and processing can be stored in georeferenced, site-specific field maps. This is necessary whenever the sensing occurs prior to the respective field operation.

The main problem with this general concept is selecting suitable sensing principles and appropriate processing methods for its signals. A vast variety of concepts and alternatives has been developed, investigated and analysed in the past for many farming operations such as *e.g.* soil cultivation, sowing, fertilizing, crop protection, irrigation and harvesting. The results of this intensive, scientific work have been published in numerous journals and proceedings of conferences. For those interested in site-specific farming, the results present themselves in a very fragmented way.

This fragmented situation may be the inevitable initial fate of any new field that is developing in science and application. Nevertheless, this new field does need a compendium, which facilitates to obtain a fast overview. This book tries to be such a compendium about site-specific precision farming.

It is well known that the general attitude of the public towards modern farming techniques differs greatly. Only part of the public views modern farming techniques in an open-minded and affirmative manner. There is a substantial part of the public, which blames high-tech, modern farming for being a burden to the environment and the society. Especially mineral fertilizers, herbicides, fungicides and insecticides that are needed for high yields are regarded as contaminants for the environment.

Fortunately, applying agricultural chemicals within single fields in a site-specific way allows to reduce the amount needed while still maintaining or even improving the yields. The efficiency in the use of farm chemicals thus can be enhanced. In general, the same can hold for the efficiency in the input of energy, seeds and water.

In short, this book intends to show that precision farming can substantially help to get high yields per unit area as well as a protected environment.

It is not within the scope of this book to deal with technical details of precision farming. Due to the abundance of alternatives available, this would be impossible within one book. Instead of this, the book aims at explaining the rationales existing between agronomical sciences, sensing principles plus its physical, chemical and biological background as well as finally possibilities in agricultural engineering and farming management. Thus the book is based on an interdisciplinary approach within several fields of the agricultural sciences and adjacent disciplines.

# Chapter 2
# Heterogeneity in Fields: Basics of Analyses

**Hermann J. Heege**

**Abstract** Sustainable and economical farming needs precise adaptation to the varying soil- and plant properties within fields. Consequently, farming operations have to be adjusted to this in a site-specific way.

An important question is, on which spatial resolution or cell size within a field these adjustments should be based. It is reasonable to expect that this depends on the spatial variations of the respective soil- or crop properties. Consequently, it is shown how the cell sizes needed can be derived from semivariances and its complement functions, the covariances.

Once thus suitable cell sizes are known, they should not be exceeded on any site-specific stage, whether this is sampling, mapping or the operations of the farm machinery.

**Keywords** Cell-sizes • Kriging • Resolution • Semivariance • Semivariogram

## 2.1 Variation and Resolution

A traveller attentive to agricultural land will notice that uniform fields are not the rule (Fig. 2.1). This becomes especially obvious to farming experts, who look more closely at soils and crops.

Many soil- and crop properties can vary within fields, such as *e.g.*

- texture (content of sand, silt, loam or clay) and pH of topsoil and subsoil
- soil content of organic matter, of water and of various minerals
- slope and orbital orientation of the soil

H.J. Heege (✉)
Department of Agricultural Systems Engineering, University of Kiel,
24098 Kiel, Germany
e-mail: hheege@ilv.uni-kiel.de

**Fig. 2.1** Aerial view of a farming area in Schleswig-Holstein, Germany

- density and morphology of crops
- crop content of water and of various minerals
- infestation of crops by different weeds and by various pests.

Nowadays, many of these soil- or crop properties can be detected and recorded within a field in a site-specific way via modern sensing techniques. Yet before these techniques are dealt with in detail in subsequent chapters, the question arises, how in principle variations of these soil- or crop properties can show up.

These variations of soil- or crop properties within a field can occur in different ways. They can show up in a complete **random pattern**, *i.e.* in a similar manner as raindrop spots within a field. All locations within the field are affected by the rain in a similar way.

Yet the variations can also show up in a **nested pattern**. This is the case, if the respective property, *e.g.* the clay content of the soil, is not uniform on the whole field, but instead there are parts in various directions where it is lower and *vice versa* higher. The respective property in this case varies with the distance.

Thus the spatial variation can be uncorrelated or correlated (Fig. 2.2), depending how it presents itself in a graph with a distance scale. However, the distance scale too can change. And if it does, the appearance of variations can be different. What looks as random arrangement or noise at one scale can be recognized as structure at another scale. This is why looking through a microscope can be so fascinating.

Therefore, we have to deal with **resolutions**. What is meant is not a resolution taken by a political assembly. Here the term resolution stands for the "resolving" or the dividing up of physical properties involved, such as the area of the field, the time or the measurement units that belong to the signals of a sensor (Fig. 2.3).

The **temporal resolution** that is required depends very much on the respective soil- or crop property. Textures and organic matter contents of soils hardly change

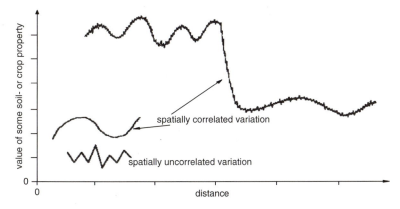

**Fig. 2.2** Types of spatial variation in a dimensionless diagram (From Oliver 1999, altered)

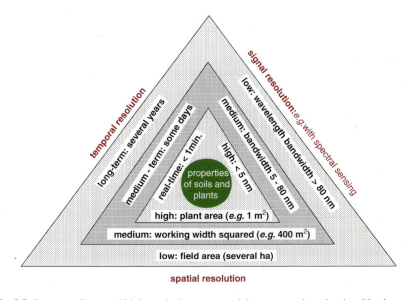

**Fig. 2.3** Low-, medium- and high resolutions on a spatial-, a temporal- and a signal basis

over time. Therefore, these properties can be recorded once on a long-term basis in field maps that can be used for several years. The situation is quite different when it comes to the water-, the nitrogen- and the pesticide supply of crops in the growing season. In these cases, the best temporal resolution would be obtained with a control system that adjusts the supply in real-time, which means immediately after sensing and "on-the-go" during the application.

The **signal resolution** refers to the physical quantities that are sensed. In case of spectral sensing, the bandwidth-ranges of the light waves in the visible- or infrared region are important and can be very different (Fig. 2.3).

Finally, precision farming aims at a high **spatial resolution**. It should be noted that instead of the term "resolution", the denotation "**cell size**" often is used. This stands for the area, for which the respective farming operations are uniformly adjusted. Therefore, a low resolution means a large cell size and *vice versa*. The traditional and still rather common cell size is the individual field. Whilst the sizes and basic operations of present farm machinery are maintained, about the smallest cell size that can be realized would be an area that corresponds to the square of the working width. This approach is derived from the assumption that the basic shape of a cell would be a square. So if a working width of 20 m for fertilizing and spraying is used, a cell size of 400 m$^2$ would result. And if fields that are controlled by small robots become a reality, a high regional resolution based on treating individual plants or even leaves might become feasible.

Yet these considerations emanate from **technical possibilities** with the respective farm machinery, provided the control components are available. A better approach is to base the resolution on the respective soils and crops and to adapt the technical solutions as well as possible to these.

If theoretically the soil- and crop properties would be completely uniform within a field, no site-specific treatments would be necessary. And if on the other hand significant variations would show up within short distances, small cell sizes would be reasonable. This leads to the question, how – based on variations existing within fields – proper **cell sizes** can be deduced. Statistical indices of soil- or crop properties like averages or standard variations are no help in this respect. This is because intrinsically these indices are independent of location. What is needed are statistical indices that rely on distances within a field. The semivariance and its graph – the semivariogram – do this.

## 2.2 Semivariance and Semivariogram

The geostatistical concept behind semivariances and semivariograms is Matheron's (1963) regionalized variable theory. It states that the differences in the values of a spatial variable – such as a soil- or crop property – between points in a field depend on the **distance** between these points. In short, the smaller the distances, the smaller the differences.

As a logical consequence, the semivariance $v$ expresses the dissimilarity of paired property values as a function of the distances between two sampling points. The general equation of the semivariance $v$ is:

$$v = \frac{1}{2N}\sum_{1}^{N}\left[f(x+h)-f(x)\right]^2$$

Here, $x$ and $x+h$ stand for the vectors of areal coordinates at two locations in the field. These locations are separated by the distance $h$. The functions $f(x)$ and $f(x+h)$ together represent thus a pair of soil- or crop properties at these places. N is the number of location pairs that are involved.

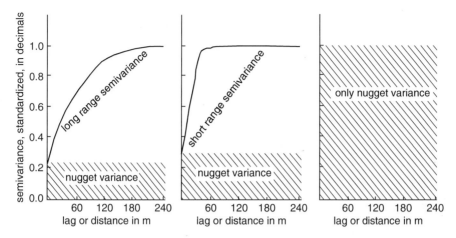

**Fig. 2.4** Semivariances, its semivariograms and pure nugget variance (From suggestions by Oliver 1999, altered and redrawn)

The equation resembles the general formula for the standard variance. However, the basic data for the standard variances are not pairs that are related by distances and orientation; they just come from a common population and not more. And because of the **pairs** that are used when calculating the semivariance, the summed result is halved. This explains the denotations semivariance for the numerical values and semivariogram for the graphical description.

Most semivariance curves – the semivariograms – are bounded, which means that they reach an asymptote (Fig. 2.4). This asymptote is called the **sill variance**. But unbounded variogram curves also can occur. Note that the semivariograms are standardized to a sill variance of 1.

The distance at which the sill variance is approached is called the **range** (Fig. 2.5). Since this is an asymptotic approach, the range is arbitrarily set to be the distance needed to get 95 % of the sill variance (Tollner et al. 2002). All points that are separated by distances smaller than the range are spatially correlated. Whenever the distances are larger than the range, the points are spatially independent. This is very important: it means that any site-specific operation that is based on squared cells with sides longer than the range is useless. This is, because with cells of this size it is not possible any more to detect or catch regional differences of the respective property. So the control distance that is used for site-specific operations must be smaller than the range of the semivariogram.

Theoretically, the semivariogram curves start with zero variance and zero distance. But in reality, this seldom occurs. In practice there always is some variance already at zero distance. It is called **nugget variance** and represents variability at distances smaller than the typical sample spacing as well as measurement errors. In rare occasions, there is only nugget variance (Fig 2.4, right), which means that any site-specific treatment – at least based on the respective property – is useless.

## 2.3 Cell Sizes

Knowing that the cell squares should have side lengths between zero and the range alone is not sufficient. Very small cells result in high costs, and rather large cells can impair the precision. And since the number of cells that must be dealt with quadruples with each halving of cell-side length, more precise information about the size that should be used is desirable. This information can be derived from the semivariance (Fig. 2.5).

An interesting approach for this was developed by Russo and Bresler (1981) as well as by Han et al. (1994). It is based on the notion that as the semivariance is an indicator of **dissimilarity** of a site-specific soil- or crop property, *vice versa* the complement function to the semivariance provides information of **similarity** or **relatedness**. For normalized situations, the semivariance plus its complement function for all respective distances or lags add up to one (1). Therefore, the complement function is the vertical mirror image of the semivariance (Fig. 2.5). It can be shown that for the pairs involved, this complement function of the semivariance is a well known statistical function – the covariance (Davis 1973; Gringarten and Deutsch 2001). In contrast to this: the sill of the semivariogram is the standard variance, which in this case stands for zero correlation. Therefore, the semivariance is standard variance minus covariance.

The area under the curve of the complement function to the semivariance can be regarded as an accumulation of all relatedness or similarity of the respective

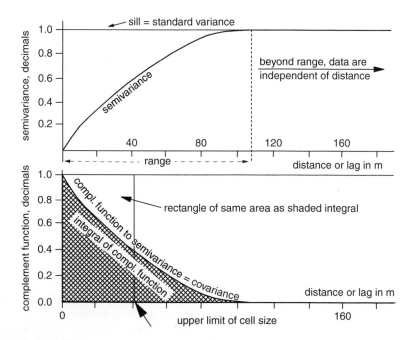

**Fig. 2.5** Semivariance, its complement function and the upper limit of cell size

property. It can be computed by integrating the equation of the complement function between the range and zero. This integral of the complement function is represented by the hatched area in Fig. 2.5.

If differences based on location exist, theoretically only cells of zero size would be completely uniform. For practical purposes, a compromise in a cell size with side-length between zero and range is necessary. The proposal of Russo and Bresler (1981) as well as of Han et al. (1994) for this compromise is based on a rectangle with a height of the sill and an area that equals the integral of the complement function. This rectangle contains all the similarity or relatedness that can be obtained from the respective semivariance from which it was derived. For this reason, the length of the rectangle side along the abscissa can be regarded as an indication of the **upper limit of the cell-size** (Fig. 2.5). Any larger cell-size overlaps into areas of pure dissimilarity. It thus deteriorates the precision in site-specific management. This upper limit of cell-size represents the largest distance for which the soil or crop property is well correlated with itself.

It is also obvious that the upper limit of the cell-size depends largely on the distance of the range. Kerry and Oliver (2004, 2008) propose to use a sampling interval – or upper limit of cell size – of less than half the range. It can be seen that less than **half the range** results in approximately the same limit of cell size.

However, these determinations of cell sizes are possible only after the semivariance has been recorded. The problem is that cell-sizes for **sampling** of soil- or crop properties must be known beforehand in order to arrive at reliable resolutions for subsequent farming operations. If the sampling is based on too large cell-sizes, no detailed computation afterwards any more can result in a reliable control for site-specific farming. The resolution that is needed depends on the respective variability of the soil or crop property. It must be met at the first site-specific operation, otherwise subsequent procedures cannot be controlled with precision. And this first operation is the sampling.

Ways out of this situation are either sampling with a very high resolution from the outset so that any site-specific requirements definitely are met or alternatively the use of standardized, default cell-sizes for sampling. The first method lends itself whenever the data are recorded automatically online and on-the-go, since many modern sensors can provide for several signals per m of travel and thus for a high resolution.

The use of standardized, default cell-sizes for sampling is advisable if manual sampling and processing is needed as with *e.g.* the conventional and traditional collection of data about soil texture, -nutrients or crop properties. Because of the high amount of labour involved, in these cases sampling with a very high resolution from the outset as with online and on-the-go methods cannot be practised. Therefore, the information about the sampling cell-sizes that are needed must be obtained from previously made semivariograms, which are based on data of soils or crops under conditions that are similar. Information about such standardized semivariograms has been published (Kerry and Oliver 2004, 2008; McBratney and Pringle 1999). These **standardized semivariograms** are based on sampling with a high resolution. They thus provide the default cell-sizes for subsequent site-specific sampling.

It should be noted that reliable semivariograms cannot be obtained from fewer than 100 data. The nugget to sill ratios of the standardized default semivariograms should match those of the respective fields (Kerry and Oliver 2008).

The term "standardized semivariogram" should not conceal the fact that in reality the curves for the semivariances are rather unique and different for every field. Yet precise sampling and site-specific management relies on the semivariograms for information about the cell-size that should be used. If the cell-size is oriented at less than half the range (Kerry and Oliver 2004, 2008) as outlined above, what ranges for soil- and crop properties do actually occur?

Actually, the **ranges** that have shown up in research with semivariograms vary immensely – as do the local conditions around the globe. In extreme cases, ranges on the one hand as short as 1 m (Solie et al. 1992) and on the other hand as long as 26 km (Cemek et al. 2007) have been recorded. However, the vast majority of ranges for soil- and crop properties is between 20 and 110 m (McBratney and Pringle 1999). So for most cell-sizes, the upper limits of side lengths should be between 10 and 55 m. This is still a rather wide span. Therefore, to define the actual need more closely, a careful deduction of required cell-sizes from suitable standardized semivariograms is necessary.

## 2.4 Processing and Adjusting the Resolution

Finding the appropriate upper limit of cell-sizes deserves some effort. This is because knowing about it is important on several stages of site-specific farming, first when sampling, then for mapping and finally when the machinery operates in the field. It is obvious that the sampling must occur within the distance limits defined by about half the range. But the same holds for techniques used to make the maps and finally for the cell-sizes, on which the farm machinery works in a site-specific way. If on any of these stages the distance limits that are defined by less than half the range of the respective semivariogram are exceeded, the precision of site-specific farming is impaired.

The question is, at which stage – sampling, mapping or machine operations – striving for small cell-sizes is most difficult. As long as sampling of soil properties and of nutrient contents is done in a manual way, this will be the sampling stage. In the long run, however, manual sampling will be more and more replaced by online and on-the-go sensing methods. Many of these methods will allow for sensing of small cell-sizes. As a consequence, then the cell-sizes that can be realized with wide farm machinery become important.

It is not recommended to directly combine fine grid spacings for sampling or sensing on the one hand with a much coarser resolution for the machinery operations on the other hand without any signal corrections. This is because the control of the machinery is less erratic and is more stable if **averages** of highly resolved signals are used. If the machinery is controlled via online and on-the-go sensing, this averaging step can easily be implemented into the processing computer program. In case a field

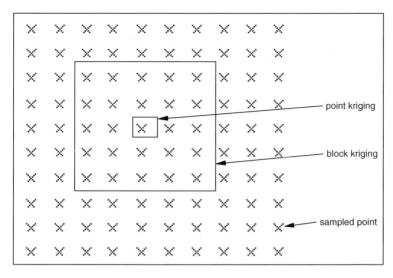

**Fig. 2.6** Principles of block-kriging and point-kriging (From Whelan et al. 2002, altered)

map for later use is made from the sampled or sensed data, the averaging can occur during mapping.

The sampling or sensing of the data usually occurs on a **punctual basis**. Hence for mapping of a continuous surface, some interpolation is needed. This **interpolation** of the punctual signals is called **kriging** in honour of D.G. Krige, a South African geologist, who was a pioneer in this processing of data for mapping. Today, many kriging methods are available. These methods are based on the assumption that geological properties in close proximity to each other are more likely to be similar than those separated by longer distances. So the basic premise – the concept of spatial dependence – is the same as for semivariances (Sect. 2.2).

Kriging can be subdivided into either block- or point methods, depending on the size of the respective map area, for which the property is estimated by interpolation. In block kriging, the average value of the property for a block within the map is calculated, whereas in point kriging the target is just a point within the area. This means that point kriging can be considered as the limiting case of block kriging when the block size approaches the sampling- or sensing size (Fig. 2.6). For details to interpolation methods and to computer programs for this see Hengl (2007), Webster and Oliver (2007) and Whelan et al. (2002).

An important feature of block-kriging is that its estimate – the block mean – may gain in reliability as the block area increases. This follows from basic features of means. This averaging advantage of **block-kriging** is effective as long as the size of the blocks does not exceed the cell-size that is effective in the machinery operations. But when the sizes of the kriged blocks get larger than the cells of the machines, this averaging gets detrimental. This is because then another averaging feature – its leveling effect – has consequences. This leveling effect deteriorates the objectives of

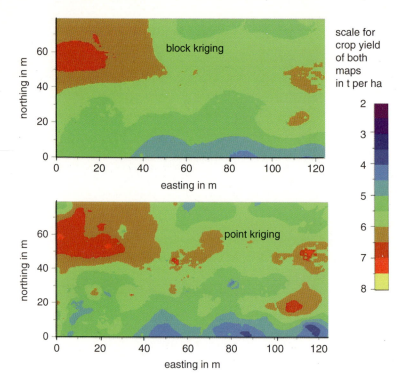

**Fig. 2.7** Maps of wheat yields from site-specific harvesting after block- and point-kriging. The smoothing effect of block-kriging is obvious (From Whelan et al. 2002, rearranged and altered)

site-specific farming, since it artificially erases local differences and thus eliminates possibilities to react on them. Yet as long as the kriged blocks are smaller than the cells of the machinery, any leveling effects of averaging are unavoidable – just because the size of the machinery does this anyway.

With **point-kriging**, there is no processing error at the actual sampling- or sensing point. However, this theoretical advantage of point-kriging can only improve the overall precision, if the farm machinery too can do punctual work. As long as this is not the case, block kriging will primarily be the choice for creating maps.

The objective is that maps should only show the spatial distribution of the respective soil- or crop property. However, maps always too reflect the influence of the sampling- or sensing techniques and in addition of the kriging method that was used (Fig. 2.7). Maps from the same basic data therefore can look very different.

The averaging and estimating that is connected with kriging should be fine-tuning of the spatial resolution as it is needed for the respective site-specific operation. A fine-tuning by averaging that is oriented at the cell-size of the machine operations should be the objective. This adjusting of the resolution can take place either during online and on-the-go control or when maps are processed.

Maps can act as precious **time-bridges** between sampling or sensing on the one hand and machine operations on the other hand. These time-bridges are essential or useful with soil- or crop properties that have a low temporal resolution, thus remain constant over a rather long time. Prime examples for this are topography, texture, organic matter and pH of the soil.

However, there are also soil- or crop properties that vary rather fast over time, thus have a high temporal resolution. Examples for this are water and nitrate in the soil or some crop properties. The temporal variance in these cases may in fact be more important than the spatial variance (McBratney and Whelan 1999), and consequently maps should then be viewed and used with caution.

## References

Cemek B, Güler M, Kilic K, Demir Y, Arslan H (2007) Assessment of spatial variability in some soil properties as related to soil salinity and alkalinity in Bafra plain in Northern Turkey. Environ Monit Assess 124:223–234
Davis JC (1973) Statistics and data analysis in geology. Wiley, New York
Gringarten E, Deutsch CV (2001) Teacher's aide. Variogram interpretation and modelling. Math Geol 33(4):507–534
Han S, Hummel JW, Goering CE, Cahn MD (1994) Cell size selection for site-specific crop management. Trans Am Soc Agric Eng 37(1):19–26
Hengl T (2007) A practical guide to geostatistical mapping of environmental variables. JRC scientific and technical reports. European Commission. JRC Ispra, Ispra. http://ies.jrc.ec.europa.eu
Kerry R, Oliver MA (2004) Average variograms to guide soil sampling. Int J Appl Earth Obs Geoinform 5:307–325
Kerry R, Oliver MA (2008) Determining nugget: sill ratios of standardized variograms from aerial photographs to krige sparse soil data. Precis Agric 9:33–56
Matheron G (1963) Principles of geostatistics. Econ Geol 58:1246–1266
McBratney AB, Pringle MJ (1999) Estimating average and proportional variograms of soil properties and their potential use in precision agriculture. Precis Agric 1:125–152
McBratney AB, Whelan BM (1999) The "null hypothesis" of precision agriculture. In: Stafford JV (ed) Precision agriculture '99. Part 2. Sheffield Academic Press, Sheffield, pp 947–957
Oliver MA (1999) Exploring soil spatial variation geostatistically. In: Stafford JV (ed) Precision agriculture '99. Part 1. Sheffield Academic Press, Sheffield, pp 3–17
Russo D, Bresler E (1981) Soil hydraulic properties as stochastic processes: I. An analysis of field spatial variability. Soil Sci Soc Am J 45:682–687
Solie JB, Raun WR, Whitney RW, Stone ML, Ringer JD (1992) Optical sensor based field element size and sensing strategy for nitrogen application. Trans Am Soc Agric Eng 39(6):1983–1992
Tollner EW, Schafer RL, Hamrita TK (2002) Sensors and controllers for primary drivers and soil engaging implements. In: Upadhyaya SK et al (eds) Advances in soil dynamics, vol 2. ASAE, St. Joseph, p 182
Webster R, Oliver MA (2007) Geostatistics for environmental scientists, 2nd edn. Wiley, Chichester
Whelan BM, McBratney AB, Minasny B (2002) VESPER 1.5 – Spatial prediction software for precision agriculture. In: Robert PC, Rust RH, Larson WE (eds) Proceedings of the 6th international conference on precision agriculture, Madison

# Chapter 3
# Sensing by Electromagnetic Radiation

Hermann J. Heege

**Abstract** Electromagnetic radiation lends itself to non-contact sensing of many soil- and crop properties. The basis for this is that theoretically any matter – including constituents of soils and crops – can be identified by an electromagnetic index that is derived from its radiation. This electromagnetic index can act as an optical fingerprint of the respective matter or constituent.

Sensing from satellites or from aerial platforms allows obtaining maps that provide an overview within approximately the same time about soil- or crop properties from fields or from wider areas for tactical inspections. Sensors that are located on farm machines never can do this, let alone because of the time it takes to cover a wide area. Yet when it comes to the control of site-specific field operations, sensors on farm machines can provide the best spatial- and temporal precision that is possible. Their excellent spatial precision results from the low distance to soils or crops. The high temporal precision is possible since the signals are recorded just in time. This is important for those soil- and crop properties that vary fast in time.

Georeferencing by positioning systems allows storing site-specific signals.

**Keywords** Absorbance • Atmopheric windows • Clouds • Emitted radiation • Georeferencing • Optical fingerprint • Radar • Reflectance • Transmittance

## 3.1 Basics in Sensing by Electromagnetic Radiation

Site-specific operations require many samples, therefore, wherever possible, manual sampling should be replaced by autonomous- or semiautonomous sensing. This sensing can be accomplished with- or without direct contact to the respective soils

---

H.J. Heege (✉)
Department of Agricultural Systems Engineering, University of Kiel,
24098 Kiel, Germany
e-mail: hheege@ilv.uni-kiel.de

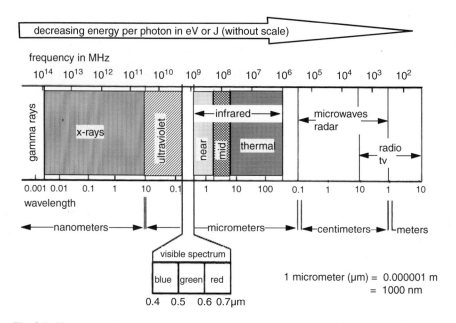

**Fig. 3.1** Electromagnetic spectrum on a contiguous wavelength scale. Please note that the wavelength units change. The exact boundaries between radiation types are not unanimously agreed upon and thus can vary somewhat. Consequently, the limits between radiation types are approximates (From Chuvieco and Huete 2010, redrawn and altered)

or crops. Hence the methods can be classified into contact- or non-contact sensing. Either of these methods can be used when sampling occurs in conjunction with farm machines, whereas satellites and aerial platforms rely solely on non- contact sensing. Since only a few contact sensing methods are available, these are dealt with in appropriate chapters later. Non-contact methods almost exclusively are based on sensing by electromagnetic radiation. This chapter concentrates on this.

Electromagnetic sensing is based on **radiation** of photons. This radiation –depending on its specific type – carries energy through space along periodic harmonic waves. There are many different types of electromagnetic radiation (Fig. 3.1). An important criterion is its **wavelength**, which can vary between a tiny fraction of a nanometer and several meters. The wavelength times the frequency is the speed of the radiation. In a vacuum and in air, this speed is the same for all types of electromagnetic radiation, namely 300,000 km per second. Therefore, the shorter the wavelengths, the higher the frequencies are and *vice versa*.

Another important item is the **energy per photon**. This energy is proportional to the frequency of the radiation type and consequently inversely proportional to the wavelength. The shorter the wavelength, the higher is the energy per photon. The energy of very short wavelengths – ultraviolet radiation and shorter – therefore can be dangerous to human health. Yet this depends on the particular situation.

The differences in energy per photon also have implications for sensing. For photons from longer wavelengths, either very sensitive sensing devices are needed or a larger area is required in order to get a sufficient amount of energy. Thus a balance between wavelengths and spatial resolutions might be necessary. If a high spatial resolution is aimed at, using radiation from a short wavelength range in principle would be advantageous. However, the spatial resolution is not the only criterion when wavelengths for sensing are selected.

Radiation may come from a natural source or may be artificially induced. The most important natural sources are the sun and the earth. The wavelengths of the radiation that is emitted by these sources depend on the respective surface temperature. Since this temperature is much higher on the sun than on the earth, the solar wavelengths are much shorter than the terrestrial ones. Practically all the solar energy flux to the earth is in the wavelength range between 0.15 and 4.0 µm; hence it consists mainly of ultra-violet-, visible- and some infrared radiation. The maximum energy flux of the solar radiation is in the visible wavelengths. On the other hand, the energy that is emitted from the surface of the earth is in the region from 3 to 100 µm, which is mainly in the thermal infrared range (Guyot 1998). In short, the earth emits long-waved, but it receives short-waved radiation. There is only a small overlap between emitted and received wavelengths.

However, this is a rather rough breakup of the energy fluxes and the wavelengths involved. It is important that the longer waves do not contribute much energy. A more precise view is obtained when considering what happens with radiation that is directed from the sun to the earth or *vice versa*.

## 3.2 Emitted, Absorbed, Reflected and Transmitted Radiation

It is important to distinguish between emitted-, reflected-, absorbed and transmitted radiation. **Emitted radiation** leaves the surface mass of the sun or the earth, as every body at a temperature above 0 K discharges photons. The higher the temperature is, the shorter the wavelengths are. Photons that hit a particle en route, rebound and change the direction. Hence these photons become **reflected radiation**. If the photons are not reflected, but instead of this provide energy for the matter that was hit – *e.g.* for heating or for photosynthesis – **absorbed radiation** is dealt with. Finally there is **transmitted radiation** that was neither reflected nor absorbed.

Instead of using the absolute values, it is often reasonable to relate the reflected, absorbed and transmitted radiation to the initial radiation. These related or normalized signals are denoted as **reflectance, absorbance** and **transmittance**. It should be noted that the initial radiation – from which the reflectance, absorbance and transmittance are obtained – can be but must not be at the stage of emission from the sun or the earth. The respective initial radiation can also be radiation that was already reflected or transmitted at an earlier stage, *e.g.* on its path from the sun to the earth. It just depends on what is regarded as the initial radiation.

The sum of the respective reflectance, absorbance and transmittance in fractions always adds up to 1 (one). So it suffices to measure only two of these radiation types, the third type can then be calculated. This is important since often the

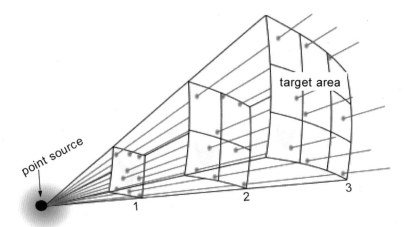

**Fig. 3.2** Schematic to the effect of the inverse square law. The farer the target area is away from the point source, the lower is the number of photons per unit area

absorbance is difficult to measure directly. Sensing in precision farming mostly relies on reflectance from soils or crops. In a few cases, transmittance is used.

In theory, any matter – including constituents of soils and crops – can be identified either by its reflected, transmitted, absorbed or emitted radiation. The identification is a matter of careful **analysis of the spectrum** – that is the wavelength distribution – of the respective matter. This might include mathematical processing of spectral data. The fundamental basis of optical sensing is that in this way for every matter or constituent of soils and crops an **optical fingerprint** can be derived.

So far the best optical fingerprints have been derived from details in the visible- and/or infrared radiation. These details depend largely on the resolution of the optical signals and on the range of the spectrum that is used. The optical fingerprint might rely on the whole range of the visible, near-infrared and mid-infrared radiation, hence might be derived from a **full spectrum approach**. However, it can also be based just on a few or even only one narrow wavelength, thus depend on a **discrete waveband approach**. For some properties of soils or crops, such optical fingerprints have been well defined. There are also cases where research still has to find the best optical fingerprints. Details to this will be dealt with in later chapters.

If electromagnetic radiation emanates from a point source at a constant rate and the distance to the target (sensor) increases, the photons will spread out over a larger area. Hence the fewer photons will land on a target area of constant size, the farer this area is away from the source. This is the effect of the well known **inverse square law**, which states that the result per unit of the target area is proportional to the inverse of the squared distance (Fig. 3.2). Since the sensor of the electromagnetic radiation can be regarded as a target, this can affect the sensing results.

The attenuating result of the inverse square law on radiation is basically independent from any effects that are caused by material barriers such as molecules, which the photons might hit in the atmosphere. But these material barriers can induce additional attenuation.

However, there is another factor that affects the results on the target area and thus the sensing records as well. This is the **sensitivity** of the target area in the sensor to the energy of the radiation. A high sensitivity can compensate for attenuated signals. The progress that has taken place in remote sensing must be attributed partly to the fact that highly sensible receivers of the radiation have made up for inevitable effects of the inverse square law.

## 3.3 Atmospheric Windows and Clouds

The solar radiation that is directed towards the earth hits molecules and aerosols in the atmosphere. The result is scattering and absorption of radiation. Hence the radiation that finally gets to the surface of the earth is filtered by the atmosphere. However, this filtering effect of the atmosphere depends very much on the type of radiation. **Atmospheric windows** show, which radiation types are transmitted to the surface of the earth or *vice versa* (Fig. 3.3).

The respective white regions show, which radiation is transmitted. Black areas indicate radiation that is either absorbed or reflected back into space. The transmittance shown is for a sky without clouds.

There are two main regions of rather unimpeded transmittance: the range of the visible light and the range of the radar-, micro- and radiowaves. The situation in the infrared region depends on the respective wavelengths. Here ranges with blocked transmittance alternate with regions with rather free penetration. Thermal infrared radiation with long waves hardly is transmitted.

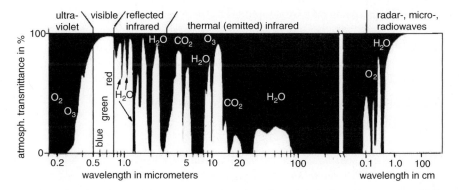

**Fig. 3.3** The atmospheric windows (*white*) show the wavelengths that penetrate the cloudless atmosphere of the earth. The gaseous molecules that can block the transmission of wavelength ranges are indicated. For the boundaries between some radiation types, see legend to Fig. 3.1 (From NASA Earth Observatory 2010, altered and redrawn)

**Table 3.1** Absorbance, reflectance and transmittance of the solar radiation spectrum by clouds (Data from Liou 1976, altered, transmittance added)

| Type of cloud | Thickness (m) | Absorbance (%) | Reflectance (%) | Transmittance (%) |
|---|---|---|---|---|
| Cumulonimbus | 6,000 | 10–20 | 80–90 | 0–10 |
| Nimbostratus | 4,000 | 10–20 | 80–90 | 0–10 |
| Altostratus | 600 | 8–15 | 57–77 | 8–35 |
| Cumulus | 450 | 4–9 | 68–85 | 6–28 |
| Stratus | 100 | 1–6 | 45–72 | 22–54 |

The gaseous molecules that block the radiation are oxygen, water and carbon dioxide. The transmittance of ultraviolet light is partly prevented by oxygen molecules, which is a blessing for human health.

The situation is quite different when **clouds** are present. Clouds are generated by water vapor near the condensation point. They consist of aerosol-sized particles of liquid water that absorb or scatter electromagnetic radiation of waves with less than about 0.1 cm length. Consequently only radar-, micro-and radiowaves are capable of penetrating clouds without being scattered, reflected or absorbed. This is a very important point for remote sensing, since in many areas of the world it is necessary to reckon with cloud covers.

So for sensing of visible- and infrared radiation by satellites, clouds can completely alter the possibilities. And on the average, clouds occupy regularly more than 50 % of the planet earth's atmosphere (Liou 1976). There are of course large regional differences in the incidence of clouds. Their attenuation of the transmittance depends on the wavelengths. Within the visible- and infrared range, the longer the waves, the more attenuation occurs. Short visible waves still have the best chance to penetrate the clouds and thus provide for some diffuse illumination of the earth's surface during an overcast day.

Even a thin stratus cloud reduces the transmittance on the average to almost one third (Table 3.1). With thick clouds, the transmittance drops to 10 % or less. The problem for recording data from satellites by visible and infrared reflectance is that indeed the terrestrial area might be regularly passed overhead, yet in regions with humid climate the actual sensing possibilities are not predictable.

## 3.4 Sensing from Satellites, Aerial Platforms and Field Machines

Electromagnetic sensing of soil- or crop properties can be achieved with passive- or active sensors. Passive sensors rely on natural electromagnetic waves that are provided either by solar energy or by radiation that is emitted from the earth. Hence passive sensing of visible light is confined to daytime. Active sensors have their own artificial radiation sources. This means that they can operate at night as well, even if visible radiation is needed for the sensing process. In case radiation outside the visible

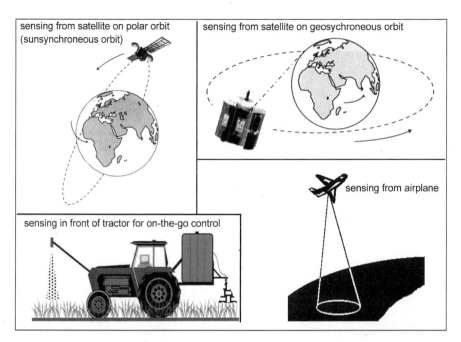

**Fig. 3.4** Sensing from satellites on different orbits, from an airplane and from a tractor (From Chuvieco and Huete 2010 and from Heege et al. 2008, altered)

region is used, operation at night might be possible with either sensing system. But an important point with active- as well as with passive sensing can be the effect of the atmosphere of the earth on the radiation.

The reflectance of the soil or the crop must be transmitted to the sensor. So the distance between the sensor and the soil or crop must be dealt with. If the sensor operates on a tractor or another farm machine, the distance is less than 3 m. As a consequence of this short distance, attenuation of the radiation by the atmosphere hardly occurs. With satellites and also with aerial platforms, because of the very much larger distances, attenuation of the radiation takes place.

Satellites operate on different **distances** from the surface of the earth, depending on whether they move on a geosynchroneous- or on a polar orbit (Fig. 3.4).

In a **geosynchroneous path**, the satellites are always in the same position with respect to the rotating earth. Since these satellites orbit at the same angular rate and in the same direction as the earth, they appear stationary from the globe. The denotion therefore often is "geostationary satellite". They orbit in an equatorial plane at an elevation of about 36,000 km and thus provide a big and constant view of the whole hemisphere in a single image. These satellite types are used for weather forecasting and for broadcasting.

Of special use for precision farming are satellites that either move on a polar orbit or those for the global positioning system (GPS). The latter will be dealt with in the next section. The satellites on a **polar orbit** circle the earth from pole to pole.

They do not rotate with the earth – as the satellites on geosynchronous orbits do. Instead many of these satellites operate on a **sunsynchronous path**, which means they pass the globe overhead at essentially the same solar time throughout the whole year while the earth rotates underneath them. So, theoretically, it is possible to get snapshots for specific places on the planet at the same solar time, which facilitates multitemporal comparisons. The elevation is between 200 and 900 km, hence much lower than with geosynchronous orbiting satellites.

For sensing from aerial platforms either a plane, a helicopter or an unmanned quadrocopter could be used. The latter has rotors like a helicopter, but four of them and can operate in an autonomous manner. With planes, the vertical distance can be several km, whereas for helicopters and especially for unmanned quadrocopters it can go down to 70 m or even less. So the attenuation of the radiation that is reflected from soils or crops can be much lower than for satellites.

The spatial- and temporal resolution that can be obtained is important. From theory, it must be expected that the **spatial resolution** decreases in the order farm machines, aerial platforms, satellites. And in fact, some years ago the spatial resolution that was obtained from satellites often did not satisfy the needs for site-specific farming. But steady advances in the sensitivity of optical instruments have improved the spatial resolutions. Today with a clear cloudless sky, satellites on polar orbits can provide spatial resolutions that make possible a terrestrial cell size of 1 $m^2$ and even less. This does not alter the general fact that it is easier to obtain a high spatial resolution when sensing occurs with smaller distances to soils or crops. Yet the situation is that with modern techniques and a clear cloudless sky, sensing from every platform can deliver the spatial resolution that is needed. Especially with optical sensors that operate from a farm machine, the resolution can be much higher than is even needed.

Concerning **temporal resolution**, sensing from satellites on polar orbits theoretically provides for the best prerequisites since data from the same field can be obtained every day, provided neither a closed atmospheric window nor clouds impede the radiation. It is practically not feasible to sense from farm machines or from aerial platforms with such a **temporal frequency**. This holds as long as farm machines and aerial platforms need drivers or pilots.

However, when sensing occurs from farm machines during a field operation, another important aspect deserves attention. Since many field operations take place just once or twice per year, the temporal resolution seems to be extremely low. But an important point is that with such proximal online and on-the-go control of farming operations, the sensing can occur exactly at the time when the information is needed. If there is temporal variation of soil- or crop properties during the growing season – and in many cases this is the situation – it can be important to sense precisely at the time when the farming operation takes place. So for these soil- and crop properties it is **temporal precision** that is needed rather than temporal resolution. A high temporal resolution might in these cases lead to a huge amount of useless data.

There are farming situations that call for a high temporal resolution or temporal frequency, *e.g.* when a crop is observed for pest infections. Yet there are also cases when temporal precision is the most important criterion, *e.g.* when in- season fertilizing of nitrogen takes place. This distinction between temporal resolution on the one hand and temporal precision on the other hand is helpful, though both might be needed.

With sensing from aerial platforms, it might be possible to avoid the transmission of signals through clouds by a low height above the surface of the earth. However, up to now unmanned observations from aerial platforms hardly occur. This limits sensing from aerial platforms. The development, permission and use of unmanned quadrocopters might alter the situation.

## 3.5 Microwaves or Radar Instead of Visible or Infrared Waves

Practically all sensing limitations that arise from atmospheric barriers (Fig. 3.3) including clouds are removed when **microwaves** are used. The name "microwaves" can be misleading, since their spectral region has the longest waves used in remote sensing. Hence the microwaves also have the lowest energy per photon. The limitations that arise from this for sensing from satellites are overcome by using **active sensors** with special antennas that provide a high sensitivity. The active sensors both emit microwave energy and detect its return from the ground. They are generally known as **radar** sensors. Radar stands for **r**adio **d**etection **a**nd **r**anging.

Modern spaceborne radar sensors work in the wavelength range of 0.1–100 cm and emit pulses of radiation in a "flashlight" manner. The signals that are reflected back to the satellite depend to a large extent on the roughness of the surface that was hit. The rougher the surface, the better the return signal is. Because from a rough surface, the radar echoes are scattered back in several directions. Hence the reflection is at least partly thrown back for recording, whereas specular reflection from a smooth target might not get back to the satellite at all. With cultivated soils, clods in the seedbed provide for a diffuse reflection (Fig. 3.5).

Yet the reflection back to the satellite depends on additional factors, especially on the wavelength and the dielectric properties of the soils or the plants. The longer the waves are, the more radiation is reflected back to the satellite and *vice versa*. Hence with long waves, a rather flat soil surface can appear as being rough, while with shorter waves it can show up as being smooth (CRISP 2010).

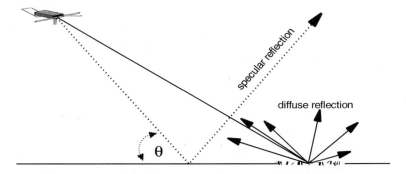

**Fig. 3.5** Reflection of radar signals from a smooth- or from a rough target surface

Formerly, the dielectric properties have been defined by a so called dielectric constant. This property, however, is not constant at all, it varies widely. Consequently, now it is denoted as "**permittivity**". This physical term generally is expressed as relative permittivity because it is related to the permittivity of a vacuum and hence is without dimensions. It defines the potential to store electric energy. Air has approximately the permittivity of a vacuum, which is 1 (one). This is the minimum. Compared with other matters or materials, water has a very high permittivity of around 80. The respective data for dry natural materials including soils and plant matter are much lower, they are in the range of 3–8 (Paul and Speckmann 2004; Lillesand and Kiefer 1979).

These large differences in the permittivities or in the dielectric properties of water on the one hand and dry soils on the other hand are the base of **moisture sensing** by radar waves in precision farming. However, a prerequisite for sensing the moisture is that effects of differences in the surface roughness do not show up. Rather long radar wavelengths can help in this respect, at least with sensing of soils. This is because long waves react less on the roughness of the soil surface. Another advantage of rather long waves is their ability to sense the moisture not exclusively on the top surface of the soil, but instead also for some vertical distance down from the surface. The moisture solely on the surface of soils is hardly important for crops, since their water is supplied by a soil layer of some thickness.

The potential of sensing by radar waves can be enhanced by **polarizing the radiation**. The normal case is that the radiation vibrates or fluctuates in all directions perpendicular to the propagation at random, even if the wavelength is uniform. Polarizing the radiation aims at controlling the direction in which the photons vibrate. So a polarizer is a device that allows only radiation with a specific angle or a specific direction of vibration to pass through. The signal is filtered by a polarizer in such a way that the wave vibrations are restricted to a single plane that is *e.g.* perpendicular or horizontal to the direction of wave propagation (Fig. 3.6). There can be additional alternatives in polarizing directions.

It should be mentioned that this polarizing in a vertical- or horizontal direction does not alter the fact that every radiation has an electrical- as well as a magnetic field. These fields incidentally also move in perpendicular planes. Yet the polarizations shown in Fig. 3.6 only refer to electric fields.

When a polarized radar radiation is transmitted to crops or soils, it generates reflectance with a variety of polarizations. So – in a simplified way – there is again a somewhat random situation. But this random radiation too can be polarized again when it is received by the radar sensor. Today, many radar sensors are designed to transmit and receive waves that are either horizontally (H) or vertically (V) polarized. With these, there can be four combinations of transmit- and receive polarizations:

- HH – for horizontal transmit to the target and also horizontal receive
- VV – for vertical transmit to the target and also vertical receive
- VH – for vertical transmit to the target but horizontal receive
- HV – for horizontal transmit to the target but vertical receive.

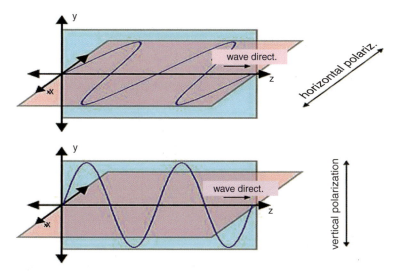

**Fig. 3.6** Polarization of radar waves in a horizontal- or in a vertical direction (From Fink et al. 2010, altered)

Because the transmit and receive situations in the first two cases are the same, they are denoted as "**like-polarized**". The last two cases are "**cross-polarized**". These detailed sensing alternatives can – when used in combination with suitable wavelengths and incidence angles $\theta$ of the radiation (Fig. 3.5) – substantially improve the results obtained (Heinzel 2007; McNairn et al. 2009; Shimoni et al. 2007).

Yet whatever results are obtained when sensing with visible, infrared or radar waves, it should always be kept in mind that the signals indicate just phenomena that are of interest because of their known relation to soil- or crop properties. The signals never directly explain the reasons for the sensed phenomena. So still a clever mind is needed for the analysis of the causes. However, a farmer has to make this analysis too when he inspects his fields visually. The difference is in the amount and kind of data available.

And the success achieved with the large pool of signals often depends very much on an intelligent processing of these. It might be necessary to create special **mathematical indices** based on the respective electromagnetic spectrum. These indices are calculated from special selected wavelength bands by algebraic or differential operations. Suitable indices for soil- and crop properties have been and still are the object of intensive research. Some of them will be dealt with in later chapters.

However, there are still fundamental differences in the sensing potential of visible and infrared radiation on the one hand or radar waves on the other hand. This potential is listed in Table 3.2. Summing up, it can be seen that sensing with visible and infrared radiation is mainly focused at **constituents** of soils and crops, whereas the applications for radar waves are more pointed towards

**Table 3.2** Feasible applications for sensing of soil-and plant properties by radiation

| Radar waves | Visible- and/or infrared waves |
|---|---|
| **Proven applications** | |
| Volume or height of crops | Plant constituents, *e.g.* chlorophyll, water and nitrogen |
| Vertical- and horizontal arrangement of plant parts | Leaf-area-index of crops |
| Roughness of soil surface | Senescence of crops |
| Moisture of soil layer of a few cm | Organic matter and water on soil surfaces |
| **Emerging applications** | |
| Classification of crop species | Classification of crop species |
| Fresh biomass of crops | Soil texture on the surface |
| Dry biomass of crops | Soil content of some nutrients on the surface |

getting information about **synoptical properties**, *e.g.* the volume of crops or the roughness of soil surfaces. Yet the items listed in Table 3.2 should not be regarded as strict limits (Kühbauch 2002).

## 3.6 Using Maps or On-The-Go Control in Real-Time

Aside from sensing limits, there are distinct differences in the domains of applications for properties that are recorded from satellites or from farm machines. Polar satellites can provide for maps that show the situation for a large area at a definite hour within a day. Machine based sensors never can do this.

There is a need for maps that provide for an **overview** of soil- and crop conditions at a definite time within a farm, a community, a county or a whole country. So overviews about *e.g.* soil water supply, fields that are fallow or cropped, crop species used, crop development, crop damage of various kinds (hail, drought, floods, diseases, insect infestation *etc.*), progress of harvesting and subsequent cultivation can be helpful. The present state of the art in sensing from satellites or aerial platforms allows not yet to provide all of these details despite the fact that the possibilities increase steadily. In many instances, combining of several radiation phenomena is needed in order to get to the desired information. Accordingly, McNairn et al. (2009) as well as Shimoni et al. (2007) have provided for methods in order to classify or identify crops that are grown in an area either by using visible- plus infrared radiation or by taking radar waves.

However, the information that is helpful differs. Governmental departments, farm agencies and agribusiness institutions need maps that provide for information over wide areas that include many farms. Farmers primarily require maps that either contain just the whole own farm or even are limited to a single field. So it is reasonable to differentiate between

- wide area maps
- farm maps and
- field maps.

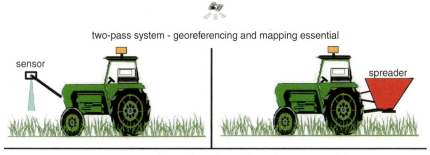

**Fig. 3.7** Tractor-based control for site-specific spreading of fertilizer with or without mapping

Wide area maps and farm maps are predominantly used for **tactical inspections** of the situation. In a similar way like a mirror might allow to see around corners, these maps make it possible to get an overview of the coverage of large areas from a central bureau, and this within a few minutes. During the growing period, these tactical inspections can be useful in time intervals ranging from several days to several weeks, *e.g.* in order to see how the crops develop. The maps can be supplied easily and at reasonable cost via internet from polar satellites that orbit the respective areas every day. Limitations can exist in some agricultural regions for maps that rely on visible- and infrared radiation as a result of the effect of clouds (Sect. 3.3). Yet for some soil- and crop properties (Table 3.2), the cloud problem can be overcome by using radar waves instead of visible- and infrared radiation. The steady advances in sensing by radar waves facilitate this.

With field maps, the situation is different. They might sometimes be used for tactical inspections as well, but this is not the most important application. The preferential use in precision farming is for the control of **site-specific field operations.** Some properties that are recorded in field maps are temporally constant, others are not constant over time at all. Maps about texture, organic matter content and contour lines of soil can be regarded as being up to date for a long time and hence be used for many years. To a somewhat lesser extent, this also applies to maps about the pH of soils. But there are many soil- and crop properties that do not allow to use the same map for several consecutive field operations or years. The plant available nitrogen- and water content in soils can change within some days. The same applies to growth stages or infestations of crops with fungi or insects.

The ideal control technique for site-specific operations when the soil- or crop properties change fast in time is online **real-time sensing** combined with on-the-go adjustment of the farm machine (Fig. 3.7). This technique allows for the best temporal precision that is possible. An imaginable alternative to this would be online transfer of site-specific soil- and crop properties from satellites or from aerial platforms in real-time to moving farm machines. But this alternative is not yet state of the art with the exception of georeferencing (Sect. 3.7).

In addition, sensing from farm machines evades the cloud problem. This is important, since visible and infrared radiation – which is needed for site-specific control of field operations – is highly affected by clouds.

Online and on-the-go controlled field operations do not rely on maps. However, field maps of the respective operations can be created as by-products that allow *posterior* studies of the situations and also make possible a joint use in the control of subsequent field operations. A prerequisite for recording these field maps is the georeferencing of the signals, hence the simultaneous use of a positioning system. The next section will deal with this.

## 3.7 Georeferencing by Positioning Systems

Precision in mapping as well as in guidance of farm machinery needs georeferencing in the fields. Global navigation satellite systems (GNSS) provide the means for this (Fig. 3.8).

The most used and universally known method is the American global positioning system (GPS). In 1995, it was supplemented by the Russian GLONASS system. The GALILEO system of the European Union will start in 2014. A Chinese-and a Japanese system will also be created.

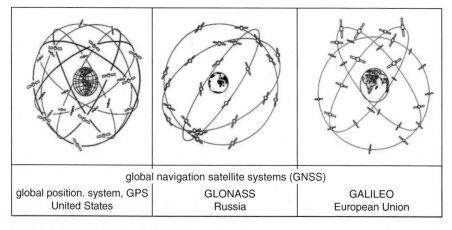

**Fig. 3.8** Orbits used for present global navigation satellite systems (Graphs from Mansfeld 2010, altered and recomposed)

The American, Russian and European satellites orbit the earth about twice per day in oblique angles to the equator. The vertical distances to the surface of the earth are between 19,000 and 23,000 km. Hence the satellites move higher than those that are on polar orbits, yet lower than those on geosynchroneous orbits (Sect. 3.4). All systems use about 24 satellites, however, the GPS satellites are on six different orbits, the GLONASS- and GALILEO systems have only three orbits.

The signals are transmitted via **microwaves**, which operate within an atmospheric window (Fig. 3.3) and penetrate clouds. So obstructions in the atmosphere do not exist.

The georeferencing is achieved by the **time interval**, within which radio signals go from the satellites to the receiver. The latter is *e.g.* on a vehicle or on a farm machine that moves in the field. The satellites carry highly accurate atomic clocks. The receivers on the ground synchronize themselves to these clocks. Hence in a simplified way, every receiver too is a highly accurate atomic clock.

Once the time-interval is known that a radio signal takes from the satellite to the receiver on earth, the calculation of the respective distance between the satellite and the target is possible. This only requires taking into account the speed of the electromagnetic radiation (Sect. 3.1). And finally, when the distances between several satellites and the receiver are known, the geometric position of the target can be found out by trigonometric means. A prerequisite for this is the knowledge about the position of the satellites. This knowledge is at hand. Thus in detail, the signals can provide the target with four dimensions:

- the time
- the geographical longitude
- the geographical latitude
- the geographical altitude.

The last three dimensions together define the respective **geometrical position**. As a first step in precision farming, geometrical positions can be used for getting the borders and exact areas of all fields. Subsequently, the position can be used as the site-specific reference for all farming operations. This reference allows to link soil- and crop properties in an intelligent way. In this respect, the position is a benchmark in precision farming. The site-specific altitude can be used as a source for mapping the contour lines of fields. Topographic maps that contain this information can be obtained as a by-product of other site-specific farming operations (Abd Aziz et al. 2009).

Important criteria in georeferencing are the availability of the satellite signals and the precision of positioning. A general prerequisite of **availability** is that the radio waves from four satellites simultaneously can get to the receiver. Clouds are no barrier since microwaves are used, but trees and buildings can reflect the signals.

Whether this prevents georeferencing, can depend on the number of satellites that are operating (Fig. 3.9). This number has been steadily increasing, not least because the global positioning system of the USA has been and still is supplemented by similar systems from other parts of the world. Different global

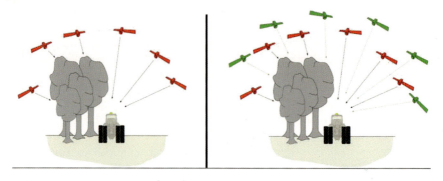

**Fig. 3.9** The position of the receiver on the tractor is the same in both drawings. In the *left* drawing, the signals from only three satellites get to the receiver, which in most cases is not sufficient for georeferencing. In the *right* drawing – because of more satellites – georeferencing is possible (From graphs by Poloni 2009)

**Table 3.3** Availability of georeferencing signals and maximal down time of receiver with varying numbers of satellites in orbit, which operate with a mask angle of 5°

| Number of satellites in orbit | Availability % of time | Maximal total down time within 30 days (min) |
| --- | --- | --- |
| 24 | 99.98 | 8 |
| 23 | 99.97 | 13 |
| 22 | 99.90 | 42 |
| 21 | 99.10 | 386 |

Compiled from Mansfeld (2010), altered

navigation satellite systems provide signals that are compatible. Accordingly, one receiver can simultaneously use signals that come from separate systems.

The drawings in Fig. 3.9 show schematically the extreme situation when a farm machine moves along high obstacles of navigation signals. But also within a normal field without patches of trees it cannot be assumed that all the navigation signals have unimpeded access to the receiver. There always is a so called "**mask angle**" between the horizontal line in the field and oblique radiation that is oriented to the receiver. Oblique radiation that is directed to the receiver at an angle to the horizontal line that is smaller than the mask angle is ineffective. Apart from the situation close to forests or hedges, mask angles between 5° and 10° must be taken into account (Mansfeld 2010).

The data in Table 3.3 hold for a mask angle of 5°, hence for favorable conditions. The deteriorating effect of a lower number of satellites on the availability in % of the time appears to be rather small at first sight. However, the availability in % of time does not indicate the conditions for georeferencing in a way that is easy to understand. The maximal total down time – that can be calculated from the availability in % of time – provides for a better insight into the situation. It is the maximal sum of

**Table 3.4** Inaccuracies and expenditures for different global positioning systems

| System | Inaccuracy | Expenditures Receiver (€) | Reference signal |
|---|---|---|---|
| Autonomous GPS, coded signal, single freq. | 2–25 m | 70–400 | None |
| Differential GPS, coded signal, single frequency | 1–3 m | 1,000–3,000 | 0–500 €/a |
| Differential GPS, carrier phase signal, dual frequ. | 10–30 cm | 5,000–10,000 | 1,000–2,000 €/a |
| Real-time kinematic differential GPS | 1–4 cm | 20,000–40,000 | None (own reference station) |

time within a period during which signals cannot be received. This maximal total down time rises fast when less than 24 satellites are available (Table 3.3). The steady increase in the number of satellites that orbit the globe has helped to avoid problems that arise from this.

While having more visible satellites can reduce the down time of receiving signals considerably, it has only a slight effect on the **precision of positioning** or of **georeferencing**. This precision depends largely on techniques that are used for correcting errors.

In a strict sense, information that is given about the precision of georeferencing in metric units does not treat accuracies. Instead it is dealing with measurement errors, hence with inaccuracies or with imprecision. Generally, the lower the measurement errors, the higher the expenditures of the respective positioning systems are. Table 3.4 shows for some systems the ranges of measurement errors as well as the expenditures. All systems have the term GPS within their denotation because of the former leading position of the US navigation system. This does not mean that receivers cannot use signals from other global navigation satellite systems (Fig. 3.8).

The autonomous GPS system with coded signals and single frequency corresponds to the devices that are million times used in cars or small handheld computers. The measuring errors of these devices can be accepted for navigation on roads, however, for many precision farming operations lower inaccuracies are needed. For this, several ways are available.

A widely-used method is **differential positioning**. It involves having two GPS receivers. One of them is stationary and called the "base" receiver. Its geographic position is in the respective area – up to 200 km from the second receiver – and is precisely known beforehand. Hence this base receiver can register errors that are involved with signals from a satellite. As a consequence, it can provide the second receiver – which is the main receiver used for controlling a moving vehicle or a farm machine – with radio signals that have correction data. This allows for substantially lower inaccuracies (Table 3.4). The correction signals can be obtained online on-the-go either for an annual fee or sometimes also free of charge.

Another significant improvement in positioning can be realized – in a simplified way – by a higher resolution of the signals that the receiver gets. This higher resolution is provided by **carrier phase signals** instead of coded signals. Details to this are dealt by Mansfeld (2010) and by van Diggelen (2009).

An amazing good georeferencing can be realized by **real-time kinematic differential GPS**, abbreviated **RTK-GPS**. This system uses all the possibilities for improvement that are mentioned above and is – in the original way – equipped with an own base receiver for corrections. This base receiver is located rather close to the moving receiver. For farm machines, it often is positioned on the headlands. With optimal conditions, the positioning error can be as low as 1–3 cm. The inaccuracies increase with the distance between the two receivers. Per 1 km distance, the increase in error is about 1 mm (Heraud and Lange 2009). So even with a distance of 4 km, the inaccuracies can be below 4 cm.

RTK-GPS technology allows a farmer to return to the exact location again later during the growing season or even in subsequent years. Hence its precision in georeferencing can be relied on not only from pass to pass during a current farm operation, but from season to season or year to year as well. This feature is important when **repeatability** in the guidance of farm machinery via positioning systems with a low error is needed. Prime examples for this are the guidance for no-till sowing into inter-row strips of the crop from the previous year (Sect. 8.4.1) or strip-till sowing when the cultivating of the strips occurs in autumn and the sowing precisely into the center of the narrow strips in spring. There are additional examples when dealing with row crops. Some farmers pour concrete pads at the headlands to ensure that the base station is returned to the exact spot for precise guiding.

The maximum distance between the base receiver station and the moving receiver with real-time kinematic differential GPS – as described above – is between 10 and 20 km. This restriction in distance with an own reference station can be avoided by using an array or **network of RTK-GPS** base receiver stations within a wide area. The distance between adjacent **network base receiver stations** can be up to 70 km (Heraud and Lange 2009). These network base stations provide for correction data that are collectively processed. The result is that despite longer distances to the moving receiver within this network, a similar low error or inaccuracy as shown in Table 3.4, bottom is possible (Edwards et al. 2008). The transmission of the correction data from the network to the user typically is via mobile phone.

Not all precision farming operations require the accuracy or low error range of RTK-GPS. In many cases, the error associated with differential GPS operating on dual frequencies and carrier phase resolution can be tolerated. This system presently is used widely, since the expenditures are much lower than for RTK-GPS.

## References

Abd Aziz S, Steward BL, Tang L, Karkee M (2009) Utilizing repeated GPS surveys from field operations for development of agricultural field DEMs. Trans ASABE 52(4):1057–1067

Chuvieco E, Huete A (2010) Fundamentals of satellite remote sensing. CRC Press, Boca Raton

CRISP, Centre of Remote Sensing, Imaging & Processing (2010) Principles of remote sensing. Research tutorial 01. http://www.crisp.nus.edu.sg/~research/tutorial/intro.htm

Edwards S, Clarke P, Goebell S, Penna N (2008) An examination of commercial network RTK GPS services in Great Britain. School of Civil Engineering and Geosciences, Newcastle University, Newcastle upon Tyne

Fink MC, Carver G, Johnson RL (2010) The polarization of light by reflection. Omega Optical Inc., Pittsfield, MA, USA. http://www.photonics.com/Article.aspx?AID=35808

Guyot G (1998) Physics of the environment and climate. Wiley, Chichester

Heege HJ, Reusch S, Thiessen E (2008) Prospects and results for optical systems for site-specific on-the-go control of nitrogen-top-dressing in Germany. Precis Agric 9(3):115–131

Heinzel V (2007) Retrieval of biophysical parameters from multi-sensoral remote sensing data assimilated into the crop-growth model CERES-Wheat. Doctoral dissertation, University of Bonn, Bonn

Heraud JA, Lange AF (2009) Agricultural automatic vehicle guidance from horses to GPS: how we got there, and where we are going. In: Distinguished lecture series, agricultural equipment technology conference, Louisville, 9–12 Feb 2009. ASABE, St. Joseph

Kühbauch W (2002) Remote sensing (in German). In: Tagungsband Precision Agriculture Tage, 13–15. März 2002 in Bonn, KTBL Sonderveröffentlichung 038, Darmstadt, pp 79–87

Lillesand TM, Kiefer RW (1979) Remote sensing and image interpretation. Wiley, New York

Liou KN (1976) On the absorption, reflection and transmission of solar radiation in cloudy atmospheres. J Atmos Sci 33:798

Mansfeld W (2010) Positioning and navigating by satellites (in German). Vieweg & Teubner, Wiesbaden

McNairn H, Shang J, Jiao X, Champagne C (2009) The contribution of ALOS PALSAR multipolarisation and polarimetric data to crop classification. IEEE Trans Geosci Remote Sens 47(12):3981–3992

NASA Earth Observatory (2010) Remote sensing. Absorption bands and atmospheric windows. http://earthobservatory.nasa.gov/Features/RemoteSensing/remote_04.php. Accessed 4 May 2013

Paul W, Speckmann H (2004) Radarsensors: new technologies for precise farming (in German). Landbauforsch Völk 54(2):73–86

Poloni D (2009) Finding position and direction by using satellites (in German). Neue Landwirtsch, Berlin 4:72–75

Shimoni M, Borghys D, Heremans R, Milisavljevic N, Derauw D, Pernel C, Orban A (2007) Land cover feature recognition by fusion of POLSAR, POLINSAR and optical data. European Space Agency. POLINSAR. In: The 3rd international workshop on science and application of SAR polarimetry and polarimetric inferometry, Esrin, Frascati, 22–26 Jan 2007

Van Diggelen F (2009) A-GPS: assisted GPS, GNNS and SBAS. (GNSS Technology and Applications) Artech House Inc., Boston

# Chapter 4
# Precision in Guidance of Farm Machinery

Hermann J. Heege

**Abstract** Georeferencing by GNSS has opened up possibilities for precise guidance of farm machinery along virtual lines in the fields. The guidance takes place either manually with the help of lightbar indications or in an automatic way. The driver concentrates on supervising the machinery.

In prior pass guidance, each run across the field follows the respective prior path offset by the operating width of the machine. Contrary to this, in fixed line guidance, the courses across the field are not defined by the respective prior path but instead solely by the first pass. Prior pass guidance is indispensable in irregularly shaped fields, whereas in rectangular fields fixed line guidance should be preferred.

On slopes, using more than one GNSS antenna allows to compensate for roll, pitch or yaw of the tractor. Downward drifting of implements on side slopes can be counteracted by passive- or active guidance corrections.

**Keywords** Automatic guidance • Fixed line guidance • Guidance on slopes • Lightbar indications • Pitch • Prior-pass guidance • Roll • Section control • Yaw

## 4.1 Principles of Guidance

Precision farming implies accurate guidance of farm machinery. The wider the width of operation is, the more difficult it is for the driver to navigate without inaccuracies between adjacent runs. Therefore, many farmers in Europe use **tramlines** in small cereals or rape (colza). These tramlines are created each year at the time of drilling by disengaging the respective seed metering rollers precisely there,

---

H.J. Heege (✉)
Department of Agricultural Systems Engineering, University of Kiel,
24098 Kiel, Germany
e-mail: hheege@ilv.uni-kiel.de

**Fig. 4.1** Tramlines in small cereals (From UK Agriculture 2013)

where during fertilizing and spraying operations the wheels of the machines go. This allows for a better aligning during in-season fertilizing and spraying, provided the operating width is an integer multiple of that of sowing. When tramlines cannot be used – as with cultivating, sowing or for all operations on permanent grasslands – either disc- or foam markers can assist in guiding the driver. With harvesting machines, mechanical feelers that slide along the still standing crop can control the guidance (Fig. 4.1).

Yet an accurate guidance for all field operations can be achieved on the basis of georeferencing by **global navigation satellite systems (GNSS)**. Via special computer programs, these systems make it possible to guide machinery precisely along **virtual tramlines** while respecting boundaries that result from headlands or grassed waterways even in irregularly shaped fields. The boundaries can be georeferenced in a prior, circuitous pass around the field. Once stored and mapped, they are available for several years.

The positioning information is provided either by differential GPS operating with carrier phase signals plus dual frequencies or by real-time kinematic differential GPS (Table 3.4). With manual- as well as with automatic steering, the start of the satellite based guidance is either along a prior pass- or along a fixed line.

With **prior pass guidance**, each run across the field follows the respective prior path offset by a given distance. Typically, this distance is the operating width of the machine. The driver starts by manually steering the machine just as usual on a first path – a so called guess row – along the field. This method is naturally used when the driver starts operating, *e.g.* along a waterway or along an irregular- or curved field boundary. Once the prior pass has been recorded, all other trips across the field follow on the basis of GNSS. The courses of all passes are georeferenced in order to get the following run. This method allows for adapting to field shapes. Yet limitations can arise because the following paths can end up with successively more bent curves.

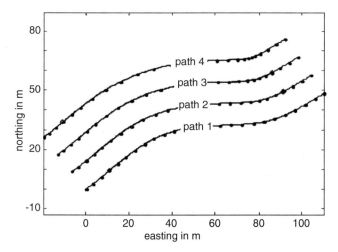

**Fig. 4.2** Modification of the curvatures of subsequent implement paths. Note the change of curvature between path 1 and 4 on the *right side* of the graph (From Heraud and Lange 2009, altered)

The geometrical explanation for this is simple: a flexible horizontal strip that is bent laterally has a steeper curve on the concave side than on the convex side. The differences in the respective **radii of curvature** between the concave- and convex side increase with the width of the strip. Hence if in the field strip by strip are joined, the radii of curvature of the paths on the concave side get shorter and shorter (Fig. 4.2). Finally when the implement cannot follow the sharp bent curve any more, farmers either have to leave a gap between passes or add a short corrective strip.

Extreme effects of this are generally known when implements are used in a circuitous manner within a rectangular field with steep, rounded edges. This inevitably creates crescent-shaped gaps in the corners that are not treated by the machine.

Contrary to this, in **fixed line guidance**, the first path follows a smooth and mathematically predefined line, commonly called A-B line. In most cases this predefined pass is a straight line. Yet for fields that are center-pivot irrigated, it can be a circle. All additional trips are defined by a given offset distance – multiplied by an integer – from the first path. Again, the offset distance typically is the operating width of the implement. But contrary to prior pass guidance, the course of each line is – in a strict sense – independent of the respective preceding pass. Instead, it is only defined by the A-B line, the offset distance and the integer of the pass number. Therefore, errors between adjacent lines do not accumulate. Once the passes are recorded, they can rather precisely be followed again in subsequent operations even if these occur in later years. This would be possible with prior pass guidance as well, provided all passes are recorded, however, the errors resulting from curvature problems (see above) remain in this case.

From this follows that generally fixed line guidance should be the choice because it isolates errors. Topography and field shapes, however, often do not allow an exclusive use of fixed line guidance. The logical consequence of this situation is that prior pass guidance is restricted to fields or parts of them where fixed line

**Fig. 4.3** Contour farming in a sloped field (From USDA NRCS 2013)

guidance is impossible. In many cases this means that one or a few passes in a circuitous pattern are necessary around a field and apart from that the main area inside these passes is handled in fixed line guidance. The number of headland turns can be minimized by choosing the longest field side respectively for the direction of the fixed lines.

In sloped regions with continental climate, often **contour farming** (Fig. 4.3) is essential since it substantially can reduce runoff of water, thus also can diminish soil erosion and improve water infiltration into the land. This implies guiding the machinery as far as possible along lines of similar elevation, hence straight paths hardly are possible and curved lines in most of the area must be dealt with. As long as the curvature is constant, this does not prevent the use of fixed line guidance. Yet when variations in curvature show up, prior pass guidance as outlined above is needed. All in all, guidance with contour farming is more complicated and often less precise.

## 4.2 Techniques of GNSS Based Guidance

Two main techniques are available, either the lightbar- or the automatic guidance system. The former often is also denoted as the manual guidance system.

Both techniques rely on GNSS signals for the indication of the cross-track errors. These are the deviations of the horizontal distance to the reference line, which can be the prior pass- or the fixed line as well as any offset line to it.

With the **lightbar guidance system**, the cross-track errors are indicated in a display by means of light emitting diodes (LEDs), which are arranged in a horizontal row (Fig. 4.4). The errors that occur from pass to pass depend on the guidance principle as well as on the driver because the steering is still manual. However, in

**Fig. 4.4** Manual guiding assisted by lightbar indications (From John Deere & Co., altered)

combination with differential GPS on carrier phase signals and dual frequencies, the cross-track-error is on the average only 22 cm (Bombien 2005). And what is important when operating with modern present-day farm machinery: this average cross-track-error in **absolute values** does not depend on the working width, which is the case with conventional navigating. Compared to conventional manual steering without using GPS and with a working width of 6 m, the cross-track-error is halved. Based on a working width of 24 m, the cross-track-error is reduced to a quarter. And guidance at night is much facilitated.

An **automatic guidance system** (Fig. 4.5) is similar to a lightbar system except that now the signals that come from the computer algorithms do not induce the driver any more to steer, instead these signals go to electric- and hydraulic steering actuators. To engage the automatic system, the driver pushes a button. This allows the system to take control and hence to lead the vehicle along the closest guidance line. And manual control is simply resumed by turning the steering wheel.

In principle, automatic guidance systems can also be used for the headland turns. However, presently many farmers still use it only for the passes across the field and steer manually at the headlands. Hence most of the time, they are relieved from steering and can concentrate on supervising and adjusting the remote-controlled implements. Shortly before the machine arrives at the headlands, a sound reminds the driver. After having completed the manual turning, the driver again pushes the button for automatic control. With real time kinematic GPS, the **cross-track-error** is as low as 2–5 cm (Bombien 2005; Reckleben 2011). So compared to manual lightbar steering based on differential GPS with carrier phased signals plus dual frequencies, the cross-track error is reduced to about one sixth. In case conventional manual steering without any GPS and without tramlines is the basis of comparison, the cross-track-error for a working width of 6 or 24 m goes down to 1/12th and 1/24th respectively. Because again – contrary to conventional steering – the

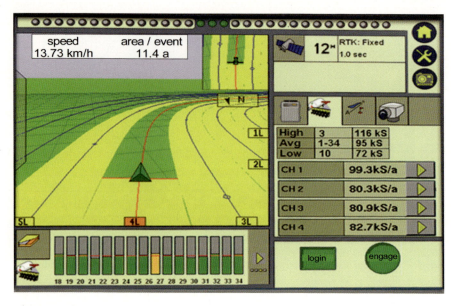

**Fig. 4.5** Display of an user interface to an automatic guidance system. The lightbar (*top*) indicates cross-track errors. The signals from two different GNSS satellite systems can be used simultaneously in order to improve availability. Site-specific section control of implements (*e.g.* sprayer) is possible (Courtesy of Trimble Agric. Div., Westminster, USA, altered)

absolute accuracy of the GPS technology is independent of the working width (see above).

The basic components needed for manual lightbar- or automatic guidance are:

- a GPS receiver
- a user interface for displaying cross-track errors and for user input, *e.g.* the working-width or the location of the first guidance line
- algorithms for path-planning that compute cross-track errors relative to a guidance line.

In addition to these items, automatic guidance always requires to install

- an actuator for vehicle steering and a detector for manual override.

In hilly fields or when very precise guidance is needed, additional components for automatic guidance can be useful, *e.g.* components for terrain compensation (Sect. 4.4) and a steering angle sensor. The still dominant practice is to **retrofit** farm machines with the guidance components. For this, a hydraulically- or electrically driven servo-motor is connected to the steering wheel in the cab. Its friction roller runs against the rim of the steering wheel and thus turns it. In many cases, the retrofitted components including the actuator for steering and the manual override for it can be moved from one tractor to another. So one set of components can be used for guiding in successive farming operations with different machines.

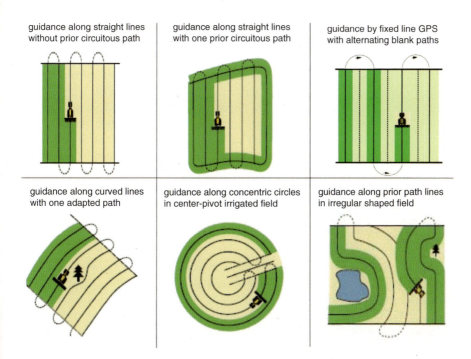

**Fig. 4.6** Driving patterns (Courtesy of Agri Con GmbH, supplemented and altered)

However, the trend is from retrofitting of used machines to **factory-fitting** of new vehicles with the components that are needed. For this, the actuator for automatic guidance replaces the conventional hydraulic steering motor and thus does not operate in the drivers cab any more. Yet this makes it more difficult to use the same actuator on several vehicles. On the other hand, factory fitting saves much time by doing away with harnessing and bothering about technical details.

Efficiency and precision of field operations depend on the **driving pattern** that is used (Fig. 4.6). In rectangular fields, circuitous driving around and around parallel to all field boundaries should be avoided whenever possible. If this driving pattern is used without disengaging the implements, crescent-shaped gaps that are not dealt with are inevitable at the corners (see text near Fig. 4.2). On the other hand, when driving around is done in a dead end pattern with a turn at each corner, time is lost because of the high number of turns per unit area. The lowest number of turns in rectangular fields with dead end driving patterns is needed when driving is done parallel to the longest field side (Fig. 4.6, top). With many machines, this can be in the pattern of adjacent strips. However, sometimes this would require a back-up for every headland turn. This can be avoided by using fixed line GPS guidance and alternating between strips that are dealt with and left blank at first (Fig. 4.6, top right).

There can be obstacles in rectangular fields for driving up and down parallel to the longest field side. With some harvesting machines, it must be taken into account that unloading onto wagons on-the-go is possible only to one side. Consequently, with these machines driving up and down along adjacent strips is not possible. Yet this limitation does not apply to harvesting machines that temporarily store in tanks – like combines – or that load onto hitched wagons.

Driving along curved lines always is more complicated. But if the curvature is constant – as with the guidance along concentric circles of a center-pivot irrigated field – an automatic fixed line GPS guidance mode still is possible. With contour farming the situation can be different, the curvature can change. The same applies to irregular shaped fields (Fig. 4.6, bottom right). Modern guidance systems allow to switch between prior pass- and fixed line guidance as well as between automatic- and manual steering and hence to adapt to different field shapes that might be within a single farm or within the area of a contractor.

## 4.3   Economics of GNSS Based Guidance

The **investment** for guidance systems can vary considerably. A main factor for this is the respective investment for the georeferencing system (Table 3.4). Normally, manual lightbar guidance systems are not combined with real-time kinematic differential GPS. This is because typical drivers are not able to guide with an error that is below 10 cm on the average (Heraud and Lange 2009). Consequently, using a very accurate positioning system that is much more precise than this is not justified. Differential GPS that operates with carrier phase signals plus dual frequencies would suffice. Based on this, the investment for a manual lightbar guidance system can be around 10,000 Euros. On the other hand, automatic guidance systems can steer so precisely that the accuracy of real time kinematic differential GPS can be utilized fully. The investment for such an automatic guidance system based on RTK-GPS can be close to 40,000 Euros.

The *costs* from the investment should be offset by savings that result from more precise guidance. Theoretically, the basis for these savings could be the economical effects of fewer gaps between passes as well as of less overlapping of strips. In practice, however, savings of a result of fewer gaps between adjacent strips hardly occur. This is because farmers hate the visual impression that gaps make and therefore by and large prevent them by a narrower alignment of passes. A consequence is that instead of gaps more **overlapping** results. Therefore, an economical analysis can concentrate on the effects of less overlapping. These effects have an impact on either the costs of machinery plus labour or of the fertilizers and pesticides that are applied. Since overlapping reduces the capacity of the machinery plus the driver, it is appropriate to assess the overlapped area with the respective costs of the operations. Undue overlapping furthermore results in wasted fertilizer and pesticides, so these costs must be considered too.

The **relative overlapping** that occurs, when no GPS based guidance techniques are used, can be different. Bombien (2005) and Reckleben (2008 as well as 2011) calculated 8 % overlapping for operations, which are carried out without some visual aid for the driver such as cultivations or operations on grassland. When these operations were done either with the manual lightbar- or with the automatic GPS guidance method, the overlapping dropped to 4.4 % or to 0.96 % respectively.

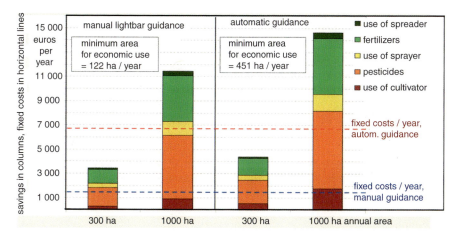

**Fig. 4.7** The columns show savings with small grains that can be realized by the reduction of overlappings when using GPS guidance methods for soil cultivation, for spreading of fertilizers and for spraying of pesticides. The operational savings for the use of the cultivator, spreader and sprayer include saved labour costs. The savings for fertilizers and pesticides are based on costs of respectively 115 and 160 Euros/ha. Any savings for seeds would be small and are not included. The *dashed horizontal lines* represent fixed costs of the guidance systems. It is obvious that the savings can be much higher than the fixed costs (From Reckleben 2008, altered)

On the other hand, the overlapping without using GPS guidance was much lower – only 4.20 % – when fertilizing- and spraying operations were carried out in **tramlines**. This cross-track error of 4.20 % includes inaccuracies that occurred when the tramlines were generated during sowing. However, the errors that result from inaccuracies in the course of tramlines often have a multiple effect. Because the tramlines are used several times during a growing period for spreading agrochemicals. Furthermore, also this error of 4.20 % was still significantly reduced when GPS guidance methods were used for all operations. With the manual lightbar method and carrier phase signals plus dual frequencies, the overlapping dropped to 0.92 %. And when the automatic guidance with RTK-GPS was used, it went down to 0.10 % (Bombien 2005; Reckleben 2008, 2011). For the respective cross-track errors in absolute- instead of relative data see Sect. 4.2.

The **savings** from less overlapping shown in Fig. 4.7 refer to operations for small grains in Germany. Those for fertilizing and spraying are based on driving in tramlines. The main savings come from reduced expenses for fertilizers and for pesticides and not from the improved use of the capacity of the machines (Fig. 4.7). Consequently, the savings will increase with the yield of crops and the respective use of agrochemicals. Any savings in expenses for seeds would be small and are not included. Yet apart from this, it is obvious that the savings from less overlapping can be much higher than the fixed costs from GPS guidance.

The general situation is that the manual lightbar system suits to medium sized farms and the automatic system to larger farms. Yet special possibilities that are associated with the precision of automatic guidance based on RTK-GPS and the repeatability of the same tracks after long time periods can be important too, *e.g.*:

- no-till sowing into the inter-row strips of the previons crop
- strip-tilling in autumn and sowing into the strips in spring
- hoeing up to a distance of 3 cm from plant rows
- applying chemicals in narrow bands precisely between narrow rows of plants
- precise operations for expensive crops as *e.g.* potatoes, beets, vegetables, asparagus *etc*.

Hence it may be that not only the use in traditional farming, but in addition such special fields of application will define the potential for modern guidance techniques in the future.

## 4.4 Problems and Solutions on Slopes

### 4.4.1 Tractors and Self-Propelled Implements

The techniques that were dealt with above work well on flat land. Yet a large portion of the global agricultural area is sloped. This makes precise guidance more difficult. Whenever soil erosion on sloped land is a problem, contour farming is essential as a corrective, even if this implies guiding along curved lines and inaccuracies that are associated with this (Sect. 4.1). However, this is not the only problem with guiding on slopes.

Generally, the GPS receiving antenna is installed on the roof of the tractor. So this is the location for which positions are received. On the one hand, the position on the roof allows rather unobstructed access to signals from the satellites. Yet on the other hand, this location is some distance above the point, at which the implements for most farm operations work.

In flat land, this vertical distance from the antenna to the control point on the soil is no problem. But on slopes, this distance results in additional errors of georeferencing (Fig. 4.8). When the antenna has a height of 3 m above the soil, 1° of side slope causes a lateral deviation from the target point on the soil surface of 5.2 cm. The deviation increases proportionally to the trigonometric tangent with the degrees of side slope. With a moderate slope gradient of 10°, the deviation is 52.1 cm. This is not trivial any more. Control algorithms that are based on the sensed slope have to correct this error.

Contour farming implies that the driving occurs as much as possible along lines of approximately the same elevation. With this driving pattern, therefore, the machines are mainly tilted sideways to the direction of travel. This side slope – often denoted as **roll** – is the main problem for which target point corrections are needed.

In a maritime climate and its drizzling rain, soil erosion is much less a problem and hence driving on the contour not essential. But if instead of contour farming

**Fig. 4.8** Shifting of the target point on side slope

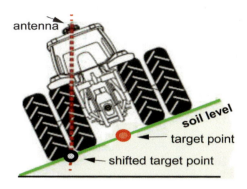

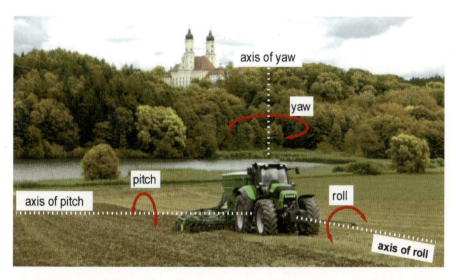

**Fig. 4.9** The three-dimensional position of a tractor on a slope depends on topography and heading. Topography defines the roll angle (side slope angle) and the pitch angle (down slope angle). Heading depends on the guidance and is defined by the yaw angle, which is the deviation between the path- and the tractor direction. This yaw angle may be close to zero on flat land, but along side slopes it may be not because the tractor crabs along the slope

fixed line guidance along straight paths is practiced, the slope directions – side slope or down slope – can vary. As a result, not only roll (side slope), but **pitch** (down slope) and yaw of the tractor as well may need attention (Fig. 4.9). **Yaw** is the heading error of the tractor that may result from crabbing and slipping, especially on side hills. Because of this, a deviation between the direction of the planned path and the heading position of the tractor can exist.

Roll, pitch and yaw can be measured via GPS in order to adapt the guidance. In case mainly the effects of **roll** are a problem – as with contour farming (Fig. 4.8) – the remedy is rather simple. If two antennas are mounted on the roof – one on the left side and the other one on the right side of the driving direction – the roll can be indicated via their relative height. And by simple trigonometric calculations concerning the target point, the path can be corrected. The **yaw** of the tractor can be obtained via the relative forward positions of these two antennas. Finally, if three antennas are used – with the third antenna mounted in front of the other two – it is possible to indicate **pitch**. Adapting the steering control to the positional data requires suitable algorithms (Heraud and Lange 2009). Yet correcting the guidance on slopes on the basis of several antennas is state of the art.

The use of GNSS guidance sometimes is supplemented by signals from **inertial sensors**. These inertial sensors measure either linear- or rotational accelerations and hence can also indicate changes in positions of vehicles (Yazdi et al. 1998). They are very tiny devices and used in huge quantities for a variety of applications, *e.g.* in automobiles for airbag employment and for dynamic stability control. When used adequately, these inertial sensors can measure translations as well as rotations in three spatial dimensions, hence with six degrees of freedom along and around the axes of roll, pitch and yaw.

Despite these abilities and their low cost, normally inertial sensors are not used to completely replace guidance by GNSS. Instead there are commercial solutions that rely on several GPS antennas plus inertial sensors for terrain compensation.

Because inherently, inertial sensors are not able to steer precisely for a long time period along fixed lines, let alone to guide along paths that were used months or years ago. This inability of inertial sensors results from the operating principle: they rely totally on **motion memory**. Their basis is that when a vehicle starts from a known position and then moves in a known direction for a known time, the final position is known. But this motion memory inherently includes the accumulation of errors along its path, often denoted as drift. Whereas with GNSS guidance, the errors that arise are corrected – constantly and automatically – by new signals that are received from satellites and base stations, with inertial sensors this does not occur. The present recommendation therefore is to use the signals from inertial sensors not primarily for controlling the **heading** or the **yaw** of vehicles, for which properly used GPS signals are superior.

The situation is different when it comes to controlling the effects of the **attitude** that are defined by roll and pitch. These attitude errors hardly can accumulate, since they do not occur in the direction of the path or of the heading of the tractor. Hence in controlling the effects of roll and pitch there is a choice: this could be on the basis of either additional GPS antennas or of inertial sensors. Li et al. (2009) as well as Kellar et al. (2008) got about the same accuracies in guidance of vehicles when either using GPS with multiple antennas or – as an alternative – employing inertial sensors for the control of roll and pitch in combination with a single antenna GPS for the control of yaw. They mention that the expenses for the latter alternative are lower.

However – even for the same positional dimension – there are reasons for combining GPS receivers and inertial sensors. GPS indications that are received have a rather long-term stability, but its signals can be unavailable or blocked by trees, buildings or a mask angle (Sect. 3.7). If both systems are combined for a fused navigation control, it is possible to bridge short-term disruptions of GPS signals by a control via inertial sensing.

## 4.4.2 Drawn- and Mounted Implements

The implement is the actual device doing the field work. But on a side-slope, it is possible for the tractor to follow the desired line, but have the implement widely off the path. This results mainly from side forces that act on the implement and drag it downhill. And crabbing of the tractor along a side-slope can have an effect too. The downhill drift on the implement occurs with towed machines as well as with those that are mounted on the three-point-hitch of the tractor. However, towed implements are particularly susceptible.

Corrections for this downhill drift can be made based on GPS signals from an antenna that is located on the implement itself. However, relying solely on signals that come from the implement without having an antenna on the tractor is not recommended. The reason for this is that then the driver would have to spend much effort for keeping the heading of the tractor in the vicinity of the path. Hence the signals from the GPS-implement-antenna are used in combination with the respective signals for the tractor.

The systems for correcting the downhill drift rely on either passive- or active steering of the implement. **Passive guidance** corrections of the implement are based on compensations via the heading of the tractor by steering somewhat away from the path in order to make up for the drift of the implement. Instead, **active guidance** corrections are made by steering devices on the implement itself. The result is that with active guidance both the tractor and the implement follow the desired path, whereas with passive steering this holds only for the implement, while the tractor is on a line that is slightly offset (Fig. 4.10).

This offset-position of the tractor can be a problem on steeper slopes, which cause more offset-distance. This can make driving and operating in inter-row strips impossible, *e.g.* of potatoes, beets, vegetables and maize. Active guidance does away with these problems, but in return requires steering tools on the implement. These steering tools vary as much as the implements do. Common devices used for this are

- steerable tires
- laterally movable hitches
- steerable frames
- steerable coulters (Fig. 4.11).

The assumption that implements on a **three point hitch** do not drift downwards does not hold well. The position of these implements is approximately parallel to the rear axle of the tractor. This means that the implement "tails out" when the attitude of the tractor changes. Tractors tend to a crab-attitude with a corresponding

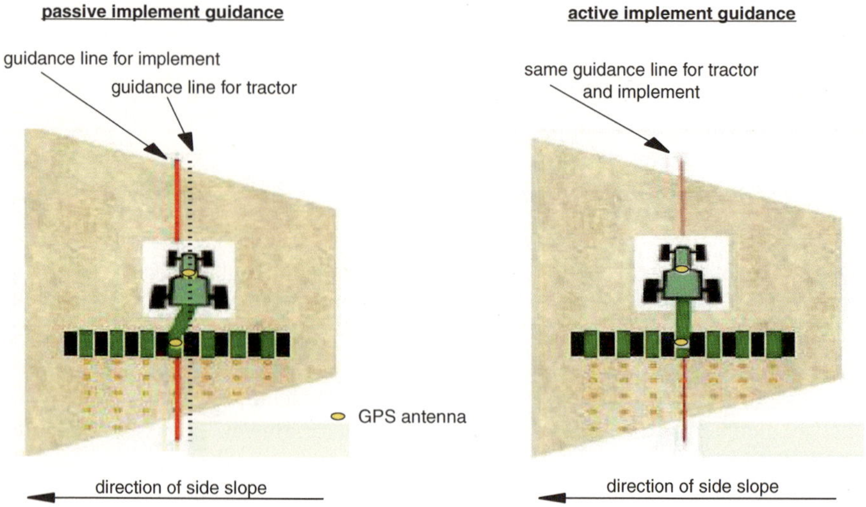

## passive implement guidance / active implement guidance

**Fig. 4.10** Passive- and active guidance of towed implement by GPS (From Deere & Co., altered)

**Fig. 4.11** Active guidance of a seeding machine on a side-slope. The GPS antenna on the roof of the tractor is hidden (From Orthman Mfg.Co., altered)

yaw angle in the direction of travel (Fig. 4.9) on side-slopes. The yaw angle varies with the side-slope, and hence with it the position of the implement. Active implement guidance that relies on GPS can prevent this uncontrolled "**tailing out**". This is achieved by GPS controlled lengthening or shortening of the respective lower arm of the three point linkage. Furthermore, devices as listed above sometimes are used for implements on three point hitches as well.

In short, solutions for precise guidance on slopes exist. The most important decision on slopes is about the best driving pattern and its effect on soil erosion.

## 4.5  Precision in Section Control of Farm Machines

The operating width of present day farm machines and irregular field shapes causes situations where overlapping cannot be avoided. But if overlapping of machinery parts cannot be prevented, twice sowing, -fertilizing and -spraying the same area should not take place. It results in waste of seeds as well as of agrochemicals and reduces yields. Section control of farm machines avoids these disadvantages.

Basis of modern section control is that all field boundaries, previously treated areas, not cultivated regions and grassed waterways within fields are geoereferenced. This prerequisite holds for all automatic guidance systems that rely on GPS. The sections refer to either **single units** of the respective machinery – *i.e.* planter units, nozzles on the sprayer boom and outlets of pneumatic spreaders – or alternatively to **several units** respectively that are grouped together. Each section has an actuating- and control device that permits switching on and off by a central controller, which has the field data and communicates with the GPS receiver. When properly used, these automatic techniques allow for precisions in irregular fields as outlined and shown in Fig. 4.12.

An economical use of section control techniques calls for frequent applications in irregularly shaped fields. The savings in seeds and agrochemicals can be as much as 25 % in very oddly-shaped fields and be almost zero in rectangular land (Stombaugh et al. 2009).

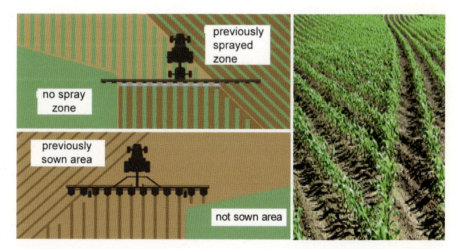

**Fig. 4.12** Schematics of section control for sowing and spraying in irregular fields (*left*, from Raven Industries Inc., altered) and a result with maize (*right*, from John Deere & Co.)

# References

Bombien M (2005) Comparison of pass by pass driving systems (in German). Rationalisierungs-Kuratorium für Landwirtschaft (RKL), Rendsburg, No. 4.1.0, pp 1203–1224

Heraud JA, Lange AF (2009) Agricultural automatic vehicle guidance from horses to GPS: how we got there, and where we are going. In: Distinguished lecture series, agricultural equipment technology conference, Louisville, 9–12 Feb 2009. ASABE, St. Joseph

Kellar W, Roberts P, Zelzer O (2008) A self calibrating attitude determination system for precision farming using multiple low-cost complementary sensors. In: Ingensand H, Stempfhuber W (eds) Proceedings of the international conference on machine control and guidance, Zurich, 24–26 June 2008

Li Y, Efatmaneshnik M, Cole A (2009) Performance evaluation of AHRS Kalman filter for MOJORTK system. In: Symposium of international global navigation Satellite Systems Society, Surfers Paradise, Queensland, 1–3 Dec 2009

Reckleben Y (2008) Electronics in agricultural engineering – example of parallel driving systems (in German). In: ISOBUS "Plugfest" Landberatung Uelzen, 6 Nov 2008. www.isobus-fuer-alle.de/.../pdf/vortrag_prof_reckleben.pdf

Reckleben Y (2011) Assistance with automatic guidance (in German). A step to more precise sowing. In: Bauernblatt für Schleswig-Holstein und Hamburg, 27 Aug 2011, pp 40–41

Stombaugh TS, Zandonadi RS, Dillon CR (2009) Assessing the potential of automatic section control. In: van Henten EJ, Goense D, Lokhorst C (eds) Precision agriculture '09. Papers presented at the 7th European conference on precision agriculture, Wageningen, 6–8 July 2009. Wageningen Academic Publishers, Wageningen, pp 759–766

UK Agriculture (2013) http://www.ukagriculture.com/multimedia/farming pictures.cfm

USDA NRCS (2013) http://www.nrcs.usda.gov/wps/portal/nrcs/main/national/newsroom/

Yazdi N, Ayazi F, Najafi K (1998) Micromachined inertial sensors. Proc IEEE 86(8):1640–1659

# Chapter 5
# Sensing of Natural Soil Properties

Hermann J. Heege

**Abstract** Site-specific sensing of varying natural soil properties is a prerequisite for an adequate control of many field operations.

Topography can be mapped rather easily as a byproduct of other farming operations by means of RTK-GPS. Information about clay, moisture and salinity of soils in a combined mode can be obtained via electric conductivity sensing. In humid areas, salinity can be left out. So here the electric conductivity is defined mainly by a combination of clay- and water content of the soil. The combined effect of these factors is well related to the yield potential of soils. Hence in humid regions, electric conductivity sensing can supply information that is needed for the control of farm operations according to yield expectations.

Electric conductivity sensing is based on soil volumes that may include the topsoil as well as the subsoil. In contrast to this, the reflectance of visible or infrared light senses only soil surfaces and thus may be less representative. Yet reflectance sensing might supply signals simultaneously about several soil properties such as texture, carbon content, cation-exchange-capacity and water content.

**Keywords** Capacitance • Electrical-conductivity • Permittivity • Reflectance • Surface-sensing • Topography • Volume-sensing

H.J. Heege (✉)
Department of Agricultural Systems Engineering, University of Kiel,
24098 Kiel, Germany
e-mail: hheege@ilv.uni-kiel.de

## 5.1 Sensing of Topography

Soil properties depend to a large extent on nature, yet partly also on human activities. Both nature as well as human activities can result in spatial variations of soil properties that should be taken into account for site-specific farming. This chapter deals with properties that mainly depend on nature such as

- topography
- texture
- organic matter content
- cation-exchange-capacity
- water content
- salinity.

Soil properties that in modern farming predominantly depend on human activities, such as the supply with nutrients, are dealt with in later chapters.

**Topography** affects farming in many aspects. Its long-term influence on run-off of water and thus on erosion results in distinct differentiation of soil qualities between uphill- and downhill locations. Short-term effects come from the fact that the inclination of fields to the sun influences the temperature of the soil. The less oblique the solar radiation hits the soil surface, the more energy is transferred per unit area and hence the higher the soil temperature is. This explains why generally fields with slope aspects that are oriented to the South are preferred in most areas of the Northern latitudes of the sphere. It is *vice versa* in areas of the Southern latitudes, here fields that are oriented to the North are more valuable in most cases.

The resulting effect of **slope orientation** on crop growth can be vast. In some areas of the Northern hemisphere, wine is only grown on slopes that are oriented to the South. Even with small cereals, the effect of slope orientation on yield can be significant. Studies of Geary (2003) with a CERES wheat model show a loss in grain yield of 1 t/ha on a slope of 10 % oriented to the North in England.

Implications for precision farming come from the influence of topography on soil qualities, on water run-off and on yield potential. When the yield potential of a field changes as a result of varying field inclinations as well as of slope orientations, both the economy and the environment ask for adapting the input of agrochemicals to this. Site-specific operations in fertilizing and crop-protection can provide for that. Variations in soil qualities, in water run-off and thus on the prerequisites for erosion within a field call for site-specific responses in cultivation intensities. Details to respective responses are dealt with in Sect 7.2.

Yet reliable site-specific data about the relief of fields are prerequisites for any responses to topography. Traditionally, these data for relief- or contour maps have been obtained via conventional manual techniques by using theodolites and level surveying. Modern techniques for sensing and recording of topography are widely automated and include sensing methods such as

- **radar interferometry** (comparing phases as well as amplitudes of outgoing- and reflected satellite radar radiation)

- **laser light** of the ultraviolet, visible or infrared range from satellites or from aerial platforms and its reflection (transit-time)
- **inertial georeferencing** by recording linear- and rotational accelerations on a moving vehicle (see Sect. 4.4.1)
- **real-time kinematic georeferencing** via Global Navigation Satellite Systems (RTK-GPS).

With adequate processing, all methods can record modules that are used in **digital elevation models** (DEM) in order to provide for contour- or topographic maps. These can then be used to control farm operations. Such maps can be obtained in some countries or areas from geological institutions. However, often the maps from geological institutions do not provide the resolution that site-specific farming operations require, and the layout may not correspond to the respective field sizes.

Probably the best method for most cases is to rely on topographic maps that are created via the georeferencing system that precision farming needs anyway. For many farms in the future, this will be RTK-GPS or at least differential GPS based on carrier phase signals and dual frequencies (Table 3.4), sometimes in combination with inertial georeferencing on slopes. This procedure of relying on the respective **farm-specific georeferencing method** does not require separate trips through all fields. Instead, the topographic maps can be obtained as a **byproduct** when driving through the fields for other purposes, which principally can be any field operation. However, in order to cover the site-specific situation within fields well and to obtain a good resolution, it might be reasonable to omit operations that are performed with very wide widths, *e.g.* spraying.

The precision that is obtained can be enhanced by repeated recording of georeferenced topography in the same field and **averaging** the results. This prospect is challenging and promising since with an adequate software this farm-specific georeferencing of the topography can be a process that goes on rather casually and automatically along with field operations (Westphalen et al. 2004). Results about the effect of such a repeated recording of topography on the precision are presented in Fig. 5.1. The standard deviation of the error that occurred when the elevation was recorded by RTK-GPS only once in a field operation was between 12 and 20 cm. This result was halved to between 6 and 10 cm when the records from four field operations were averaged. The respective spans in the results were due to different speeds of the operations. A halving of the standard deviation was also obtained when during one field operation, both RTK-GPS and inertial sensing were used simultaneously and the results were combined. Recording the elevation in a "stop and go" mode instead of an "on-the-go" method did not change the results much.

However, a question is how with repeated recordings of the elevation the averaging should be done. An easy and obvious approach is to use **arithmetic averages** of the elevation data per grid element within the field. The results for the repeated measurements in Fig. 5.1 are based on such a procedure. An alternative to this procedure is the use of **weighted averages** per grid element. The theoretical background for this is the concept that the data from one field operation may contain more errors than those from another run. But on which basis should such weighted averages be calculated?

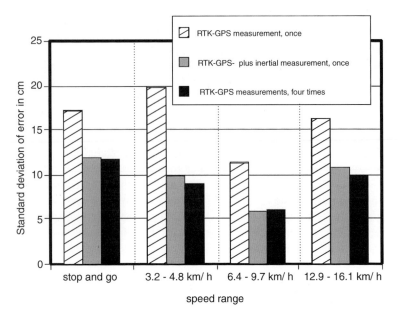

**Fig. 5.1** Standard deviation of error when recording digital elevation models (From Westphalen et al. 2004, altered)

Abd Aziz et al. (2009) concluded that from each field operation the intermediate results of the subsequent krigings (see Sect. 2.4) can provide an indication for the weighing of the data. The kriging – if it occurs for blocks or grid elements – is associated with a variance of the respective data per run within each grid element. This implies that several data per block or grid element accumulate. The smaller the variance of the data within a grid element is, the better the respective elevation estimate is supported by the measured data and *vice versa*. Consequently, the averaging function weighted the elevation estimate based on the variance within the respective grid element. For details to this see Abd Aziz et al. (2009).

This processing of the data by weighted averaging instead of simple averaging reduces the errors in digital elevation models (DEMs), especially when a higher number of surveys is combined (Fig. 5.2). Since such topographic elevation maps can be generated as a byproduct, including a higher number of surveys hardly affects the costs.

Among the various methods that principally are available for generating digital elevation models or topographic maps, **RTK-GPS georeferencing** can be regarded as a favourite, possibly in combination with inertial georeferencing. This technology is needed anyway for guidance- or control purposes in precision farming. So investments on additional hardware are not needed, solely adequate software must be obtained. This procedure makes it possible to get digital elevation models and maps that precisely have the resolution and layout that fits to individual fields. These digital elevation models or maps can be used for several decades, since the topography of fields hardly changes over time.

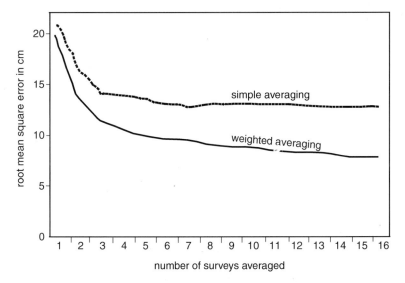

**Fig. 5.2** Errors with averaging methods used for field elevation data. Effect of combining repeated elevation surveys of RTK-GPS by either simple averaging or by weighted averaging of the error data (From Abd Aziz et al. 2009, altered and simplified)

## 5.2 Sensing of Soil Properties on a Volume Basis

Natural soil properties such as texture, water, organic matter and salinity influence the yield potential. Georeferenced knowledge about these properties can assist in pinpointing the site-specific control of cultivation-, sowing-, fertilizing- and crop protection operations.

Spatial variations within fields are frequent for all these soil properties. Compared to this, **temporal variations** are important only for soil moisture. This has implications for mapping practices. Maps about texture and organic matter content are useful and once properly created can be up to date to assist farming for decades. But because of the varying weather, maps about the site-specific water content can be obsolete within a few days and in rain-fed areas therefore seldom are of interest except for research purposes. However, the situation for soil water is completely different when online and on-the-go control of farm operations and hence instant reaction is possible. This control can be very useful for site-specific adjustments of irrigation operations, of the depth of cultivation or the depth of sowing.

In the past, the **spatial resolutions** in the recording of the data were insufficient. The costs of manual soil sampling in the field and subsequent expensive analysing in laboratories for texture, organic matter and water did not allow for high resolutions. Often data from only one site per ha were analysed, which in most cases did not at all provide a reliable basis for precision in site-specific operations. Handheld measuring instruments can save sending the samples to laboratories, yet for high spatial resolutions still imply laborious manual sampling procedures. Either online and on-the-go procedures for proximal sensing or alternatively remote sensing methods are needed, at least for site-specific farming on large areas.

The sensing concepts that are available for this rely almost exclusively on **electromagnetic radiation** or on **electric current**. The wide spectrum of electromagnetic radiation provides for a huge variety of sensing alternatives. The interest is focussed on using visible as well as infrared light, micro- or radarwaves and finally also on electric current as indicators for soil properties. It is important to realize that in all cases no direct indications of soil properties are possible. The indications always are indirect phenomena. The respective soil properties can at best influence details of the radiation or current such as amplitude, frequency, speed of propagation *etc*. Via changes in these details it is possible to make deductions or inferences for soil properties. In short, the estimations of the soil properties are based on correlations to physical details of electromagnetic radiation or electric current. These correlations allow to use intermediates that can easily be sensed or recorded online and hence applied for a control.

It is very important to realize for which part of the soil the sensing should take place. Should the information about the respective property be obtained on a soil **volume basis** or on a soil **area basis**? The interest in most cases is focussed on a soil volume basis. The crops rely on texture, organic matter and water for a soil depth that approximately equals the vertical root length. Hence the respective properties in the topsoil as well as in an adjacent part of the subsoil count. So if the maximal vertical root length is about 1 m, it might be appropriate to get site-specific signal averages that are related to this soil layer.

But there are exceptions for which such averaging procedures do not comply to the needs. A special case for which such coarse averaging does not fit is when sensing **water in the seedbed zone** solely is required. In this special situation it might be reasonable to sense at which depth a dry topsoil zone ends and a subjacent moist zone precisely begins, where then the seeds should be placed. Hence in this special case the objective would be more a method of water sensing on a two dimensional basis. Or if the sensor scans along a **vertical cross section** within the soil, it could be defined as **sensing for a water line**. In addition, there may be reasons for controlling the application of soil-herbicides according to the organic matter content of the soil just at its surface, hence also for sensing this soil property on an area basis. There may be additional cases where sensing on an area basis or on a line basis conform to the needs.

If sensing properties on a volume basis is the objective, getting the signals from the surface would suffice only if the soil constituents were uniformly distributed within the volume. This usually is the case for the topsoil that is cultivated and thus mixed. But this does not hold for the subsoil. So if information about soil properties is needed within the layer that the roots of the crops penetrate, signals that are obtained from the soil surface alone do not suffice. One- or two dimensional sensing of soil water or soil organic matter can be reasonable for the special cases of controlling the sowing depth or the application of soil herbicides as explained above. This does not alter the general need for sensing texture, organic matter and water on a three-dimensional or on a volume basis.

**Table 5.1** Methods of sensing soil properties on volume bases via electricity or via radiation

| Frequencies | Wavelengths[a] | Sensing objectives | On-the-go use |
|---|---|---|---|
| **Electrical conductivity, contact methods** | | | |
| 0–1 kHz | Infinite – 300 km | Texture, water, salinity, soil-layers | State of the art |
| **Electrical conductivity, electromagnetic induction methods** | | | |
| 0.4–40 kHz | 750–7.5 km | Texture, water, salinity, soil-layers | State of the art |
| **Electrical capacitance** | | | |
| 40–175 MHz | 790–200 cm | Water | Possible |
| **Time domain reflectometry** | | | |
| 50–5,000 MHz | 600–6 cm | Water | Not yet possible |
| **Soil penetrating radar, surface reflection mode** | | | |
| 0.5–30 GHz | 0.60–1 cm | Water | Possible |
| **Micro- or radarwaves, mainly satellite based sensing** | | | |
| 0.3–30 GHz | 100–1 cm | Water, roughness of soil surface | Does not apply |

Compiled from data by Allred et al. (2008); Corwin (2008); Lesch et al. (2005) and Lueck et al. (2009)
[a]From higher range- to lower range limits, thus corresponding to frequencies

This distinction between volume sensing on the one hand and surface- or area sensing on the other hand is necessary since not all electromagnetic waves are able to penetrate the soil. Especially visible and infrared light only provide signals that are based on the soil surface that was hit. The situation is different for microwaves, radar waves as well as for electric currents and its electromagnetic waves. Some criteria that refer to **volume sensing techniques** are presented in Table 5.1.

Among the alternatives listed there are two techniques that can operate online as well as on-the-go and already have been introduced widely into practical farming: **electrical conductivity**- and **electromagnetic induction** sensing. These methods operate either with direct current or with alternating current on the lowest frequency end and consequently with long waves. The soil properties that can be derived from the signals of these methods are not without ambiguity. This will be dealt with later. **Electrical capacitance** sensing operates in medium ranges of frequencies and wavelengths (Table 5.1) and depends on the soil water content. This method can be used online and on-the-go as well, but up to now seldom is state of the art in practical farming. **Time domain reflectometry** is based on radar- or radio frequency signals that are guided along a transmission line or cable that is embedded in the soil. Therefore the denotion often is "cable radar". The velocity of wave propagation depends on dielectric soil properties and thus can indicate the soil water content, but not yet for on-the-go operations. The situation is different if **soil penetrating radar** is used in a surface reflection mode for water sensing. This method in principle is suited for on-the-go sensing, but despite this up to now hardly is used in farming. The same applies to soil water sensing by **micro- or radarwaves from satellites**, a method that at present can be excluded for online and on-the-go control of farm operations, but might become very useful for tactical inspections of wide areas.

## 5.2.1 Methods for Sensing of Electrical Conductivities

Electrical conductivity is a measure of the ease with which an electric current flows through a substance, in this case through soil. It is indicated in units of Siemens per m (S/m). Occasionally the electrical resistivity – the reciprocal value – is used instead of the conductivity.

The electric current is introduced into the soil either by direct **galvanic contact** or by **electromagnetic induction** between the measuring instrument and the ground. Hence the sensing occurs either in an intrusive or in a non-intrusive way. In both cases the reactions of soils to electricity are sensed.

Soil is a very heterogeneous matter for an electric current since it consists of solids, gases and liquids. All these components can vary immensely. The solids include both mineral- and organic matter. If rocks and unbound organic matter are excluded, they can be broken down by particle diameters into sand (2.00–0.05 mm), silt (0.05–0.002 mm) and clay (less than 0.002 mm). Sand is primarily quartz and – if dry – can be considered as an electrical insulator. The clay size fraction is made up not only of clay minerals, but in addition of organic matter that is bonded to the minerals. These clay-humus bonds contribute considerably to current flow in soils, especially under wet conditions. Silt has an intermediate position. While the air in the soil too is a good insulator, the liquids can be regarded as an electrolytic aqueous solution with ions that are dissolved in it. The ions in the liquids as well as on the surface of clay-humus bonds are mainly responsible for current flows in soils.

#### 5.2.1.1 Methods Based on Galvanic Contact with the Soil

Theoretically, either direct current or alternating current up to a frequency of 1 kHz can be used (Table 5.1). *Ceteris paribus*, direct current senses deeper. But the electrodes that introduce the direct current into the soil can get polarized by ions and thus can loose electrical contact. This problem is alleviated by employing low frequency alternating current (Allred et al. 2008), which some geologists still denote as direct current. The present commercial implements mostly run on alternating current with frequencies between 150 and 220 Hz (Lueck et al. 2009).

The sensing process is a rather simple procedure. The current flow occurs between a rolling coulter at the left and right side of the machine (Fig. 5.3). The conductivity of the soil is sensed by one or more pairs of voltage coulters that roll between the current coulters.

A principally still simpler procedure would be to use only two electrodes for sensing both the current flow and the voltage between them. Although this configuration can be used, it is more unstable (Corwin 2008). The method of using an outer pair of current electrodes and at least one pair of separate voltage electrodes between them that provides the data about the soil properties goes back to Wenner (1915). It is consequently denoted as a "**Wenner array**". This array has proven to supply more reliable results.

The sensing implement can be pulled by a vehicle with speeds up to 15 km/h. The distance between measurement passes should be adapted to local soil variations, it usually ranges from 6 to 20 m. Consequently between 60 and 200 ha can be

# 5 Sensing of Natural Soil Properties

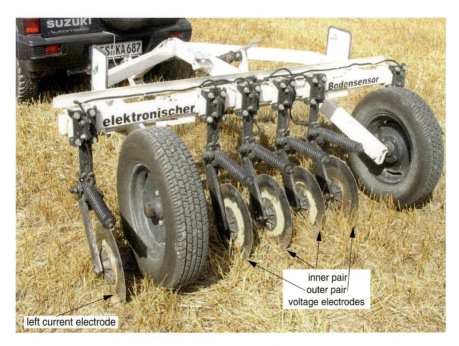

**Fig. 5.3** Online and on-the-go sensing of soil electrical conductivity by a contact method, System Veris. The right current electrode is concealed by the wheel (Photo from Lorenz, Lufa Nord-West, Oldenburg, Germany, altered and supplemented)

sensed, georeferenced, logged and mapped per day. About 120 readings are obtained per ha, so the spatial resolution is very much better than with conventional soil testing methods. Yet soils that are frozen, very dry, stony or covered with much residues can prevent the application.

The volume of soil that is sensed with the Wenner array can be adjusted. It includes all the soil between the respective pair of voltage electrodes from the soil surface to a depth that equals approximately the horizontal distance between the voltage electrodes. Thus, taking the signals from the outer pair of electrodes instead of the inner pair (Fig. 5.3) allows increasing the depth of sensing.

## 5.2.1.2 Methods Based on Electromagnetic Induction

Electromagnetic induction occurs when a magnetic field crosses a conductor or *vice versa*. In this case, the soil is the conductor. The implement that generates the primary magnetic field just is moved at a defined distance above the soil. Travel speed and area capacity can be about the same as with contact methods (see previous section).

However, whereas the contact methods might use current that can be quasi or almost direct current (see above), electromagnetic induction methods rely on alternating current with a frequency well in the kilohertz range (Table 5.1). This is because the induction process needs alternating current.

**Fig. 5.4** Online and on-the-go sensing of soil electrical conductivity by electromagnetic induction. The sensing instrument (EM 38 of Geonics LTD, Canada) can be moved on a sled, on a cart or even be carried by hand. In order to prevent interference from metal, some distance to the vehicle is needed (Photo from Agri Con GmbH, Jahna, Germany, altered)

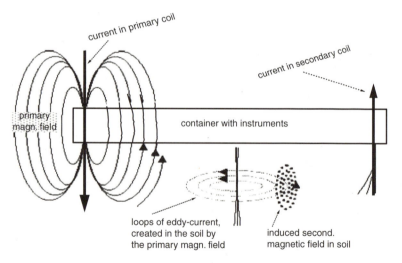

**Fig. 5.5** Operating principle of soil sensing by electromagnetic induction (From Lesch et al. 2005, altered and supplemented)

The sensing implement (Figs. 5.4 and 5.5) has two wire coils, a primary- or transmitter coil plus a secondary- or receiver coil. The current that flows through the transmitter coil produces a magnetic field around it. This magnetic field also extends into the soil and thus produces there a **primary magnetic field**. While the implement moves, this primary magnetic field induces an eddy current within the soil. This eddy current in turn causes a **secondary magnetic field**. Finally, both magnetic fields induce currents in the second wire coil. This current varies with the soil properties. The measurement units are the same as with conductivity methods that rely on direct contact, namely Siemens per m (S/m). So actually, conductivity units are sensed, but the conductivity is created by electromagnetic induction.

Because of the indirect measuring approach, the induction sensing is more difficult to calibrate than contact methods that intrude the soil. The respective adjustments in the field need more time (Sudduth et al. 2003). On the other hand, the soil penetrating methods rely on good electrical contact between the coulters and the soil, which can be a problem on dry or stony soils. This problem does not exist with induction methods.

### 5.2.1.3 Depth of Sensing and Soil Layers

The sensing depth should fit to the maximal vertical soil penetration by the roots. For many crops, this vertical penetration by the roots is in the range of 70–150 cm. The penetration of the conductivity sensing can in a simplification be perceived as a vertical cross section that starts at the soil surface. It is not precisely known whether this *vertical cross section* perpendicular to the direction of travel should be oriented at a rectangle – whose vertical side equals the root penetration depth – or whether such a schematic approach is too fussy. Since the root density with many crops decreases beyond a medium depth, it might be reasonable to aim for a sensing density that too gradually tapers off with depth. The present sensing methods actually follow this approach.

Principally, several possibilities exist for increasing the depth of sensing. For a contact method (Fig. 5.3) this can be done by

- extending the distance between the voltage electrodes (Sect. 5.2.1.1)
- using lower frequencies.

With electromagnetic induction, an increase in sensing depth can be obtained by

- changing the coil orientation from a horizontal mode to a vertical mode
- lowering the height of the sensing implement above the soil
- increasing the lateral distance between the coils
- using lower frequencies.

Most sensing implements that are used commercially at present allow only for one or two of these adjustments. The contact sensing method of Fig. 5.3 is presently widely applied in the USA. It permits only to choose or alternate between two different distances of the voltage electrodes. The electromagnetic induction method of Fig. 5.4 is dominating in Europe and Canada. It allows for changing the coil orientations and for adjusting the height of the implement above the soil.

The measured reading for a soil layer of a given conductivity depends on the distance from this layer to the instrument. An important point is how in detail the vertical distances to the soil affect the results. **Response curves** that compare the presently used techniques under *ceteris paribus* conditions show the depth effects (Fig. 5.6).

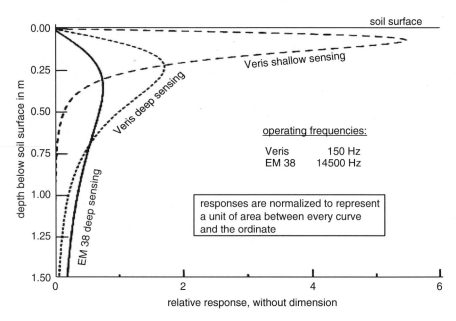

**Fig. 5.6** Relative response of conductivity sensors as a function of depth. For details to the implements Veris and EM 38 see Figs. 5.3 and 5.4 (From Sudduth et al. 2005, altered)

The objective should be to have a depth weighted sensed area that approximately incorporates the maximal root expansion during the growing season of all crops within a rotation. Since principally several possibilities exist for adjusting the sensed depth, it should at least roughly be tried to get the sensed vertical cross-section adapted to this. However, going beyond the maximal root depth probably is less disadvantageous than falling behind it, since during dry periods some water is sucked from deeper horizons to the roots.

Important is also that sensing along a defined response curve pretends a uniform soil electrical conductivity within the respective vertical cross section for the whole field because the sensed signals are **integrals**. A soil with uniform properties within the vertical cross section on the one hand and another one that is composed of different layers within the sensed section on the other hand can result in the same integrated signal.

So sensing along a single defined response curve is a sensible and target-oriented procedure only with soils that have uniform properties within the rooted depth. This method can be misleading with **layered soils** since no depth resolution is delivered. And there are many soils that have special layers or horizons within the subsoil. One the one hand, these layers may prevent the drainage of water when they consist of dense claypans or similar gleysolic, hydromorphic horizons. On the other hand, there may exist layers that let seep the water too fast and hence do not store enough of it because they are made up from coarse sand and gravel. And all layers may not be positioned parallel to the soil surface. So in a three-dimensional analysis, there can be a variety of different soil conditions.

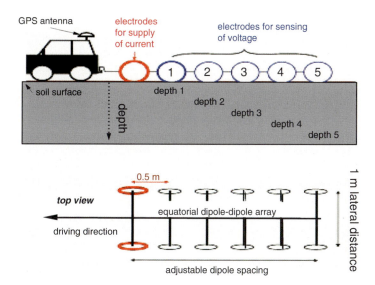

**Fig. 5.7** Sensing electrical conductivity in a contact mode by the Geophylus electricus via rolling electrodes that are spaced in the direction of travel (From Lueck et al. 2009, altered)

A response to this problem for layered soils is successive or simultaneous sensing along response curves that have different **gradients with depth**. It has been shown (Mertens et al. 2008; Saey et al. 2009; Sudduth et al. 2010) that special software and differences existing in the response curves of the present sensing implements allow for a rough delineation of soil layers. In case the soil within a field consists of only two layers with distinct differences in electrical conductivities, this vertical horizon sensing can be quite successful.

A special data processing technique has been developed – called **inversion of electrical conductivity** measurements – that aims at mapping special layers or horizons. This technique is not simple since the sensed volumes of different response curves (Fig. 5.6) overlap and because it must be prevented that different combinations of layers with their respective conductivities generate the same final result (Gebbers et al. 2007; Sudduth et al. 2013). It is obvious that the precision in the detection of soil layers can be enhanced if signals from more than two response curves within the respective soil depth can be obtained.

The "**Geophilus electricus**" – a development of the University of Potsdam and the Technical University of Berlin – delivers signals along five response curves or even more simultaneously on-the-go (Lueck et al. 2009; Lueck and Ruehlmann 2013; Radic 2008). The implement is based on rolling electrodes, so it operates in the contact mode as the instrument in Fig. 5.3. The rolling electrodes are spiked at the circumference. However, the current electrodes on the one hand and the voltage electrodes on the other hand are not arranged on one axis perpendicular to the direction of travel (Fig. 5.3), but instead they run in pairs that are separated in the direction of travel (Fig. 5.7). Five voltage electrode pairs – spaced at successive distances in the direction of travel behind the current electrode pair – independently sense electrical conductivity. With increasing distance between the current pair of electrodes and a respective voltage pair, signals from deeper soil are delivered. The increased number of response curves allows a more accurate resolution of depth signals. Hence principally the prerequisites for precise three-dimensional recording and mapping of electrical conductivities via inversed signals are better.

Further enhancing of the depth resolution is possible by varying **electrical frequencies** within the range of 1 mHz to 1 kHz and thus increasing the depth of sensing by lower frequencies or *vice versa*. Four different frequencies can be sensed simultaneously, which multiplied with five voltage electrode spacings allow to record 20 response curves almost simultaneously on-the-go. And when operating at the higher frequency ranges even more than 20 response curves can be obtained. This is because current with higher frequencies is at least partly transferred by induction via its magnetic field and not solely via conduction. And the transfer via induction instead of conduction causes phase shifts in alternating current. Since these phase shifts diminish the power that is transmitted, they too can be recorded and allow to obtain additional response curves.

Operating modes such as travel speed, field capacity and number of local readings per ha come up to those mentioned in the latter part of Sect. 5.2.1.1.

This method is not yet used commercially and more experimental experience might be necessary (Lueck et al. 2009). And more response curves alone do not alter the fact that there still is overlapping for all response curves near the soil surface. This is because all response curves – even if these successively go deeper and hence differ in shape – still originate at the soil surface similar to the gradients shown in Fig. 5.6 from other sensing instruments. By elaborate post-processing that involves **inversion** of all electrical conductivity measurements, mapping of rather thin layers or horizons might be possible – as mentioned above (Gebbers et al. 2007). In the future, powerful computers might even allow to do this on-the-go.

Another approach for a more detailed depth resolution and hence better sensing of soil horizons is based on **varying the coil orientation** of the electromagnetic induction system (Figs. 5.4 and 5.5) continuously on-the-go between the vertical and the horizontal position (Adamchuk et al. 2011). Hence the vertical horizon is scanned in modes that alternate between deep and shallow readings. This concept allows to sense in varying depths with a very compact implement. Postprocessing of the response curves via inversion procedures here too is necessary.

The prospects of thus sensing soils in three dimensions with a high resolution deserve attention. It is not the layers or horizons of soils alone that are of interest. Equally important is what happens to the water that is moving through the soil. The water travels through the soil to water tables on top of the saturated soil zone and from there into rivers and might carry nutrients from mineral- or organic fertilizers as well as even pesticides with it (Schepers 2008). Hence there is increasing concern about the **preferential water flow** routes that bypass most of the soil and can be regarded as traffic lanes for an unwanted transport of these components into the environment. Soil electrical conductivity is related to the water content and thus principally also to preferential water flow routes. However, there exist substantial

**Table 5.2** Soil types and electrical conductivities

| Soil texture class or influence of salt | Electrical conductivity in mS/m |
|---|---|
| Sand | 0.1–1.0 |
| Loamy sand | 1.0–5.0 |
| Loam | 5.0–12.5 |
| Silt | 12.5–25.0 |
| Clay | 25.0–100 |
| Saline soil | >100 |

From Bevan (1998), simplified and altered

temporal differences for the situation of soil layers on the one hand and water flow on the other hand. The soil layers hardly change over time, but the water flow does. Hence sensing the preferential water flow will depend on **change detection** by means of repeated recordings over time. And reliable signals probably will need specific electrical frequencies as well.

Future experience will have to show, how much vertical **horizon sensing** is needed and for which cases simpler procedures of uniform **volume sensing** with a depth that approximates the maximal vertical root length is sufficient.

## 5.2.2 Electrical Conductivities, Soil Properties and Yields

The sensing results in units of electrical conductivity partly depend on the temperature of the soil and of the measuring devices. To eliminate this effect, an adjustment according to the prevailing temperature is necessary. This applies especially when conductivity is sensed via electrical induction. Provided the implements are well calibrated, properly adjusted and are sensing the same soil depths, the results for the common systems – the contact method (Fig. 5.3) as well as the induction method (Fig. 5.4) – are very similar (Sudduth et al. 2003, 2005). Therefore, the results will be dealt with *in cumulo*.

The sensed soil properties can be divided up on the one hand in fairly **static properties** such as texture and organic matter content and on the other hand in **dynamic properties** that vary in time as *e.g.* water content. Generally, soil texture can be regarded as the most static property. Its influence on the electrical conductivity is listed in Table 5.2.

Within the texture classes, **clay** exerts a dominating influence. Contrary to this, the effect of sand theoretically is close to zero. But in reality it is not since an autocorrelation exists between the contents of sand and clay in a soil. Because the higher the content of sand is, the lower the shares of clay and other soil constituents become.

The pre-eminent influence of clay results from several facts. Firstly, the clay fraction has the highest ion exchange capacity of all texture classes. In the absence of electrical conductors like metallic materials with movable electrons, the exchangeable ions are the carriers of electricity. But secondly, the ions need a moist medium – sufficient water – to fulfill this function. The clay also supplies this prerequisite. This is obvious when the **field capacity**, which is the moisture held in soil after excess water has drained away and the downward movement has stopped, is compared. At field capacity, a pure sandy soil would have about 15 % volumetric water, whereas the content for a clay soil can go up to 50 % (Lueck and Eisenreich 2000). And a third factor contributes to the electrical conductivity of clay soils. In most cases, the soil content of organic matter increases with its clay content. This can be the long-term result of the higher water content of these soils, since this decreases the decomposition of the organic matter. In addition, many clay constituents can form special bonds with decomposed products from organic matter. The clay-humus bonds that thus are created still further enhance the water holding capacity of soils. So in essence, a positive interaction between clay, water and organic matter can further enhance the high electrical conductivity of the clay fraction. The clay constituents – left alone – cause only part of this effect. And the interaction between clay and water means that on a temporary basis neither static properties nor dynamic properties completely dominate (see above). Yet this situation exists in reality for all texture classes as a result of varying moisture in the soil.

Excessive electrical conductivities exist in **saline soils** (Table 5.2). These soils are found in dry, arid regions when hardly water moves downward in soils, but instead water is sucked to the surface where it evaporates. This water transports **dissolvable salts** towards the surface plus topsoil and leaves them there. The resulting salinity limits water uptake by plants because it reduces the osmotic potential for this.

The dominating effect of salinity explains why historically the sensing of soil conductivities started in arid regions (Corwin 2008) and from there later spread out into humid areas as well. And it must be mentioned that Table 5.2 conceals the fact that the electrical conductivity always is defined by interactions of several factors. If the influence of temperature on the signals is eliminated by means of careful calibrations, adjustments or post-processing, there remain three important factors or parameters in **arid regions**, namely the concentration of salt ions in the soil water, soil texture and water content. In more **humid areas** where no accumulation of salts near the soil surface has taken place, the dominating factors that affect the conductivities are just texture and water content of the soil. Less important factors such as the bulk density of soil here can be left out.

### 5.2.2.1 Electric Conductivities and Soil Properties in Humid Areas

Figure 5.8 shows an example of the influence of texture classes on electrical soil conductivity for humid conditions. The sensing was done in two fields in an EM 38 vertical mode, thus with a depth up to about 1.5 m (Fig. 5.6).

For both fields, the effects of clay were in the same direction and similar, therefore, the signals were pooled. So the data from both fields appear on the same regression, but they stand for different ranges of clay content.

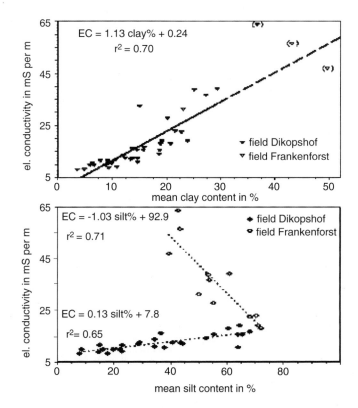

**Fig. 5.8** Electrical conductivities depending on the clay- and the silt contents of two soils near Bonn, Germany. For both soils, the effects of the clay on the conductivities were similar. Therefore, the results are presented in a common regression (*top*). Data points in *brackets* were excluded from the calculation. The effects of the silt were quite different (*bottom*). A comment to this is in the text (From Mertens et al. 2008, altered)

However, the sensing method does not provide for signals about different layers within the depth response curves, it supplies averaged data for the whole sensed depth. The topsoils of both fields are mainly loams that originated from loess with varying thicknesses between some cm and 1.5 m. The subsoil of both fields is quite different, for the field Dikopshof it is sandy and for the field Frankenforst it is clayey. These differences in the subsoil mainly cause the shifting of the clay points for Frankenforst to a much higher range. So the textures of both soils were different not only horizontally but especially in vertical directions.

Theoretically, the silt content should have only a small- and the sand content almost no influence, since the conductivity is mainly defined by ions of the soil constituents. However, **autocorrelation** with the clay content must be considered. Sand content is defined when clay- and silt content are known since these three texture classes add up to about 100 %. And apart from sand, even the silt content alone can depend on the clay content. When the clay content is very high, there is less space left for silt. So if total autocorrelation for the sand content holds when

**Table 5.3** Correlation between soil properties and electrical conductivities

| Soil properties | Corr. coeff. squared ($r^2$) to surface-layer soil | Corr. coeff. squared ($r^2$) to soil profile-averages |
|---|---|---|
| Soil moisture | 0.50 | 0.24 |
| Clay | 0.66 | 0.72 |
| Silt | 0.30 | 0.28 |
| Cation-exch. cap. | 0.70 | 0.70 |

The horizontal surface-layers and the vertical soil profiles denote the places where the reference samples were taken (averages for 12 fields in the North-Central USA, extracted from Sudduth et al. 2005)

clay- and silt content are defined, at least partial autocorrelation for the silt still applies, when the effect of a high clay content is known.

It is this partial autocorrelation to the effect of clay that probably explains the contradictory results about the influence of the silt content on the electrical conductivities (Fig. 5.8, bottom). The depth weighted clay contents in the field Frankenforst are much higher than those in the field Dikopshof. The misleading effect of autocorrelation thus probably was much higher in Frankenforst than in Dikopshof. In short, the results shown in Fig. 5.8, bottom, might be uncertain.

An important question is the respective impact of clay on tbe one hand and water on the other hand on electrical conductivities. Table 5.3 shows summarized results of extensive sensing with the presently dominating systems either by contact methods (Fig. 5.3) or by induction (Fig. 5.4). The depth of sensing corresponded to the **deep response curves** in Fig. 5.6. Since with well adjusted implements the records for the contact- or induction methods are similar if the depth that is sensed is about the same, such results were pooled. Effects of salinities on the signals probably can be ruled out, since these are based on areas with humid climates. The results are presented separately for reference samples taken from surface-layers (topsoils) and for samples that came from vertical soil profiles that include the effects of respective subsoils.

For both cases, the influence of the soil moisture on the electrical conductivity is lower than the effect of the clay content (Table 5.3). The influence of the silt too is rather low. The correlation to the site-specific cation-exchange capacity (CEC) is on a similar level with the clay content. This is in line with basic expectations since it is the clay particles in combination with organic matter bonded to them that mainly provide the cation-exchange capacity. Moreover, the ions are the carriers of the conductivity.

Yet the **ions need water** to function as carriers. Whereas the texture and organic matter in the soil might vary spatially but remain temporally constant, the soil moisture changes on a time basis as well. Hence the question arises, should the sensing be done when the moisture is at a low-, at a medium- or at a high level. Whereas a low level might reduce the temporary variations of the signals since it decreases the influence of a transient factor, a high level principally promotes the current flow.

The electrical conductivities in Fig. 5.9 are based on sensing of several loamy soils in a vertical induction mode, thus with a depth of approximately 1.5 m (Fig. 5.6).

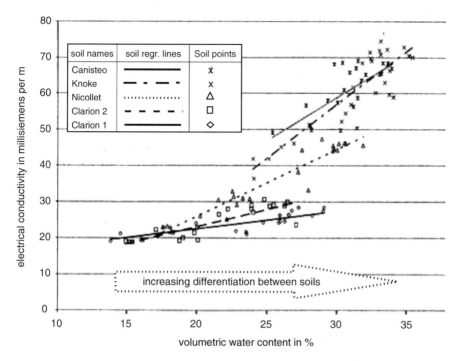

**Fig. 5.9** Electric conductivity depending on the volumetric water content for loamy soils in Iowa, USA (From Brevik et al. 2006, altered)

The varying site-specific effects of texture and organic matter have been eliminated completely by the design of the experiments, namely by sensing at exactly the same locations at different times, and hence when the water content was not the same. So the regressions represent only **temporal variations** that were caused by the respective soil water content.

The results indicate that electrical conductivity has the greatest potential to differentiate between soils when these are moist (Fig. 5.9). Hence obtaining reliable signals about **soil texture** is more difficult in areas where low soil moistures prevail – *e.g.* in dryland regions without irrigation. But instead of this, the perspectives of getting suitable information about **soil moisture** from conductivity data improve in these areas. This is in line with results that were obtained by McCutcheon et al. (2006), Neudecker et al. (2001) as well as Padhi and Misra (2011).

However, it can be questioned whether conductivity signals are the best choice for sensing water. Methods that rely on electrical capacitance, on reflectance or on radar waves might be better suited. For details on this see later chapters.

### 5.2.2.2 Electric Conductivities and Soil Properties in Arid Areas

In arid areas, **soil salinity** deserves attention. Its negative effects can be diminished or even disposed of by site-specific, georeferenced melioration or reclamation, *e.g.* by removing the salinity through site-specific irrigation and leaching out of salts in combination with adequate drainage. In case the salinity is mainly based on high levels of exchangeable **sodium**, it is not only crop growth that is impaired, since in this case soil tilth deteriorates as well. This is because sodium ions induce the soil particles to deflocculate or disperse and thus promote soil crusts. The aim is to replace the sodium with calcium and then to leach the sodium out. This can be enhanced by applying calcium-sulfate, sulfur or sulfuric acid. The latter two chemicals help if free lime is present in the soil, which then reacts with sulfuric acid to calcium-sulfate.

Irrespective of the particular situation, the first step is sensing the salinity in general. The traditional method for this has been to measure the electrical conductivity of a current, which passes through a **soil solution** that was extracted from a saturated soil sample. This method is precise since it focuses on salinity and eliminates or neutralizes the effects of texture or moisture within the soil. It is still used as a reference method. Yet up to now this method cannot be applied in an on-the-go manner, it is just used for soil samples in the laboratory. Hence for practical farming, this method can be ruled out when it comes to site-specific sensing with a high spatial resolution on larger fields.

However, it has been shown at several places (Hendrickx et al. 1992; Lesch et al. 2005; McKenzie et al. 1997; Rhoades et al. 1997) that for practical purposes the sensing of electrical conductivities of soil volumes is a suitable surrogate of solution sensing in laboratories. This volume sensing can be done either by methods that use contact electrodes (Sect. 5.2.1.1) or via systems that employ electromagnetic induction (Sect. 5.2.1.2). These techniques allow for on-the-go sensing with high spatial resolutions. Correlations with varying texture and with changing moisture of soils exist, but for many situations these do not alter much the indicated results in terms of general salinity (Hendrickx et al. 1992). This is because in most cases the salinity has an overriding influence on the respective electrical conductivity (Table 5.2).

Whether crops suffer from soil salinity depends on the respective species. The yields of most crops are not much affected when salt levels in terms of electrical conductivities are below 200 mS/m. Levels above 400 mS/m hurt many crops and above 800 mS/m all but the very tolerant plants are affected (Cardon et al. 2010). This means that the field shown in Fig 5.10 presents serious problems of salinity for most crops, although the distinct local differences call for site-specific ameliorations.

It must be expected that differentiating between the effects of salinity, texture and water content on electrical conductivity gets more difficult with low salinity levels. And even with higher salinity levels the signals needed for site-specific ameliorations might be more precise if **separating the effects** of soil constituents were possible. A concept in this direction has been developed by Zhang et al. (2004) as well as by Lee et al. (2007) and Lee and Zhang (2007). It is based on the fact that the amount of total current flowing in a soil can depend on conductive- as well as on capacitative behaviours of the soil. The methods of sensing by electrical conductivity as used hitherto use either direct current or alternating current with frequencies well below 40 kHz. Electric current with these properties ensures that the conductive behaviour of soils dominates.

# 5 Sensing of Natural Soil Properties

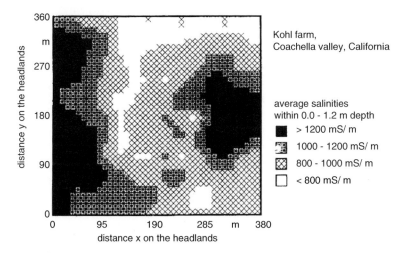

**Fig. 5.10** Map of soil salinity based on electrical conductivities in the rootzone of a field in California, USA (From Rhoades et al. 1997, altered)

The situation is quite different if alternating current with frequencies in the MHz range is used. Under these circumstances, the capacitative behaviour of soils becomes more prominent. Hence, if the sensing is done successively by different frequencies within a sufficiently wide range, signals of both the conductive- and the capacitative characteristics of soils can be obtained. The electric conductivity primarily provides signals about the salinity, whereas the capacitative behavior of soils predominantly depends on the respective water content. Thus principally differentiating between salinity and moisture of soils is possible. For further details about soil moisture sensing see Sect. 5.2.3.

### 5.2.2.3 Electric Conductivities and Crop Yields in Humid Areas

The economics of farming depend largely on crop yields. These can be regarded in a retrospective or in a perspective. Site-specific techniques for retrospective views on yields are dealt with in Chap. 12. Both views have to deal with a multitude of factors that influence the yields, such as the weather in various stages of crop development, soil and crop properties as well as techniques and practices used for irrigation, cultivation, sowing, crop protection and harvesting. This itemization shows that the indication of crop yields by soil electrical conductivities can – at most – be a partial one. But since maps about soil conductivities are becoming standard facilities for site-specific farming, their usefulness for predicting yield potentials should be known. After all, the local yield potential – if it can be derived *a priori* – can help to define the adequate site-specific input of seeds, irrigation water and agrochemicals.

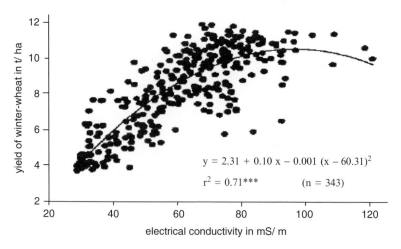

**Fig. 5.11** Electrical conductivity sensed in an induction mode with a depth of about 1.5 m and yields of wheat in Germany (From Neudecker et al. 2001, altered)

Traditionally, the yield potential of soils is attributed to its texture, particularly to its clay content. This holds especially for the topsoil. The clay and its bonds with organic matter affect the water holding capacity as well as the hydraulic properties and provide for the cation-exchange-capacity that is needed to store plant minerals. However, there are limits to the **agronomic benefits of clay**. Very high clay contents can be detrimental because this can reduce water permeability, inhibit deep drainage and consequently lead to waterlogged soil conditions. This applies in general for the whole soil depth that the plants roots penetrate and even some depth below this. Yet this holds particularly for the subsoil, since the latter hardly is cultivated and in addition is less penetrated and thus not loosened by the roots of crops.

It should be realized that sensing of the water situation in soils by means of electrical conductivity is done in a twofold manner, *i.e.* directly since water is a carrier of ions and indirectly via the effect of clay on the water regime.

When yields and electrical conductivities are compared on a site-specific basis, rather unambiguous results can be expected provided the conductivity varies distinctly within the field and the sensed subsoil has no hydromorphic layers or claypans.

Figure 5.11 shows the situation for a field, in which the electrical conductivities vary along a rather wide range. The diagram shows two typical criteria for the relation between site-specific electrical conductivites and yields: firstly a rather wide spread of the data and secondly on the average a change from a positive- to a negative influence on the yield at a high level of conductivity. The first criterion is the result of the many factors that affect yields. And the second characteristic can be explained by the fact that beyond a certain clay level in the soil, its effect on the yield is reversed (see above).

# 5 Sensing of Natural Soil Properties

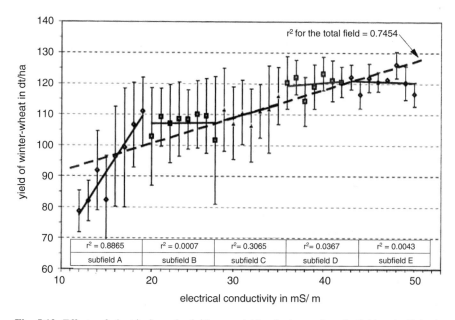

**Fig. 5.12** Effects of electrical conductivities on yields of winter-wheat in Schleswig-Holstein, Germany for a field and its subfields. Regressions are shown for the total field as well as for its subfields. The regression for subfield C is on the same line as the main regression for the whole field and hence is difficult to depict. The *vertical lines* represent the respective standard deviations (From Reckleben 2004 and Schwark and Reckleben 2006, altered)

The change from a positive- to a negative effect on yield occurs in Fig. 5.11 at a rather high level of conductivity, namely about 80 mS/m. It is reasonable to suspect that the transition from rising to declining yields depends on the water supply during the growing season.

In humid areas, the transition from a positive to a negative effect can already be attained with an electrical conductivity of 40 mS/m or less (Domsch et al. 2003; Lueck et al. 2002). And where a rather uniform soil prevails, the effects of its properties on site-specific yields may fail to appear at all.

The regressions in Fig. 5.12 indicate how different the relation between electrical conductivities and site-specific yields within a single field can be. The regression for the total field suggests a distinct positive effect of conductivities in the range from 12 to 49 mS/m on wheat yields. However, the division into subfields shows that this is largely due to the increasing yields between 12 and 19 mS/m and much less to rising yields above this level. There are ranges above this level where no increases occur. In this case, the decrease of yields probably starts somewhere between 44 and 49 mS/m. The standard deviations of the site-specific yields decrease with high yield levels (Reckleben 2004).

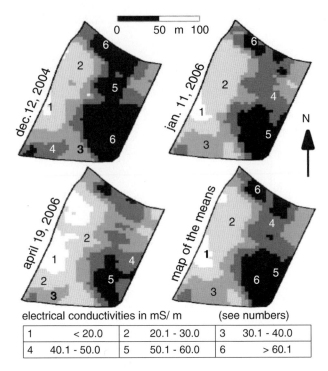

**Fig. 5.13** Maps of three repeated recordings and of the mean signals that were obtained from these (From Mertens et al. 2008, altered)

#### 5.2.2.4 Perspectives for the Use of Electrical Conductivities

For **humid regions**, clay- and water content of the soil are the main factors that define the site-specific electrical conductivities. Whereas the clay content is constant on a time basis, the moisture changes steadily. The temporally varying site-specific signals hence are mainly the result of the changes in water content. An effective method to even out these temporal variations resulting from the transient water factor is to record signals online and on-the-go at different dates and then to create maps that rely on mean data per cell. A prerequisite for this method is precise georeferencing that provides accurate spatial matching of the signals.

The original maps in Fig. 5.13 were taken from the same area (field Frankenforst in Fig. 5.8) at different times. They show distinct similarities. But they reveal also some differences, which probably are mainly due to temporal variations in the soil moisture. The **map of the means** neutralizes the temporal water effect at least partly. This procedure of averaging the results from repeated mappings must not be expensive. A single map causes an investment of about 6 Euros per ha on a contractor basis. So the contracting expenses for three maps add up to 12 Euros per ha. The expenditures for processing to maps of the means hardly count and hence can be ignored. Yet relevant is that the maps can be used for many years. If the use is

10 years, the costs per year resulting from the contracting expenses would only be between 1 and 2 Euros per ha.

An interesting perspective would be the creation of conductivity maps from repeated readings as a **byproduct** of other farm operations. With the increasing use of GNSS- or GPS based precision guidance, it might become feasible to record signals about soil conductivities as well as about topography while moving across the fields for other farming purposes. Hence more reliable maps for yield predictions could be obtained by averaging of signals from the same spot (Figs. 5.1 and 5.2).

It is general experience that the soil's potential for high yields depends largely on its ability to store and provide water for plant growth. The objective of averaging conductivity signals would be not to eliminate the water effect, but instead to remove the influence of its variation. The water tension in soils differs greatly, depending on whether the water is absorbed by soil particles, is in soil capillaries or moves freely. Apart from the weather or irrigation, it is the soil texture that determines the amount of water that is absorbed, stored and available for plant growth. So in short, it is an **interaction between soil and water** that defines largely the development of crops – aside from the application of agrochemicals.

Sensing of electrical conductivities in humid areas relies to a large extent on this interaction between soil texture and water. The conductivity signals quasi integrate the effects of clay and moisture. And fortunately, the effects of increasing clay- and water content on electrical conductivities on the one hand and on yields on the other hand point in the same direction, if soils with very high clay contents are excluded. Consequently, if by mapping of the means the temporally varying water effect can be removed, the signals obtained from conductivity sensing principally might be better suited for estimating yield potentials than the traditional analyses of only soil textures in laboratories. Hence maps of the mean electrical conductivity might become **yield-predicting-maps**. And traditional maps about soil texture instead probably will only serve as supplementing references in the future.

So sensing of electrical conductivities instead of the traditional analyzing of textures has distinct advantages:

- The signals are easily mapped online and on-the go instead of taking samples in the field and analyzing them in laboratories. Thus data from hundreds of sites per ha instead of one or two sites per ha can be afforded, which is a prerequisite for subsequent site-specific operations.
- The sensing criterion is not only texture, but in addition the water situation in soils. In principle, this complies with agronomic needs.

However, any attempt to derive site-specific yield potentials from electrical conductivities must take into consideration that – without being aware of this – sensing of **soil layers** may be included. In case soil layers in the subsoil are sensed that exist as pans, restrict the drainage of water and are not penetrated by the roots, the results can be very misleading. This is because the response curves of the presently dominating conductivity sensing systems (Figs. 5.3, 5.4, 5.5, and 5.6) pretend uniform soil properties within the vertical area that is sensed. Generally, an increasing clay content of soils increases the yield potentials – within limits. But this advantage can turn into the opposite if the clay sensed is concentrated in a dense pan that restricts drainage as well as root development. More sensing and processing techniques for precise site-specific mapping of such layers and pans within the rooted soil volume should be developed. This could help to precisely evaluate the agronomic effects of uneven texture and water situations within the sensed soil volumes in a vertical direction.

In **arid regions**, the interpretion of the signals that are obtained is somewhat simpler as a result of the overriding influence of salinity (Table 5.2). And because of this, maps of the means might not be necessary. Yet the benefits from salinity mapping in arid regions can be huge, since this is the starting basis for precise site-specific soil reclamation and soil fertility.

Presently less evident are the perspectives for electrical conductivity sensing in **dryland regions** without irrigation, where effects of salinity and of soil water are difficult to assess (McCutcheon et al. 2006). Yet the perspectives for the use in these regions might get better when precise separating of the effects that these soil constituents exert on the signals becomes state of the art (see Sect. 5.2.2.2, bottom).

## 5.2.3 Water Sensing Based on Permittivity and Capacitance

### 5.2.3.1 Basics

Soil sensing via electrical conductivity presently is mainly used for getting site-specific information in an integrated and summarizing way about the factors texture, organic matter and water. This integrated and summarizing information helps to assess the yield potential *a priori*. But there exist situations when information about the site-specific situation for a single factor alone – water – is needed.

The water content can be regarded as the most transient soil property. It can increase drastically within a few hours because of rain and decrease again within some days during dry spells. From this follows that a map about the respective water situation might help to explain the yield of a crop *ex post*, but the use of the same map for site-specific control of farming operations hardly ever can be extended over long time periods. Yet there can be occasions where a more precise knowledge about the temporal soil water content can help substantially, *e.g.*

- when **cultivations** must be scheduled since the breakup of clods that is needed as well as the prevention of compaction due to the weight of farm machines for many soils can depend on the water content.
- for **sowing- and planting** in the defining of the best time and in the control of the site-specific depth since the seeds need water for emergence.
- in the scheduling and site-specific control of **irrigation**. Agricultural crops extract most – but not all – of their water requirements from the top 30 cm soil

layer (Sharma and Gupta 2010). Irrigation water that ends up in deeper soil layers remains largely unconsumed by crops and might simply seep down, taking with it dissolved agrochemicals that pollute underground water. Consequently, a continuous monitoring of moisture in the top soil layer combined with controlled irrigation can save water and avoid unwanted leaching of minerals and pesticides.

Presently feasible methods for site-specific and on-the-go sensing of the water situation in bare fields rely either on electrical capacitance, on electrical permittivity or on infrared radiation. The latter method is based on **soil surface** sensing whereas electrical capacitance and -permittivity are measured within **soil volumes**. These volumes may be restricted, yet principally provide information from three dimensions instead of only from the two dimensions of surfaces. Capacitance methods use signals obtained from electric current flow, hence from electrons. And permittivity methods rely on recording of electromagnetic radiation – thus photons – from microwaves or radar waves.

Principally, **electrical capacitance** is the ability of a body to hold an electric charge It is a measure of the amount of electrical energy stored or separated for a given electric potential, *e.g.* in a parallel-plate capacitor or in a given soil volume. Since electrical charges are expressed in units of coulombs and electric potentials in units of volts, the capacitance has the SI unit of a coulomb per volt, which is defined as a unit of farad (F).

In contrast to capacitance, the **electrical permittivity** is the ability to resist the formation of an electromagnetic field, in this case in the soil. In other words, it is a measure of how an electromagnetic field affects the surrounding – and is affected by it. Thus permittivity relates to the ability of a material to "permit" an electromagnetic field. Important is that the permittivity $\varepsilon$ is the sum of a real part and an imaginary part.

The **real part of the permittivity** is associated with storage of electrical energy and thus with the capacitance of a material when an alternating electrical field is applied. In fact, the real part of the permittivity can be obtained from the capacitance in farad by dividing it by the overlap area of the capacitor plates and by multiplying it by the separating distance of these plates. Consequently, from the dimensions involved, it follows that permittivity is expressed in farad per m. However, this is the absolute permittivity of a material. In most cases, this absolute permittivity is replaced by the relative permittivity. This relative permittivity represents the absolute permittivity divided by the permittivity of a vacuum or of air, which equals 1 (one). This means that the numerical values for the absolute- and relative permittivities are identical. The difference is that the relative permittivity is dimensionless and because of this, it often is denoted as the **dielectric constant,** although it is not a real "constant".

The **imaginary part of the permittivity** is associated with energy dissipation, it is therefore often denoted as dielectric loss. There are applications where this energy dissipation is the main objective, *e.g.* when foods are heated in a microwave oven.

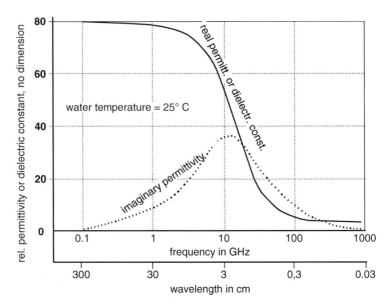

**Fig. 5.14** Permittivities of water with a temperature of 25 °C as depending on frequencies or lengths of microwaves (Compiled from data by Komarov et al. 2005)

For sensing of soil moisture, it is the real part of the relative permittivity or the dielectric constant that is used. Because regarding this physical criterion, water has outstanding properties. Within a wide range of frequencies, the real part of the relative permittivity or the dielectric constant is

- around 80 for free water
- between 3 and 7 for dry soil minerals (sand, clay *etc*.)
- between 2 and 5 for dry organic constituents of soils
- around 1 for air or for a vacuum.

Hence in principle, both the **dielectric constant** and the **capacitance** – that depends on it – offer good prerequisites for differentiating between water and other soil constituents. Yet there are still some important details to consider.

Both the real part and the imaginary part of the permittivity of moist materials depend strongly on the frequency of an electromagnetic field. This can be explained by interactions with water molecules. Because of its dipole structure, the water molecules get polarized in an electromagnetic field. And the alternating electromagnetic field causes the polarized water molecules to vibrate. As a result of inertial forces, these vibrations can get out of line with the respective frequency of the electromagnetic field.

The courses of the permittivity curves – as shown in Fig. 5.14 over frequencies and corresponding wavelengths – probably are mainly due to such effects on molecular vibrations. Regarding signals from the real part of the permittivity or the dielectric constants, the frequencies used for water sensing usually are below 10 GHz. Hence the respective wavelengths are above 3 cm.

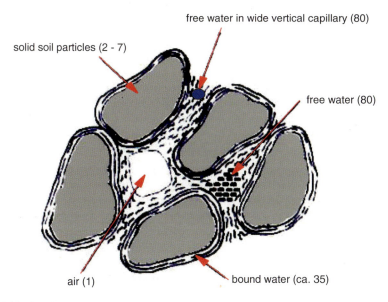

**Fig. 5.15** Constituents of a soil in a horizontal cross-section and – *in brackets* – its real relative permittivities (From Scheuermann et al. 2002, altered)

Water is held in soils either as "**free water**" or as "**bound water**". Free water drains readily and is easily available for uptake by crops. Contrary to this, bound water molecules are attached to soil particles by means of capillary- or colloidal forces. The finer the soil particles are and thus the higher the clay content is, the more water is bound. And since at least some of the bound water might not be available for uptake by crops, agronomists preferably use the tension that is needed to extract water from the soil as a criterion for the supply of plants. This tension can be expressed in units of Pascal or in mm of water column.

The problem is that for direct sensing of the **water tension**, presently no on-the-go methods are available or in sight, neither for operating from farm vehicles, nor from aerial platforms or from satellites. However, in an indirect way it is possible to sense whether moisture can be drawn from the soil, namely during the growing season by means of site-specific signals about the water-transpiration of crops. This will be dealt with in Sect. 6.5.2.

Yet fortunately the sensing of soil moisture by way of real permittivity too takes into account mainly free water and leaves out bound water – at least approximately. This is because bound water has lower permittivities than free water as a result of surface tensions that act on it (Fig. 5.15). However, a standard method for defining the bound water by means of permittivities does not exist. In a laboratory, bound water can be determined by first removing it in a drying oven using a temperature of 105 °C and 24 h time. After this, the soil is subjected to air with 50–60 % relative humidity. It takes up moisture from this air because of its hygroscopic properties. The bound water is equivalent to this hygrocopic equilibrium moisture that can easily be recorded by weighing before and after uptake (Robinson et al. 2002).

After having sensed the real relative permittivity or dielectric constant, the soil water content excluding bound water can be estimated using **Topp's equation** (Topp et al. 1980):

Volumetric soil water content $= -0.053 + 0.029\,\varepsilon - 0.00055\,\varepsilon^2 + 0.0000043\,\varepsilon^3$

$\varepsilon$ is the real relative permittivity or dielectric constant.

This equation provides for reliable results (Stoffregen et al. 2002) under the premise that $\varepsilon$ was sensed precisely for the respective soil.

Sensing of permittivity is possible either on the basis of velocities or of reflections of waves. For wave velocities, time signals are recorded. Thus the sensing occurs in the time domain. A well established method of this kind is the **time domain reflectometry (TDR)**. Electromagnetic waves are guided along transmission lines or cables within the soil. The time that is needed depends on the permittivity of the soil and thus indicates its water content. This method provides for rather reliable results and therefore often is used as a reference. Topp's equation was obtained using this method.

But unfortunately, methods that sense on a time domain up to now are not yet suited for on-the-go site-specific operations. These methods therefore are left out here. Instead, methods that rely on signals from **reflected radiation** will be dealt with. These methods operate either from satellites or on-the-go from terrestrial vehicles. Yet it will be shown that using reflected radiation can make it difficult to obtain accurate dielectric constants of soils.

### 5.2.3.2 Water Sensing from Satellites by Permittivity

Water sensing from satellites is used extensively for observing the earth's atmosphere with the objective of weather forecasting within large areas. But contrary to this, sensing of soil moisture from satellites within single fields and thus for site-specific farming still is not state of the art. In the past, neither the spatial- nor the temporal resolutions did correspond to the needs.

Yet both drawbacks are slowly disappearing. The spatial resolution for data from some modern **radar satellites** that operate in an **active mode** now even is going down to 1–100 m$^2$, which is sufficient. In addition, providing for information on a daily basis may become feasible because of more satellites. And the capability of micro- or radar waves to penetrate the space between the satellites and the earth never has been a problem. These waves have almost **all weather capabilities** – contrary to those from the visible- and infrared range (Sect. 3.3). An exception from this all weather capabilitiy may hold solely at times of heavy rainfall. Hence some general perspectives for signals about soil moisture from satellites by means of micro- or radar waves are encouraging.

But what about the ability of micro- or radar waves to provide sufficiently accurate signals about soil water? The relations between soil moisture and real

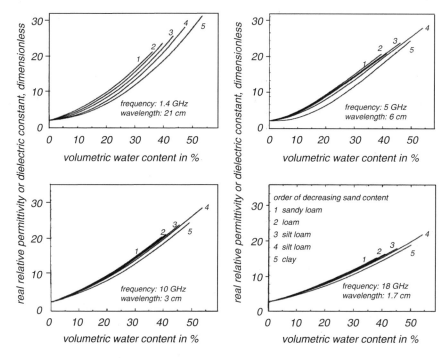

**Fig. 5.16** Real permittivites or dielectric constants of five soils sensed with different frequencies (Compiled from Hallikainen et al. 1985)

permittivities presented in Fig. 5.16 are based on experiments in a laboratory and hence refer to well controlled conditions. The respective volumetric water contents were obtained by adding water to soils that had been completely dehydrated in a drying oven before. This means that the moistures indicated include also bound water. And since bound water has lower real permittivities or dielectric constants than free water (Fig. 5.15), this should show up when soils that differ in their texture and thus in the sand- or clay content are compared.

This assumption is supported by the curves for the soils with different textures. For all frequencies and wavelengths that were used, it shows that the sandy soils had the highest real permittivities and the clay soils the lowest (Fig. 5.16).

However, the effect of soil texture depends on the micowave properties. The differences between the soil types are quite apparent when low frequencies of 1.4 GHz and correspondingly long waves were used, but get stepwise smaller with higher frequencies and hence shorter waves. So in order to account for the bound water – which is not available to plants – long microwaves are needed.

Very important is the **depth of sensing**. In case the signals are reflected only from the soil surface, which is the case with visible- and NIR radiation, a few raindrops or even dew can cause misleading results. Fortunately, microwaves do

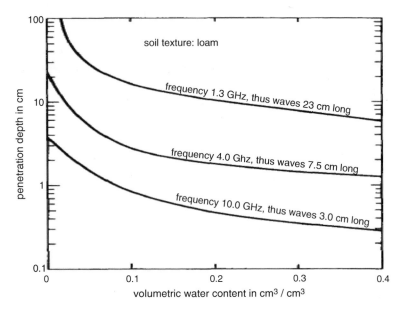

**Fig. 5.17** Depths of sensing by microwaves and soil moisture. The y axis has a logarithmic scale (From Ulaby et al. 1996, altered)

penetrate into soil. However, the depth depends on wavelength and on soil moisture (Fig. 5.17). The longer the waves and the lower the soil water content, the deeper can be sensed.

Yet the farmer does not know exactly *a priori* about the soil moisture content, especially not on a site-specific basis. This is, what he wants to know. A general rule – which disregards the actual water content and relies on average situations – is that the depth of moisture sensing is one fifth of the wavelength (Paul and Speckmann 2004).

Hence the depth of sensing

- for a frequency of 30 GHz and a wavelength of 1 cm is only 0.2 cm
- for a frequency of 10 GHz and a wavelength of 3 cm is about 0.6 cm
- for a frequency of 1 GHz and a wavelength of 30 cm is about 6 cm
- for a frequency of 0.5 GHz and a wavelength of 60 cm goes up to 12 cm.

The latter depth is about the maximum which can be achieved with present day technology operating on a surface reflection mode. This maximum depth suffices for controlling the sowing depth of crops on a site specific basis. It does not quite suffice, if for irrigation purposes, information about the water available for a growing crop is needed. The roots of most crops take up water far beyond this maximum depth. In short, regarding the depth of sensing there are limits for radar waves that operate in the surface reflection mode.

However, there are additional limits for water sensing by active radar waves. This method relies on reflection of radiation and this depends not only on the frequencies and the real permittivities or dielectric constants of the respective soils. Equally important are factors that have to do with the site-specific **surface** that is hit by the radiation such as

- the slope of the soil relative to the incident radar wave
- the roughness of the soil surface at scales relative to the wavelength
- the structure of any vegetation.

The last factor will be dealt with in Chap. 6 since it is important for sensing of crops by means of radar waves. When it comes to water sensing in bare fields, only the first two factors apply (Fig. 3.5). But the effect of these factors on the signals that are obtained with active radar sensing can be equal of even greater than the effects of soil moisture (Engman and Chauhan 1995). So methods are needed that make it possible to eliminate these disturbing effects.

Two approaches to cope with these problems seem to be feasible. The first approach is based on the fact that the error or noise caused by varying slopes or by changing roughness of the soil surface depends largely on the **incidence angle** with which the radiation coming from the satellite hits the earth. This angle varies between satellites and can often be adjusted. By systematically using different incidence angles, the effects of varying slopes as well as different surface roughnesses can be detected and be taken care of while the sensed data are processed (Baghdadi et al. 2008 and Srivastava et al. 2003). However, up to now neither signals nor maps that are corrected in such a way are available commercially.

The second approach relies on "**change detection**". Its theoretical background is that the factors which define the permittivity change differently on a **time basis**. The soil moisture varies almost constantly and sometimes even rather fast. Contrary to this, the slope of the soil always is about the same. And the roughness of the soil surface as well as the structure of the vegetation do change, however, in most cases much more slowly than the soil water content does. Hence repeated sensing within defined time spans combined with sophisticated processing of the signals allows to separate the effects of soil water from those of site-specific variations in slope, surface roughness and even vegetation (Moran et al. 2000; Kim and van Zyl 2009). But this method of "change detection by means of multi-temporal sensing" too is not yet state of the art. So the future will have to show whether using several incidence angles or change detection will provide a breakthrough for more precision in the sensing of soil moisture.

However, if bare areas are flat and have smooth surfaces, it is even possible to do without these special sensing and processing techniques as shown in Fig. 5.18 for fields from a farming region in Western Turkey.

Summing up the outlook for soil water sensing by radar satellites: the perspectives are encouraging, especially when taking into consideration that in the future operating with various incidence angles and change detection might remove still existing obstacles. **Low frequencies** and hence **long waves** will be needed in order to obtain sufficient sensing depth. Perhaps it might be feasible in the future to provide farm machinery that is operating in fields in an online and on-the-go manner with signals or maps about the respective site-specific soil water situation in a

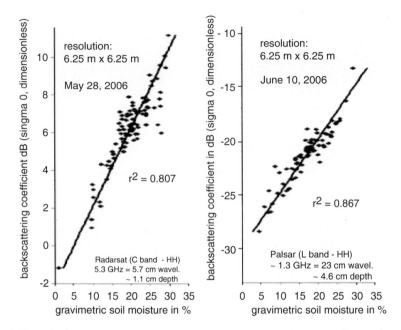

**Fig. 5.18** Reflection of radar microwaves back to the emitting satellites depending on the moisture content of bare soils. The backscattering coefficient on the y axis indicates the ratio between the wave power that is backscattered to the satellite and the incident wave power. This ratio of wave power is usually expressed in decibel (dB). The latter is ten times the common logarithm of this power ratio. Any negative values result from common logarithms below 1 (From Balik Sanli et al. 2008, altered)

similar way as georeferencing from satellites today is state of the art. However, such continuous signal transfer from the satellites to farm machines is not yet possible. Since radar satellites are not operating in a geostationary mode but instead orbit in a polar mode, this would require the availability of many more satellites in order to provide a continuous signal transfer. Furthermore, a network that links satellites, processing and farmers would be a prerequisite.

### 5.2.3.3 Water Sensing from Field Vehicles by Permittivity

This mode of water sensing would be attractive in precision farming if it could occur during usual farming operations and thus be used for simultaneous on-the-go control. Such a procedure is not yet state of the art, however, not beyond feasibility.

The emitting and sensing of radar reflectance from a land vehicle – as shown in Fig. 5.19 – is based on a frequency of 500 MHz. This frequency is below the usual range that is used by satellites. The lower frequencies and hence longer waves result in deeper sensing. A sensing depth of about 12 cm can be obtained (see Sect. 5.2.3.2).

**Fig. 5.19** Sensing surface reflectance of ground penetrating radar (From Redman et al. 2003, altered)

This is probably the deepest sensing that can be realized in the radar surface reflectance mode presently. Some farming operations and irrigation methods might benefit from control procedures that were based on such a depth. However, the equipment that is used for this mode of land based on-the-go surface reflectance sensing still is rather clumsy and not available commercially.

Principally there is no reason why sensing of radar reflectance – if it occurs from land vehicles instead of from satellites – should not have to cope with the influence of soil surface roughness on the signals. However, a general experience is that the effect of **surface roughness** on the signals decreases when the wavelength increases (Paul and Speckmann 2004). This would imply that as a result of the longer wavelengths that are used, the land based sensing as outlined above should have less to deal with surface roughness.

Yet results that have been obtained so long do not indicate that this problem therefore has gone. The raw signals that were recorded via Topp's equation (Sect. 5.2.3.1, last part) when operating in a grassy field and silt loam soil reveal more variability than would be expected from the water situation (Fig. 5.20). It is suspected that despite the longer wavelengths, at least part of this excessive variability is due to surface scattering of the reflected radiation. In order to make up for this, means of ten adjacent signals respectively were generated by applying a simple moving averaging filter. The thus obtained curve of the means seems to provide reasonable estimates although the question of biased results arises.

The black points represent results from time domain reflectometry. The estimates obtained with this method are generally presumed to be accurate and can hardly be influenced by surface roughness, which is also suggested by their course (Fig. 5.20).

### 5.2.3.4 Water Sensing from Farm Machinery by Capacitance

This method needs galvanic contact with the soil, similar to the electrical conductivity method in Fig. 5.3. But contrary to this conductivity sensing method, the positive and negative electrode are positioned very close to each other, *e.g.* within the same cultivator tine.

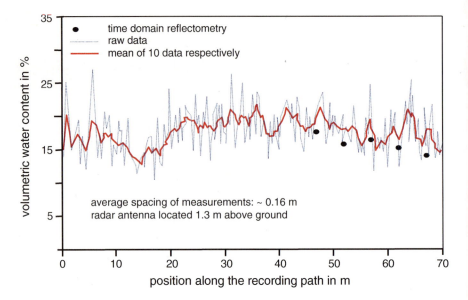

**Fig. 5.20** Soil water content sensed from a land-based vehicle by means of ground penetrating radar signals that were obtained via surface reflectance. The data were processed by applying Topp's equation to the signals (From Redman et al. 2003, altered)

The two electrodes – a brass cone and a metallic ring – make up the tip of a cultivator tine. They are separated by an insulator (Fig. 5.21). The soil that surrounds the tine is part of the capacitor. The sensing is done by measuring the "impedance" that exists for the current flow. This electro-physical criterion defines – in a rather simplified description – the resistance to current flow within a capacitor. The design of the sensor allows simultaneous recording of soil moisture via impedance as well as of penetration resistance, which depends heavily on soil water. Of course, this does not prevent the system to be used solely for moisture sensing.

An important factor is the **electrical frequency**. The present implements operate with frequencies between 40 and 175 MHz. Low frequencies make it difficult to sense precisely. This is because with low electrical frequencies, the signals that are obtained depend not only on the real permittivity or the dielectric constant, but on electrical conductivity as well (see Table 5.1 and Sect. 5.2.2.2 last part). The effect is that texture and perhaps also the ion concentration in the soil water influence the sensor output and thus the moisture that is indicated. On the other hand, higher frequencies increase the expenses for the electronic devices (Kizito et al. 2008).

A way out of this situation can be using the results from conductivity sensing for a **site-specific correction** of the signals from capacitance sensing in order to arrive at precise water mapping. Kelleners et al. (2009) indicate that such a procedure could improve the accuracy of water sensing. The instrumental solution for this could be sensing of capacitance and conductivity in one operation by employing **several frequencies** simultaneously with combined processing of the signals for the

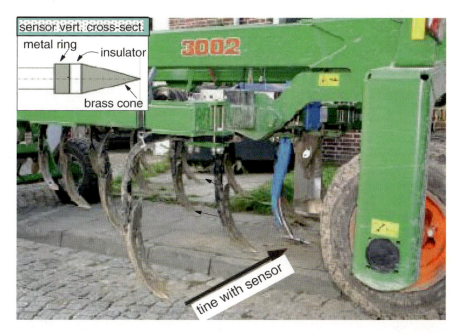

**Fig. 5.21** Capacitance sensing in the tip of a cultivator-tine. The *insert – top left –* is a vertical cross-section of the sensor. The electrical frequency is 100 GHz (From Drücker et al. 2009 and Sun et al. 2004, altered)

desired output criterions. An interesting alternative to this solution would be sensing of capacitance in real-time on-the-go and converting previously made maps of the mean conductivity by the results. Not only sensing for the respective water situation, but also the generating of "yield-predicting-maps" might benefit from such procedures to enhance the precision.

However, the temporal implications of the respective sensing objectives should be considered. Sensing for water always is a **short-term matter** because of the influence of the weather. In contrast to this, sensing for soil properties mainly aims at yield indicating maps and these – if created properly – can be used for decades. This fundamental temporal difference will have consequences for the practical management.

## 5.3 Sensing of Soil Properties on a Surface Basis by Reflectance

When soil sensing is done by reflected visible or infrared radiation, properties of the soil's skin and not of a defined volume are indicated. The signals originate from a surface that can be part of the **top-surface** of a field or of a surface within a **vertical**

**cross-section** of a soil, which *e.g.* is disclosed by a cultivating tool. Whatever the situation is, an important difference to volume sensing is that the signals are based on two dimensions and not on three dimensions in distance. And this generally means that – *ceteris paribus* – the signals come from less soil and hence might be less representative.

Yet what matters really: is it the amount of soil or is it the respective geometrical place within the soil?

When it comes to **sensing of water**, the depth of the water carrying soil layer from the surface is important for crop growth. Sensing the top-surface of a field does not supply any indication about the depth where a soil stores water. Immediately after a short rain, just 2 mm of the top-surface may be wetted. Following a longterm precipitation, 100 mm down from the top-surface might store water. Yet the signals from the top-surface might be the same. So these signals hardly help.

However, the situation for water sensing can be different if instead of the horizontal top-surface a vertical cross-section within the soil is scanned. This requires some scraping or plowing aside of soil that can be done during soil cultivation or during sowing. Thus an adequate sensing perspective for the control of sowing-, cultivation- or irrigation operations might be created.

It can be relevant that the topsoil is occasionally mixed when cultivations take place. Because of this mixing, the soil constituents that respectively are at the surface will vary. The soil constituents change their position within the cultivation depth. So even if only constituents that just were at the surface were recorded during the mapping, previous mixing within the topsoil provides actually for some volume sensing. This is especially the case when the sensing of the same field is repeated several times in the course of some years and thus **maps of the means** are obtained. These maps of the means then are estimates of the situation within the volume of the cultivated topsoil although the sensing occurred in two-dimensional patterns. However, this **cultivation induced change** from top-surface dimensions to volume dimensions does not make sense with transient soil properties such as water content. This method solely is well suited for soil properties that not at all or hardly all change over time. Thus obtainable maps of the means could be reasonable for texture, organic matter content and cation-exchange-capacity.

The situation is different when site-specific signals about soil conditions for the application of pre-emergence herbicides are needed. The application rate for these herbicides should be adapted to the site-specific **organic matter** content of the soil top-surface. Because the pre-emergence herbicides are partly absorbed by the organic matter of the soil surface and this makes them ineffective. In order to make up for the growing absorbance with increasing organic matter content, it is generally recommended to adapt the rates of application to this (Fig. 5.22).

This does not necessarily imply a higher burden to the environment. Because higher organic matter contents in soils also result in faster decomposition of herbicides (Hance 1973). So if a higher herbicide rate is absorbed, it is also faster decomposed.

Because these herbicides mainly get in contact with the top-surface of the soil, on-the-go information about the organic matter within this top-layer is precisely what is needed for controlling the application.

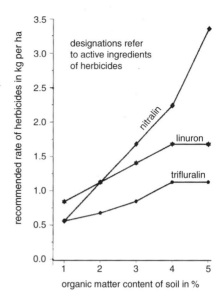

**Fig. 5.22** Recommended application-rates for pre-emergence herbicides and soil organic matter content (Compiled from specifications by Krishnan et al. 1981)

A simple – though not precise – indicator of the organic matter content is the **color of soils**. This is because the black color of humus removes all other colors of soil constituents. It especially replaces the red color of oxidized iron components and the yellow colors of aluminum in soils. Hence generally, the darker a soil appears, the more organic matter it has. But this is a very rough indication, because the water content too influences the color of a soil. The dryer a soil is, the brighter it looks.

Sensing methods that rely on details of colors have been used occasionally for purposes of precision farming in the past, however, they have largely been replaced by spectroscopic methods. Since colors are caused by wavelengths, it is reasonable to avoid the detours that inherently are connected with color measurements and to sense the wavelengths directly. This is dealt with in the next section.

## 5.3.1 Basics of Surface Sensing

Soil surface sensing can be based on diffuse reflectance in various wavelength ranges: the **visible** region (400–700 nm), the **near-infrared** domain (700–2,500 nm) and the **mid-infrared** range (2,500–25,000 nm). Traditionally, the visible- and the near-infrared region have been used, however, results with the mid-infrared range are promising.

The sensing approach can be oriented at indicating a **single soil property** – e.g. solely water or organic matter – or at recording and mapping **multiple soil properties** in one operation. The single property approach might make sense for online and on-the-go control of field operations for which just one soil property is very significant. Examples for this are controlling the sowing depth according to soil moisture or the application of pre-emergence herbicides in proportion to organic matter in the soil. All first attempts of online and on-the-go controls for site-specific operations have been based on single soil property sensing concepts.

However, the better the knowledge about the agronomic effects of soil properties is and the more the technology of spectroscopic sensing advances, the more favorable the terms get for multiple soil property sensing. This concept corresponds to the notion that some field operations probably require site-specific adjustments that simultaneously are based on several factors. For example: controlling the sowing depth could be reasonable not only according to soil moisture in the vertical soil cross-section, but also proportional to texture. And for the application of pre-emergence herbicides, the control could be based not only on organic matter, but again on texture as well.

There is another point that supports the multiple soil property concept. It is the fact that several soil properties do not exist independently from each other. In many cases, the organic matter content too depends on the texture. For instance, the water content firstly depends on the weather. Yet following precipitation, what remains in the topsoil and is available to a crop depends on texture and organic matter as well. The cation-exchange-capacity relies heavily on texture and organic matter too. So there are many interdependencies among soil properties.

An important question is, on which wavelengths the sensing should be based on. There are two general approaches for this: the full spectrum approach or the discrete waveband approach.

A **full spectrum approach** means that within the spectrum chosen (*e.g.* visible and near-infrared range) practically all wavelengths are included in the sensing process by recording in steps of for instance 5 nm or even less. Modern spectroradiometers can do this within a fraction of a second. Hence at every site-specific spot, many signals are sensed. These signals are subjected to sophisticated statistical evaluation processes like partial least squares regressions or others in order to obtain information about soil properties.

Since a full spectrum can include the important ranges or wavebands of several soil characteristics with their specific "fingerprints", this method is principally suited for **multiple soil property sensing**. Such multiple soil property sensing and mapping is close to becoming a reality for a variety of site-specific farming operations (Lee et al. 2009; Viscarra Rossel et al. 2006).

The **discrete waveband approach** dispenses with signals from a wide spectral range and just is confined to using narrow key wavelength bands. Much effort has been and still is devoted to detecting these **key narrow bands**. This sensing method can be reasonable for estimating just one soil property, *e.g.* water or organic matter. In many cases, interdependencies among soil properties are not taken into account. This might not be necessary, if *a priori* the correlations between the respective properties and the key reflectances are high, as it is the case with water and organic matter.

Experience will show, which properties can successfully be sensed via full spectrum approaches or when discrete waveband approaches are reasonable. But whatever method is used, the accurate **calibration** of the sensing equipment deserves attention. The calibration is an adjusting procedure. It aims at getting the right scale or relation between property data on the one hand and spectral signals on the other hand. The best data basis for an accurate calibration can be provided by the respective field, for which the sensing of properties is pending. But theoretically this is impossible, since it implies direct converting of output to input. Hence the input for the calibration must be supplied from another field that has a similar soil. And since worldwide a huge variety of different soils exist, this means that a reservoir for site-specific spectral calibration data is needed. Brown et al. (2006) have started to develop such spectral soil reflectance libraries.

### 5.3.2 Results of Surface Sensing in Laboratories

Figure 5.23 shows effects of moisture, organic matter and texture plus iron on visible and near-infrared reflectance of soils. Increasing contents of water, organic matter or clay result in decreasing reflectance. However, in the same order of soil properties, this effect becomes smaller. As for texture, this effect is ambiguous in the visible- and in the near-infrared range that is adjacent to it. The iron content of the soil may be important as well.

For a **discrete waveband approach**, the best wavelengths (fingerprints) are important. Within the visible and near-infrared range these are (Lee et al. 2009; Mouazen et al. 2007; Shonk et al. 1991; Zhu et al. 2010):

**for water**
  970 nm; 1,200 nm; 1,400 nm; **1,450 nm**; 1,820 nm; **1,940 nm**; 2,000 nm; 2,250 nm,
**for organic matter**
  660 nm; **1,772 nm**; **1,871 nm**; 2,070 nm; 2,177 nm; 2,246 nm; 2,351 nm; 2,483 nm,
**for clay**
  1,877 nm; 1,904 nm; 2,177 nm; 2,192 nm; **2,201 nm**; 2,220 nm; 2,492 nm and
**for cation-exchange-capacity**
  1,772 nm; 1,805 nm; **1,877 nm**; 2,090 nm; 2,276 nm; 2,306 nm; 2,498 nm.

Key wavelengths are indicated in **bold**.

The wavelengths belong – with one exception – to the near-infrared range, which is defined here as extending between 700 and 2,500 nm.

Coefficients of determination or squared correlation coefficients ($r^2$) that are based on **full spectrum** sensing of numerous soil samples via reflectance in laboratories are listed in Table 5.4. The respective full spectra were used to indicate various soil properties simultaneously. Only some of the sensed soil properties are shown. The data were processed by partial least squares regressions.

The summarized result is that the accuracy of sensing for the soil properties listed increases in the order visible-, near-infrared- and finally mid-infrared reflectance, hence with the wavelengths. In most cases, the results for organic carbon excelled those for all other properties. For the silt- and sand content, it must be taken

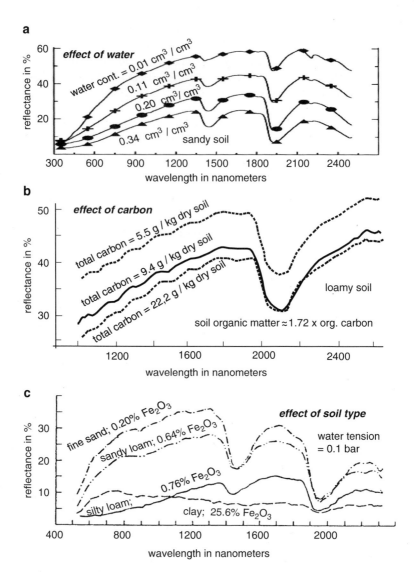

**Fig. 5.23** Soil Reflectance as affected by water, carbon and texture (*top*: from Zhu et al. 2010; *center*: from Huang et al. 2007; *bottom*: from Baumgardner et al. 1986, all altered)

into account that these properties together with the clay content add up to 100 %. Therefore, autocorrelation with the clay content must exist. And finally, soil water content was not included in these investigations.

It is generally known that soil water can be sensed as precisely or even more accurately than soil organic matter by near-infrared radiation (Mouazen et al. 2007; Viscarra Rossel et al. 2006; Zhu et al. 2010; Slaughter et al. 2001). Actually, for water, the sensing process is no problem. The question is whether the sensing for

**Table 5.4** Correlation between soil properties in laboratories and ranges of reflectance

| Soil properties | Correlation coefficients squared ($r^2$) | | |
| --- | --- | --- | --- |
| | Range of reflectance[a] | | |
| | Visible | Near-infrared | Mid-infrared |
| **From Viscarra Rossel et al. (2006), summary of extensive literature review** | | | |
| Organic carbon content | 0.78 | 0.81 | 0.96 |
| Clay content | | 0.71 combined | 0.82 |
| Cation-exchange-capacity | | 0.73 combined | 0.88 |
| **From experiments of Viscarra Rossel et al. (2006)** | | | |
| Organic carbon content | 0.60 | 0.60 | 0.73 |
| Clay content | 0.43 | 0.60 | 0.67 |
| Silt content | 0.31 | 0.41 | 0.49 |
| Sand content | 0.47 | 0.59 | 0.74 |
| Cation-exchange-capacity | 0.16 | 0.13 | 0.34 |
| **From experiments of McCarty et al. (2002)** | | | |
| Organic carbon content, 1.series | – | 0.82 | 0.94 |
| Organic carbon content, 2.series | – | 0.98 | 0.98 |

[a]The ranges are for visible reflectance 400–700 nm, for near-infrared reflectance 700–2,500 nm and for mid-infrared reflectance 2,500–25,000 nm

this soil property makes "sense" via reflectance due to the transient situation and the limitation to the soil surface. An essential point for soil water sensing is whether this is done on the top-surface or along surfaces of vertical cross-sections within the soil (see above).

It is expected that for sensing in laboratories, mid-infrared radiation will replace the hitherto dominating near-infrared reflectance due to the more precise indications. The situation is different for sensing in fields in an on-the-go mode. Because when using mid-infrared radiation, the soil has to be rather dry – a prerequisite that hardly can always be met in fields. Mid-infrared radiation is absorbed very strongly by moist soil and consequently not enough of it is reflected for sensing (Christy 2008). Visible and near-infrared radiation is less absorbed by moist soils. This allows measurements from moist field samples – a prerequisite of on-the-go signals for simultaneous online control of field machinery or for mapping. An important point is also the investment for the sensing instruments. The longer the wavelengths, the more expensive the spectroscopic implements are. Yet technological progress is more and more reducing the differences in investment.

Figure 5.24 shows similar results from full spectrum sensing of multiple soil properties with processing of the data by partial least squares regression. The soil samples came from ten fields in various regions of the Midwestern United States and were either taken in a segmented manner from the pedogenic horizons along **vertical soil profiles** or from **top surfaces** at various sites within each field. From the various soil properties that were included in the investigation, only the results for the most important natural properties – organic matter and clay content – are shown. These properties provided the best correlations between the traditional analyses and the reflectance sensing in the order organic carbon, clay.

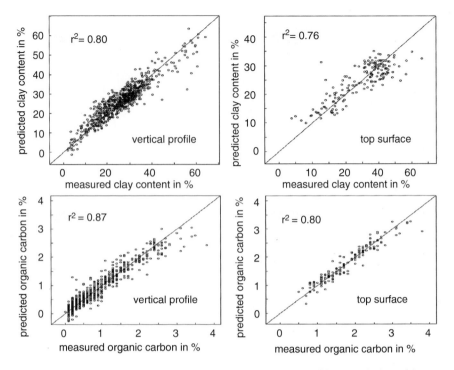

**Fig. 5.24** Comparison of predicted soil properties – obtained by traditional analysis – with measured values from full spectra of visible plus near infrared reflectance sensing in laboratories. The soil samples were taken either from vertical soil profiles (1.2 m deep) or from top surfaces. For wavelength ranges see Table 5.4 (From Lee et al. 2009, altered)

Since the vertical profiles represented not only the topsoil but the subsoil as well, its properties varied more than those that were taken solely from the top surfaces. The sensing results for the samples from the vertical profiles were better than those that were based on the top surfaces (Fig. 5.24).

The wavelength range that should be used for a full spectrum approach might depend on the soil properties needed. Lee et al. (2009) found that the soil property estimates from a reduced wavelength range of 1,770–2,500 nm were of similar accuracy than those obtained from the complete visible plus near-infrared range. The most important information obviously originates from the upper part of the near-infrared range. However, Lee et al. (2009) did not include mid-infrared reflectance and soil water in their investigations. When it comes to water, near-infrared radiation below the range mentioned above can provide important information.

Yet apart from this, single- as well as multiple soil property **sensing in laboratories** is possible with remarkable accuracies and is becoming "state of the art". For on-the-go **sensing in fields**, this stage of development largely still is a matter of the future.

For **remote sensing** from satellites or from aerial platforms, limitations that result from atmospheric barriers and from clouds must be considered. Atmospheric barriers that exist outside atmospheric windows (Fig. 3.3) can affect near- and mid-infrared radiation. In addition, clouds can block the transmission of all visible and all infrared waves. And regarding crop- or plant sensing, the unique effect of visible light on the photosynthetic process must be considered.

### 5.3.3   Concepts and Results for Surface Sensing in Fields

The ultimate goal is site-specific control of field operations by means of local soil properties. Proximal on-the-go sensing from farm machines lends itself for doing this either by simultaneous online control in real-time or by subsequent control via mapping. When remote sensing from satellites or from aerial platforms is used, up to now online control in real-time is not possible, so in these cases subsequent control via mapping is the choice.

Soil moisture hardly is suitable for mapping because of its transient feature. Its use for online control of the sowing depth is dealt with in Sect. 8.3.1.3. Other natural soil properties that can be recorded by reflectance such as texture, organic matter and cation-exchange-capacity are rather constant over time and hence well suited for control via mapping.

Sensing soils in fields occurs under much less controlled conditions than in laboratories, where dried and sieved samples in an accurately fixed position are subjected to the radiation. In fields, the soil moisture changes, and the soil particles vary from dust to crumbs, clods or residue pieces. Furthermore, on-the-go sensing excludes any fixed position. Hence a lower accuracy must be expected.

All concepts that have been used so far for proximal sensing by field machinery sense from a **flattened soil surface**. This allows to keep the distance to the soil rather constant. The flattened surface is obtained by sensing the soil at the bottom of a cultivator sweep (Fig. 5.25). Hence the information about the soil properties does not originate precisely from the top surface but instead from an area a few cm below this depending on the depth adjustment of the cultivator shank. This system of **sensing underneath a cultivator sweep** sometimes simply operates in the space between the upper flanks of the sweep and the soil below it (see Sect. 8.3.1.3). However, the concept that is outlined in Fig. 5.25 uses a closed bottom of the sweep. The radiation passes a sapphire window that is mounted along the bottom of the sweep. Contact between the passing soil and this window is supposed to create a self-cleaning effect and to prevent contamination of the optical path by dust or mud (Christy 2008).

Such soil property sensing by means of reflectance lends itself for combining with suitable field operations. It seems reasonable to **group the sensing** according to the time spans for which the respective soil properties can be used. Properties like *e.g.* conductivity, organic matter, clay and cation-exchange-capacity of soils are valid over a long time. Hence it is sensible to record these properties with simultaneous georeferencing in the same map or map-series and thus to combine the respective sensors into one machine (Fig. 5.26).

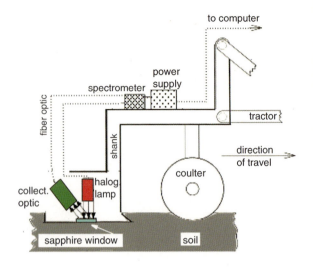

**Fig. 5.25** Operating principle of on-the-go soil property sensing by near-infrared reflectance (From Christy 2008, altered and not to scale)

**Fig. 5.26** Simultaneous mapping of both near-infrared reflectances underneath a cultivator sweep and electrical conductivities in the contact mode (Photo from Veris Technologies, Salina, USA, altered)

The situation is quite different for the short-timed water content. Mapping might be useless, but instead immediate control of sowing depth, cultivation depth or irrigation intensity might be needed. So for this, the combination with sowing, cultivation or irrigation would be reasonable.

But what about the **accuracy** of sensing in on-the-go field operations compared to recording in laboratories? Unfortunately, a direct and unbiased comparison of the

**Table 5.5** Accuracies of on-the-go and full spectrum carbon sensing via reflectance for defined point estimations depending on the variables considered in a stepwise regression

| Variable | Corr.coeff. squared ($r^2$), without dimension | Root mean squared error in 10 g/kg |
|---|---|---|
| Carbon only | 0.70 | 0.189 |
| Carbon + topography | 0.81 | 0.156 |
| Carbon + topography + water | 0.83 | 0.144 |
| Carbon + topography + water + clay | 0.88 | 0.127 |

Extract from Huang et al. (2007)
Legend: field size = 50 ha. 85 geo-referenced and defined points for traditional soil analyses. 3,700 points for mean reflectance readings. Carbon negatively correlated to topography and positively correlated to water or clay
Range of total carbon: 5.5–28.9 g/kg. Range of elevation: 290–303 m

accuracies is not possible since this would imply the same spatial resolution for records from laboratories and from on-the-go sensing in fields. This prerequisite does not hold. The technique shown in Figs. 5.25 and 5.26 easily can provide a spectrum for every 8 cm of travel. Present practice is to use 50 readings respectively for an average spectrum. Thus for every 4 m of travel, a signal is available. And with a swath-width of 20 m, about 125 readings per ha result. Such a high resolution is completely beyond any real possibilities when the recording is based on analyses of samples in laboratories. Hence from the **spatial resolution** that can be obtained, online and on-the-go sensing inherently is much better suited for site-specific farming than analyzing in laboratories. It could have a lower **accuracy per signal** than laboratory techniques yet still provide a benefit as a result of the much better spatial coverage

But disregarding any spatial resolution, online and on-the-go reflectance sensing can be compared with traditional analysing on the basis how well for defined points within a field it supplies soil property data that agree with those that are obtained with conventional state of the art laboratory methods. Such a **defined point estimation** relies on the assumption that conventional laboratory analyses are accurate. This assumption probably is reasonable because of the long experiences with these conventional laboratory methods and the well developed procedures when using them.

Among the various natural soil properties, much interest goes to organic matter or the **carbon** content, particularly since water is a special case as a result of its transient character and clay can be roughly detected by electrical conductivity – except for arid regions. However, the natural soil properties are interdependent. A question is how much can be gained in accuracy for the **sensing of carbon** if the **dependence on other soil properties** is taken into account.

The results in Table 5.5 are based on defined point estimations of total carbon sensing by the on-the-go technique of Figs. 5.25 and 5.26 in a sandy loam of glacial origin in Michigan, USA, hence in a humid, moderate climate. The reflectance signals were recorded within a range of 900–1,700 nm with a spectral resolution of 6 nm. A multiple soil property sensing approach and a sophisticated processing of

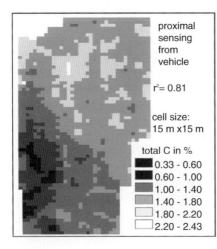 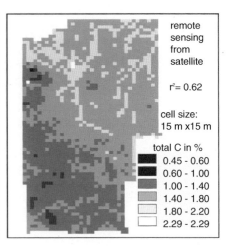

**Fig. 5.27** Carbon maps obtained by reflectance sensing either from a land based vehicle or from a satellite (From Huang et al. 2007, altered)

the reflectance signals by stepwise multiple regression with principal component analysis allowed to obtain the effects of various soil properties on the prediction of carbon sensing. The processing included site-specific topographic signals that were obtained via RTK-GPS (Sect. 5.1).

For all properties or variables listed, it holds that their incorporation in the sensing and processing improved the estimation of soil carbon (Table 5.5). The most important effect on the coefficient of determination ($r^2$) as well as on the root mean squared error had the topography (slope and inclination).

However, this topographical effect is – at least partly – an indirect one. Because site-specific moisture and texture too depend on topography. And since topography with present day technology is easy to sense and map, it should be the first choice for improving the carbon sensing on a multiple property basis.

The carbon maps in Fig. 5.27 are based on such multiple soil property sensing. However, besides carbon only topography was taken into account by the processing program. So actually, **dual soil property sensing** with the aim of carbon recording took place. The wavelength range for the proximal sensing in the left map extended from 900 to 1,700 nm – as with the correlations in Table 5.5. For the remote sensing in the right map, the range went from 450 to 2,350 nm, however, with some interruptions in the infrared part due to transmission barriers outside the atmospheric windows. Although generally longer wavelengths allow for more precision, the results for remote sensing were less accurate than those for proximal sensing. Its coefficient of determination ($r^2$) – based on defined point estimations – is lower. Explanations for this may be – among others – the interruptions caused outside the atmospheric windows and the much longer sensing distances (see Sect. 3.2). Yet apart from this, the maps from proximal and remote sensing look similar.

It should be noted that the variation of the carbon content within the field was rather high (Fig. 5.27), which facilitates the sensing. On-the-go sensing with the same implement in fields with a smaller carbon variation resulted in much lower coefficients of determination and hence less useful results (Bricklemyer and Brown 2010). So far, the best results with on-the-go carbon sensing have been obtained in humid, moderate areas and with weathered soils. In arid and semiarid regions, the organic carbon content of soils is much lower and the fraction of **inorganic carbon** on the total carbon content generally is higher. This makes it much more difficult to get reliable maps about the site-specific situation of organic carbon.

## 5.4 Summary

Among the many concepts that basically can be used to sense soil properties, a few are about mature for on-the-go applications in site-specific farming such as:

- Sensing topography and mapping of digital elevation by means of RTK-GPS
- Sensing electric conductivities in humid areas for maps that inform about the site-specific yield potential
- Sensing salinity in arid areas by means of electric conductivities
- Sensing water via optical reflectance from horizontal and vertical soil surfaces for the control of sowing depths and irrigation
- Sensing organic matter in humid, moderate areas by means of optical reflectance.

Additional techniques for sensing soil properties are developing. These are *e.g.* techniques for sensing soil water in a separate mode via radar waves or via electric capacitance and especially techniques for sensing multiple soil properties simultaneously by means of infrared reflectance.

## References

Abd Aziz S, Steward BL, Tang L, Karkee M (2009) Utilizing repeated GPS surveys from field operations for development of agricultural field DEMs. Trans ASABE 52(4):1057–1067

Adamchuk VI, Mat Su AS, Eigenberg RA, Ferguson RB (2011) Development of an angular scanning system for sensing vertical profiles of soil electrical conductivity. Trans ASABE 54(3):757–767

Allred BJ, Daniels JJ, Ehsani MR (2008) Handbook of agricultural geophysics. CRC Press, Boca Raton

Baghdadi N, Cerdan O, Zribi M, Auzet V, Darboux F, El Hajj M, Bou Kheir R (2008) Operational performance of current synthetic aperture radar sensors in mapping soil surface characteristics in agricultural environments: application to hydrological and erosion modelling. Hydrol Process 22:9–20

Balik Sanli F, Kurucu Y, Esetili MT, Abdikan S (2008) Soil moisture estimation from Radarsat-1, Asar and Palsar data in agricultural fields of Menemen plane of Western Turkey. The international archives of the photogrammetry, Remote sensing and spatial information sciences, Beijing, vol. XXXVII, Part B7, pp 75–81

Baumgardner MF, Silva LF, Biehl LL, Stoner ER (1986) Reflectance properties of soils. Adv Agron 38:1–44

Bevan B (1998) Geophysical exploration for archaeology: an introduction to geophysical exploration. Midwest Archaeological Center. National Park Service, US Department of the Interior. Special Report No. 1:13

Brevik EC, Fenton TE, Lazari A (2006) Soil electrical conductivity as a function of soil water and implications for soil mapping. Precis Agric 7:393–404

Bricklemyer RS, Brown DJ (2010) On-the-go VisNIR: potential and limitations for mapping soil clay and organic matter. Comput Electron Agric 70:209–216

Brown DJ, Shepherd KD, Walsh MG, Mays MD, Reinsch TG (2006) Global soil characterization with VNIR diffuse reflectance spectroscopy. Geoderma 132:273–290

Cardon GE, Davis JG, Bauder TH, Waskom RM (2010) Managing saline soils. Ext. paper no. 0.503. Colorado State University, Fort Collins. www.ext.colostate.edu

Christy CD (2008) Real-time measurement of soil attributes using on-the-go near-infrared reflectance spectroscopy. Comput Electron Agric 61:10–19

Corwin DL (2008) Past, present and future trends of soil electrical conductivity measurement using geophysical methods. In: Allred BJ, Daniels JJ, Ehsani MR (eds) Handbook of agricultural geophysics. CRC Press, Boca Raton, pp 19–44

Domsch H, Kaiser T, Witzke K, Giebel A (2003) Precision farming – the indirect way (in German). Neue Landwirtsch 5:48–50

Drücker H, Zeng Q, Sun Y, Roller O, Schulze Lammers P, Hartung E (2009) Sensory recording of soil moisture during tillage near the ground level (inEnglish and German). Landtechnik 64(4):272–275

Engman ET, Chauhan N (1995) Status of microwave soil moisture measurements with remote sensing. Remote Sens Environ 51:189–198

Geary P (2003) Evaluation of field surface topography for improved precision in farming. Ph.D. thesis, Cranfield University at Silsoe, Silsoe, Bedford

Gebbers R, Lück E, Heil K (2007) Depth sounding with the EM 38 – detection of soil layering by apparent electrical conductivity measurements. In: Stafford JV (ed) Precision agriculture '07. Wageningen Academic Publishers, Wageningen, pp 95–102

Hallikainen MT, Ulaby FT, Dobson MC, El-Rayes ME, Wu LK (1985) Microwave dielectric behaviour of wet soil – part I: empirical models and experimental observations. IEEE Trans Geosci Remote Sens GE-23(1):25–35

Hance RJ (1973) Soil organic matter and the adsorption and decomposition of the herbicides Atrazin and Linoron. Soil Biol Biochem 6:39–42

Hendrickx JMH, Baerends B, Raza I, Sadig M, Akram Chaudhry M (1992) Soil salinity assessment by electromagnetic induction of irrigated land. Soil Sci Soc Am J 56:1933–1941

Huang X, Senthilkumar S, Kravchenko A, Thelen K, Qi J (2007) Total carbon mapping in glacial till soils using near-infrared spectroscopy. Landsat imagery and topographical information. Geoderma 141:34–42

Kelleners TJ, Paige GB, Gray ST (2009) Measurement of the dielectric properties of Wyoming soils using electromagnetic sensors. Soil Sci Soc Am J 73(5):1626–1637

Kim Y, van Zyl JJ (2009) A time-series approach to estimate soil moisture using polarimetric radar data. IEEE Trans Geosci Remote Sens 47(8):2519–2527

Kizito F, Campbell CS, Campbell GS, Cobos DR, Teare BL, Carter B, Hopmans JW (2008) Frequency, electrical conductivity and temperature analysis of a low-cost capacitance soil moisture sensor. J Hydrol 352:367–378

Komarov V, Wang S, Tang J (2005) Permittivity and measurement. In: Chang K (ed) Encyclopedia of RF and microwave engineering, 6 vol set, Wiley, Hoboken, 3696

Krishnan P, Butler BJ, Hummel J (1981) Close-range sensing of soil organic matter. Trans Am Soc Agric Eng 24(2):306

Lee KH, Zhang N (2007) A frequency response permittivity sensor for simultaneous measurement of multiple soil properties: part II. Calibration model tests. Trans ASABE 50(6):2327–2336

Lee KH, Zhang N, Kuhn WB, Kluitenberg GJ (2007) A frequency response permittivity sensor for simultaneous measurement of multiple soil properties: part I. The frequency-response method. Trans ASABE 50(6):2315–2326

Lee KS, Lee DH, Sudduth KA, Chung SO, Kitchen NR, Drummond ST (2009) Wavelength identification and diffuse reflectance estimation for surface and profile soil properties. Trans ASABE 52(3):683–695

Lesch SM, Corwin DL, Robinson DA (2005) Apparent soil electrical conductivity mapping as an agricultural management tool in arid zone soils. Comput Electron Agric 46:351–378

Lueck E, Eisenreich M (2000). Introduction into geophysical sensing to record soil properties for site-specific farming (in German). IX. Arbeitsseminar "Hochauflösende Geoelektrik", Bucha, Sachsen

Lueck E, Ruehlmann J (2013) Resistivity mapping with GEOPHILUS ELECTRICUS – Information about lateral and vertical soil heterogeneity. Geoderma 199:2–11. http://dx.doi.org/10.1016/j.geoderma.2012.11.009

Lueck E, Eisenreich M, Domsch H (2002) Innovative methods for precision agriculture. In: Blumenstein O, Schachtzabel H (eds) Dynamics of matters in geosystems. University of Potsdam, Potsdam

Lueck E, Spangenberg U, Ruehlman J (2009) Comparison of different EC-mapping sensors. In: van Henten EJ et al (eds) Precision agriculture '09. Wageningen Academic Publishers, Wageningen, pp 445–451

McCarty GW, Reeves JB III, Reeves VB, Follett RF, Kimble JM (2002) Mid-infrared and near-infrared diffuse reflectance spectroscopy for soil carbon measurement. Soil Sci So Am J 66:640–646

McCutcheon MC, Farahani HJ, Stednick JD, Buchleiter GW, Green TR (2006) Effect of soil water on apparent soil electrical conductivity and texture relationships in a dryland field. Biosyst Eng 94(1):19–32

McKenzie RC, George RJ, Woods SA, Cannon MA, Bennett DL (1997) Use of electromagnetic-induction meter (EM38) as a tool in managing salinization. Hydrogeol J 5(1):37–50

Mertens FM, Paetzold S, Welp G (2008) Spatial heterogeneity of soil properties and its mapping with apparent electrical conductivity. J Plant Nutr Soil Sci 171:148–154

Moran MS, Hymer DC, Qi J, Sano EE (2000) Soil moisture evaluation using multi-temporal synthetic aperture radar (SAR) in semiarid rangeland. Agric For Meteorol 105:69–80

Mouazen AM, Maleki MR, De Baerdemaeker J, Ramon H (2007) On-line measurement of some selected soil properties using a VIS-NIR sensor. Soil Till Res 93:13–27

Neudecker E, Schmidhalter U, Sperl C, Selige T (2001) Site-specific mapping by electromagnetic induction. In: Grenier G, Blackmore S (eds) 3rd European conference precision agriculture, Montpellier, pp 271–276

Pahdi J, Misra RK (2011) Sensitivity of EM38 in determining soil water in an irrigated wheat field. Soil Till Res 117:93–102

Paul W, Speckmann H (2004) Radarsensors: new technologies for precise farming. Part 1: Basics and measuring of soil moisture (in German). Landbauforsch Voelkenrode 54(2):73–86

Radic T (2008) Instrumental and processing investigations on geoelectrics (in German) Ph.D. thesis, Technical University of Berlin, Berlin

Reckleben Y (2004) Innovative and real-time sensor technology for recording and control of product quality of small grains during combining (in German). Doctoral dissertation, University of Kiel, Forschungsbericht Agrartechnik des Arbeitskreises Forschung und Lehre der Max Eyth Gesellschaft im VDI (VDI-MEG) 424

Redman JD, Galagedara L, Parkin G (2003) Measuring soil water content with the ground penetrating radar surface reflectivity method: effects of spatial variability. ASAE paper no. 032279, 2003 ASAE annual international meeting, Las Vegas

Rhoades JD, Lesch SM, LeMert RD, Alves WJ (1997) Assessing irrigation/drainage/salinity management using spatially referenced salinity measurements. Agric Water Manag 35:147–165

Robinson DA, Cooper JD, Gardner CMK (2002) Modelling the relative permittivity of soils using soil hygroscopic water content. J Hydrol 255:39–49

Saey T, Simpson D, Vermeersch H, Cockx L, Van Meirvenne M (2009) Comparing the EM38DD and DUALEM-21S sensors for depth-to-clay mapping. Soil Sci Soc Am J 73(1):7–13

Schepers JS (2008) Potential of precision agriculture to protect water bodies from negative impacts of agriculture. Landbauforschung – vTI Agric For Res 58(3):199–206

Scheuermann A, Schlaeger S, Becker R, Schädel W, Schuhmann R (2002) Advantages of TDR-measuring techniques for evaluating unsaturated soils in geotechnics (in German). In: BAW-Kolloquium: Der Einfluss von Lufteinschlüssen auf die Strömungs- und Druckdynamik in Erdbauwerken. University of Karlsruhe, Karlsruhe

Schwark A, Reckleben Y (2006) The EM 38-system as soil-sensor for practical farming (in German). In: Rationalisierungs-Kuratorium für Landwirtschaft (RKL), Rendsburg, file 4.1.0:1225–1245

Sharma RK, Gupta AK (2010) Continuous wave acoustic method for determination of moisture content in agricultural soils. Comput Electron Agric 73:105–111

Shonk JL, Gaultney LD, Schulze DG, Van Scoyoc GE (1991) Spectroscopic sensing of soil organic matter. Trans Am Soc Agric Eng 34(5):1978–1984

Slaughter DC, Pelletier MG, Upadhyaya SK (2001) Sensing soil moisture using NIR spectroscopy. Appl Eng Agric 17(2):241–247

Srivastava HS, Patel P, Manchanda ML, Adiga S (2003) Use of multiincidence angle RADARSAT-1 SAR data to incorporate the effect of surface roughness in soil moisture estimation. IEEE Trans Geosci Remote Sens 41(7):1638–1640

Stoffregen H, Yaramanei U, Zenker T, Wessolek G (2002) Accuracy of soil water content measurements using ground penetrating radar; comparison of ground penetrating radar and lysimeter data. J Hydrol 267:201–206

Sudduth KA, Kitchen NR, Bollero GA, Bullock DG, Wiebold WJ (2003) Comparison of electromagnetic induction and direct sensing of soil electrical conductivity. Agron J 95:472–482

Sudduth KA, Kitchen NR, Wiebold WJ, Batchelor WD, Bollero WA, Bullock DG, Clay DE, Palm HL, Pierce FJ, Schuler RT, Thelen KD (2005) Relating apparent electrical conductivity to soil properties across the North-Central USA. Comput Electron Agric 46:263–283

Sudduth KA, Kitchen NR, Myers DB, Drummond ST (2010) Estimating depth to argillic soil horizons using apparent electrical conductivity. J Environ Eng Geophys 15(3):135–146

Sudduth KA, Myers DB, Kitchen NR, Drummond ST (2013) Modeling soil electrical conductivity-depth relationships with data from proximal and penetrating ECa sensors. Geoderma 199:12–21. http://dx.doi.org/10.1016/j.geoderma.2012.10.006

Sun Y, Schulze Lammers P, Ma D (2004) Evaluation of a combined penetrometer for simultaneous measurement of penetration resistance and soil water content. J Plant Nutr Soil Sci 167:745–751

Topp GC, Davis JL, Annan AP (1980) Electromagnetic determination of water content: measurements in coaxial transmission lines. Water Resour Res 16:574–584

Ulaby FT, Dubois PC, van Zyl J (1996) Radar mapping of surface soil moisture. J Hydrol 184:57–84

Viscarra Rossel RA, Walvoort DJJ, McBratney AB, Janik LJ, Skjemstal JO (2006) Visible, near infrared, mid infrared or combined diffuse reflectance spectroscopy for simultaneous assessment of various soil properties. Geoderma 131:59–75

Wenner F (1915). A method of measuring earth resistivity. Sci. Paper No. 258. US Department of Commerce Bureau of Standards, NIST, Gaithersburg

Westphalen ML, Steward BL, Han S (2004) Topographic mapping through measurement of vehicle attitude and elevation. Trans ASABE 47(5):1841–1849

Zhang N, Fan G, Lee KH, Kluitenberg GJ, Loughin TM (2004) Simultaneous measurement of soil water content and salinity using a frequency-response method. Soil Sci Soc Am J 68:1515–1525

Zhu Y, Weindorf DC, Chakraborty S, Haggard B, Johnson S, Bakr N (2010) Characterizing surface soil water with field portable diffuse reflectance spectroscopy. J Hydrol 391:133–140

# Chapter 6
# Sensing of Crop Properties

Hermann J. Heege and Eiko Thiessen

**Abstract** Sensing of crops by visible and infrared reflectance allows estimating the chlorophyll concentration within leaves as well as the leaf-area-index. The product of the chlorophyll concentration within leaves and the leaf-area-index supplies the chlorophyll content per unit field area. Recording this criterion repeatedly during the season provides reliable estimates of the site-specific yield potential as based on past growing conditions.

Fluorescent light too can sense the chlorophyll concentration within leaves or the functioning of the photosynthetic apparatus of crops. Infrared reflectance as well as thermal radiation can be used to get information about the site-specific water supply of crops. From the backscatter of radar waves, information about the biomass, the leaf-area-index and especially about the crop species for vegetation classification within large agricultural areas can be obtained.

Proximal sensing from farm machines allows direct site-specific control of farm operations in real-time. On the other hand, remote sensing from satellites lends itself for repeated recording of fields or larger areas during the growing season. Yet remote sensing needs radiation that can penetrate the atmosphere.

**Keywords** Chlorophyll sensing • Fluorescence sensing • Infrared reflectance • Sensing by microwaves • Visible reflectance • Water sensing • Yield sensing

---

H.J. Heege (✉) • E. Thiessen
Department of Agricultural Systems Engineering, University of Kiel,
24098 Kiel, Germany
e-mail: hheege@ilv.uni-kiel.de

## 6.1 Basics of Sensing by Visible and Infrared Reflectance

Crop properties often vary even more than soil properties do. This is, because they depend not only on the respective soils but on many additional factors as well such as microclimate, species, variety, growth stage, farming operations, nutrient supply, weed competitions and pest infestations. It certainly is useful to know, which factors are responsible for differences in crop development. Yet before such questions can be pursued, temporal and spatial differences in crop growth must be identified. And for doing this on a site-specific basis with a sufficiently high resolution – similar to the situation with soil properties – sensing via electromagnetic radiation offers challenging opportunities.

There exist fundamental differences in the **temporal variation** or stability of crop properties compared to most soil properties. The most important soil properties such as texture, organic matter content and cation-exchange-capacity often vary spatially, yet remain rather constant during a growing period. The sole exception to this is the water content of soils. But the main crop characteristics – such as biomass, structure of plants and ingredients – change steadily while a crop develops. Consequently for crop properties, the respective time of sensing is very important as it is for soil water.

Sensing of crop properties by means of electromagnetic radiation predominantly relies on **reflectance**. Theoretically, it could be based on transmittance or even absorbance as well. However, both the transmitted- and the absorbed radiation cannot easily or not at all be recorded on-the-go via proximal sensing by farm vehicles or via remote sensing from aerial platforms or satellites.

But it should be realized that the reflectance itself depends heavily on the absorbance as well as on the transmittance. This is because irradiance (= incident radiation) that is absorbed or transmitted is not available any more for reflection.

From visible radiation that hits healthy and growing plants, the dominant part is absorbed for **photosynthesis** (Fig. 6.1, left). Consequently, both the reflectance and the transmittance are rather small in this part of the spectrum. However, there are differences within the visible region. The reflectance is smaller in the blue and in the red part than in the green range. Because the blue and especially the red light are better suited for photosynthesis. The higher reflectance in the green range causes the color of growing crops.

When dealing with crops, the **near-infrared range** from 700 to 2,500 nm is subdivided. Because the lower part of this range – 700 to 1,300 nm – has an extremely low absorbance and consequently a very high reflectance and transmittance. And contrary to this, the upper part of this range has regions of very high absorbance by water and thus low reflectance and low transmittance (Figs. 6.1 and 6.2). Therefore with crops only the lower part from 700 to 1,300 nm is denoted as near-infrared, whereas the upper part from 1,300 to 2,500 nm is named **shortwave-infrared**. The inconsistency in definition when dealing with soils (Table 5.4) or crops might be confusing. The respective wavelength range must be noted.

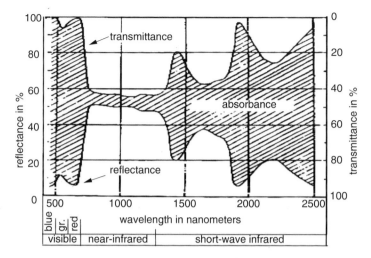

**Fig. 6.1** Reflectance and transmittance spectra of wheat leaves (variety: Talent) in the visible and infrared range. The *shaded part* is the absorbance. The latter plus the reflectance and transmittance add up to respectively 100 % (From Guyot 1998, altered)

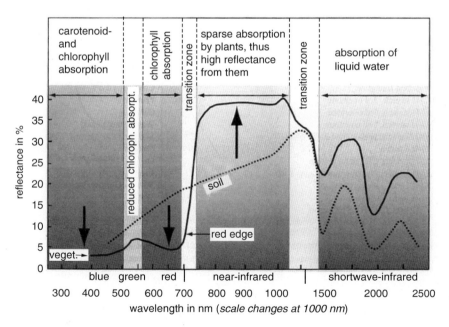

**Fig. 6.2** Reflectance spectra of vegetation and soil in the visible-, near-infrared and shortwave-infrared range (From Jensen 2007, altered)

When regarding the general course of the spectra that are reflected from soil or from vegetation, some distinct differences show up (Fig. 6.2). The reflectance from soils increases rather steadily and uniformly with the wavelength in the visible and near-infrared range. Contrary to this, the reflectance of plants in the visible regions is defined by the absorbance for photosynthesis and hence is below that of soil. Yet within the near-infrared range, the absence of absorbance for photosynthesis allows a multiplication of the reflectance from crops. The result of this is a steep rise of the reflectance in the transition zone from the visible to the near-infrared radiation. This steep rise in the reflectance is generally known as the "**red edge**" because it is located between the visible red- and the near-infrared radiation.

However, the curves of the spectra in Fig. 6.2 represent **average situations** just in order to show the principal differences in reflectance between soil and vegetation. In detail, the reflectance spectra of soils differ according to texture, organic matter and chemical constituents (Fig. 5.23). And those of vegetation too depend on type, ingredients and chemical composition. The sensing possibilities by means of reflectance rely on these differences.

The most important differences for vegetation result from biomass and chlorophyll. The **biomass** always has been a very significant criterion for defining the development of a crop. For forage crops, the total aboveground plant mass is utilized, hence the biomass in t per ha as well as its ingredients are of interest. With grain crops, the main objective is not the total aboveground biomass but mainly the reproductive part of it. Yet for all crops, the leaf area is important since photosynthesis takes place in the leaves. Consequently, the **leaf-area-index** of crops is a significant criterion in order to assess the development. It is defined as the relation between the photo-chemically active, one-sided leaf area and the ground surface. In a simplified way, the leaf-area-index shows how much "factory space" for photosynthesis a crop supplies. Depending on the development stage, species, variety and growing conditions, the leaf-area-index can vary widely. Starting with 0 (zero) before emergence, it can go up to 8 (eight) for well developed, lush grain crops before the ripening process begins.

Within the leaves, **chlorophyll** is the essential driver of photosynthesis and hence of plant production. Hence information about the leaf-area-index must be supplemented by data about the chlorophyll concentration within the leaves, which can be defined by the mass of chlorophyll per unit of leaf area.

So knowledge about the separate effects of chlorophyll on the one hand and the leaf-area-index on the other hand on reflectance spectra is helpful. These effects can be shown separately by using simulation models that have been developed by Verhoef (1984) and Jacquemoud and Baret (1990). Looking at the results from these simulation models (Fig. 6.3), it is obvious that the effect of the chlorophyll per unit leaf area is restricted to wavelengths below the red edge. Above this edge, the chlorophyll per unit leaf area does not affect the reflectance. But in the visible region, the reflectance is the lower, the higher the chlorophyll concentration per unit leaf area is.

Contrary to this, the effect of the leaf-area-index mainly is above the red edge inflection point in the near-infrared range. The reflectance here increases very

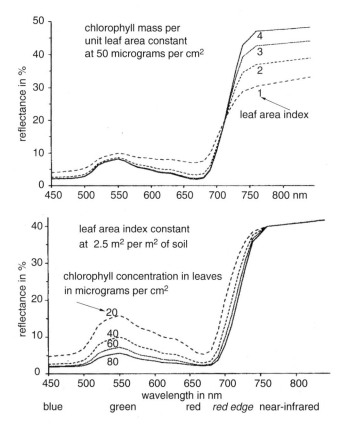

**Fig. 6.3** Reflectance of plants depending on either the leaf-area-index (*top*) or on the chlorophyll concentration in the leaves (*bottom*). The curves are based on simulation models. For the *top graph*, the chlorophyll mass per unit leaf area is constant. Vice versa, for the *bottom graph*, the leaf-area-index is constant (From Reusch 1997, altered)

clearly with the leaf-area-index. In the visible region, more leaves generally result in some decrease of reflectance because – even if the chlorophyll content within single leaves is constant – this improves the absorbance. However, the effects of either more leaves or of more chlorophyll within the leaves in the visible range are not the same: more leaves mainly reduce the reflectance in the red part, whereas more chlorophyll lowers it in the green region (Fig. 6.3).

The visible and near-infrared reflectance from both sides of the red edge lends itself for differentiating between soil and plants. Because within this range, the reflectance for soil increases slowly and steadily, however, for vegetation or crops it rises drastically (Fig. 6.2).

Hence the relation between red and infrared reflectance can be used as an **indicator of a vegetation cover** within a field (Fig. 6.4). This holds despite the fact that in a strict sense the reflectance of a bare soil within a field might not be constant as a

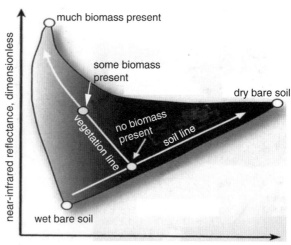

**Fig. 6.4** The site-specific significance of red and infra-red reflection within a heterogeneous field. The *black* and *gray shaded* field is bare along the *lower side* but covered with plants in the *upper region*. The soil line extends within bare soil from a wet area into a dry region. The vegetation line goes from bare soil into areas that are progressively vegetated (From Jensen 2007, altered)

result of varying water content, texture, organic matter content and iron content (Fig. 5.23) and that the reflectance of vegetation too can vary (Fig. 6.3). Because in most cases, the general effect of vegetation on the reflectance is much more pronounced than the differences that result from varying soil- or plant properties. The **soil line** and **vegetation line** concept as defined by the red- and near-infrared reflectance in Fig. 6.4 result from this. However, the soil line and vegetation line concept can only supply for a very rough survey in order to differentiate between bare ground and plants.

A more detailed explanation about crop properties requires an analysis of the canopy reflectance curve either while relying on a **full spectrum** approach or on the basis of **discrete wavebands**. A full spectrum analysis might be reasonable if sensing of several crop properties simultaneously is aimed at, *e.g.* biomass, water and nutrients. The techniques that are needed for such multiple crop property sensing are still in a developmental stage. This holds especially for the data processing. On the other hand, discrete waveband sensing lends itself mainly for single or dual crop property recording. Applications for this are in a feasible stage and dealt with in the next section.

## 6.2 Defining the Reflectance by Indices

The easiest method to get information about vegetation is to create a **simple ratio** of the near infrared to red reflectance, *e.g.* with centres of the respective wavebands at 800 and 670 nm (Figs. 6.2, 6.3 and 6.4). Such **near infrared to red indices** with the formula $R\,800/R\,670$ can be used for differentiating between soil and vegetation within a field by **"green seeking"** in order to obtain **spot spraying** instead of treating the whole area including bare soil. Thus for chemical weed control in a fallow field, a site-specific application of herbicides only in areas – where really weeds stand – is possible. Regions that are free of weeds are not treated.

A similar control system that can detect the weed infestation within a vegetated field relies on sensing the vegetation within the tracks of tramlines. This concept is based on the assumption that the site-specific weed infestation within the tracks of tramlines will extend also into the adjacent area. It can be employed on fields that are bare as well as on those that are fully vegetated by a crop excluding the tracks. But then instead of "spot spraying" – as in fallow fields – the control now results in site-specific **strip spraying**. The respective strip is oriented perpendicular to the direction of travel and its length corresponds to the width of the sprayer. This method inherently cannot be very accurate, since the weed infestation may not correspond to rectangular strips.

A "green seeking" effect can also be realized on the basis of the **Normalized Difference Vegetation Index**, abbreviated **NDVI**. In most cases, this index uses reflectance bands that are centered at 800 and 670 nm too as with the above mentioned simple ratio. But instead of a simple relation, a quotient consisting of the difference and the sum of these bands is calculated. Hence the equation is:

$$\text{Normalized Difference Vegetation Index} = \frac{R\,800 - R\,670}{R\,800 + R\,670}$$

The difference between the wavelengths is "normalized" by relating it to its sum. This is done because the reflectance spectra can be on different levels.

The Normalized Difference Vegetation Index (NDVI) is a widely used standard measure that has been employed since many years for sensing from satellites, from aerial platforms and occasionally also from farm machines. Because of the "normalization", it might supply more reliable results than the simple ratio of the same wavelengths. Common values of the NDVI are between 0 and 0.8. Very low values (0.1 and below) represent sand, snow or barren areas of rock. Medium measurements (0.2–0.3) correspond to grasslands and shrubs. And higher values (0.5 to nearly 1.0) indicate lush crops as well as temperate and tropical forests. So the NDVI can be used to roughly classify the earth's surface via satellite data.

In case of sensing from farm machines, the NDVI instead of simple ratios is presently employed in "**green-seeking**" devices for differentiating between soil and vegetation and thus for spot spraying (Fig. 6.5). This technique can be useful for weed control in dryland farming regions where lack of water necessitates rotations with long fallow periods during which weeds must be eliminated. Compared to spraying whole fields, it allows for substantial savings in herbicides and thus reduces their environmental impact. In order to compensate for the additional investment in the green-seeking devices, the technique should be used on adequately large annual areas.

An index that increasingly is used for sensing vegetation is the **Red Edge Inflection Point** (Fig. 6.6). Basically, this index is not defined by a discrete wavelength. Instead, it is located at the precise point where the concave and convex part

**Fig. 6.5** Sprayer boom with optical sensors for spot spraying of weeds in a fallow field (Courtesy of Trimble Agric. Div., Westminster, USA, altered)

**Fig. 6.6** Red edge inflection points

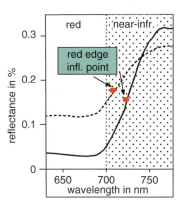

of the respective reflectance curve along the red edge meet. So the red edge inflection point changes its position along the wavelength scale with the course of the spectrum. If the algebraic formula of the reflectance within the red and adjacent near-infrared range is known, this point can be calculated by differentiating twice and putting the result of this equal to zero.

This method of calculating the red edge inflection point requires that the reflectance data in the red and the adjacent near-infrared range are available with a high spectral resolution. Therefore, many bands are needed and the method may become expensive. For this reason, several indices that estimate the position of the inflection

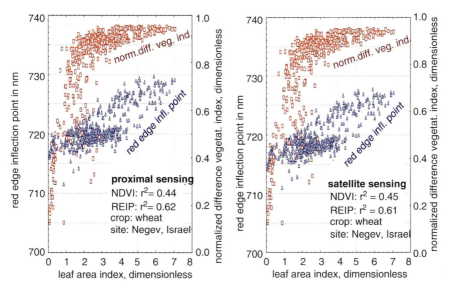

**Fig. 6.7** Sensing the leaf-area-index by using the normalized difference vegetation index (NDVI) or the red edge inflection point (From Herrmann et al. 2010, altered)

point from reflectances at just three or four wavelengths in the red edge region have been developed (Dash and Curran 2007; Guyot et al. 1988; Herrmann et al. 2010). Details to these reflectance indices used for approximating the position of red edge inflection point are in the literature cited.

The prime crop property that farmers are interested in is the plant productivity per unit of field area. For this, valuable control signals can be expected from crops itself, since its properties reflect the environment and human activities. Important is the time span between the recording of crop properties and respective control operations. Signals that are derived from real yield monitoring at harvest time imply long time spans and represent conditions that existed within the whole preceding growing season.

In case the signals are from the still growing crop that is to be controlled, the question is, which plant properties deliver suitable data. Prime candidates for such a control system that is oriented at the yield and takes place within the growing season are the **leaf-area-index** and the **chlorophyll concentration**.

The leaf-area-index as well as the chlorophyll concentration within the leaves can be estimated by means of reflectance sensing. When doing this for lush crops, it is important to use a reflectance index that can differentiate between different leaf-area-indices at high levels. The normalized difference vegetation index generally is not able to do this. As can be seen in Fig. 6.7, the estimation by this reflectance index is admittedly very distinct at the start of the growing period, but then flattens out when the leaf-area-index exceeds the value 2. The **saturation** of this present day standard vegetation index at about this crop growth stage has been observed at several places (Serrano et al. 2000; Mistele et al. 2004; Sticksel et al. 2004). This limitation is serious when taking into account that a well developed small grain crop can attain a leaf-area-index of 8. But in areas where the climate does not allow for lush crops, the use of the NDVI may be a satisfactory approach.

When instead of the NDVI the red edge inflection point was used, this saturation effect did not show up (Fig. 6.7). Furthermore, the correlation ($r^2$) along the whole range of leaf-area-indices was better.

It should be noted that the results of proximal recording with a height of 1.5 m above the ground on the one hand and of sensing from a satellite are very similar (Fig. 6.7, left and right). However, these results were obtained in an arid climate with normally clear skies. In areas with humid climate, such results often cannot be expected because of limitations in the radiation transfer.

### 6.2.1 Precision in Sensing of Chlorophyll

The higher the chlorophyll content of the leaves, the smaller their reflectance is (Fig. 6.3, bottom). Therefore, if for a particular wavelength $\lambda$ instead of the straight reflectance $R_\lambda$, its inversion or its reciprocal is used, a more convenient relation is obtained. Because this inversion $(R_\lambda)^{-1}$ just rises and falls as the chlorophyll content does.

Indicating the chlorophyll effect can be further improved if any interfering with the leaf-area-index is taken care of. Because tendentially, a high leaf-area-index affects the reflectance in the visible range in a similar way as a high chlorophyll content within leaves. However, the result of the leaf- area-index on the spectrum mainly shows up in variations of the near-infrared reflectance (Fig. 6.3, top). Consequently, a further logical step is to subtract the inverted reflection of this near-infrared range – abbreviated $(R_{NIR})^{-1}$ – from the inversion of the respective reflection in the visible range. The thus created difference $(R_\lambda)^{-1} - (R_{NIR})^{-1}$ supplies an optical index that is linearly proportional to the chlorophyll content in spectral bands of the green and red edge range (Gitelson et al. 2003).

It is finally suggested to correct for variations in the leaf structure of crops (Gitelson et al. 2003). These variations are taken care of by multiplying the difference $(R_\lambda)^{-1} - (R_{NIR})^{-1}$ with the non inverted reflectance in the near-infrared range $R_{NIR}$. This product can be simplified:

$$\{(R_\lambda)^{-1} - (R_{NIR})^{-1}\} \times R_{NIR} = (R_{NIR}/R_\lambda) - 1$$

So from these logical deductions, there resulted three indices for estimating the chlorophyll effect, namely

- the simple **inverted reflectance**: $(R_\lambda)^{-1}$
- the **difference of the inversions**: $(R_\lambda)^{-1} - (R_{NIR})^{-1}$
- the **reflectance ratio minus one**: $(R_{NIR}/R_\lambda) - 1$.

The ability of these three indices to estimate the chlorophyll in leaves is shown in Fig. 6.8 by means of their coefficients of determination ($r^2$) from experiments in

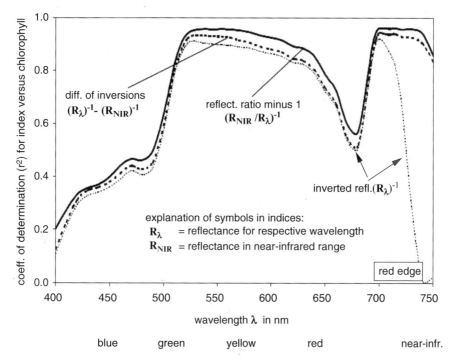

**Fig. 6.8** Coefficient of determination ($r^2$) of three reflectance indices versus chlorophyll content of leaves (From Gitelson et al. 2003, altered)

laboratories. Along the whole range of visible wavelengths plus the red edge, the prediction is slightly improved in the order of firstly the simple inverted reflectance $(R_\lambda)^{-1}$, secondly the difference of inverted reflectances $(R_\lambda)^{-1} - (R_{NIR})^{-1}$ and thirdly the reflectance ratio minus one $(R_{NIR}/R_\lambda) - 1$. So the logical expectations regarding the corrections or enhancing the predictions have been confirmed.

Yet within the visible radiation, the differences between these three indices on the reliability of the estimates are very small compared to the effect of the wavelengths $\lambda$. For the best index – the **reflectance ratio minus one** – the $r^2$ for the wavelengths $\lambda$ involved goes from 0.20 to about 0.95 (Fig. 6.8). Hence getting reliable predictions about the chlorophyll in the leaves is primarily a matter of selecting the most suitable wavelength bands. The bands should be located either in the green range or in the red edge range and perhaps also in both of them.

Based on the reflectance ratio minus one, the recommended indices and wavelengths for sensing of chlorophyll in leaves are either (Gitelson et al. 2003)

- the green chlorophyll index with the formula $(R_{NIR}/R_{520-585}) - 1$, or
- the red edge chlorophyll index with the formula $(R_{NIR}/R_{695-740}) - 1$.

The rather wide wavelength ranges in the green region (520–585 nm) and in the red edge region (695–740 nm) leave some tolerances for narrower bands within and hence for adapting radiometers.

The question may arise why in the main absorption regions – the blue and especially in the red range – the chlorophyll prediction is so poor (Fig. 6.8). The explanation for this just is the high absorption of the incoming light in these ranges. As a result of the very high absorption in these ranges, the **depth of light penetration** into the leaf or leaves is rather low. Because even low amounts of chlorophyll suffice to saturate the absorption. And when saturation is attained, a further increase in pigment content influences neither absorbance nor reflectance. In the green and especially in the red edge region, the absorption of the light by chlorophyll is much lower. Therefore the light penetrates deeper (Ciganda et al. 2012) and the sensitivity of absorbance as well as reflectance to the chlorophyll content of the canopy volume is much higher.

This context also explains to a large extent, why the normalized difference vegetation index (NDVI) saturates when it is used for sensing the leaf-area-index (Fig. 6.7). Because the NDVI too relies on red reflectance. Logically the same disadvantage applies for simple ratios of near-infrared to red radiation.

## 6.3 Sensing Yield Potential of Crops by Reflectance

A key criterion is the crop productivity potential per unit of field area. If this potential can be sensed during the growing season on a site-specific basis, it can supply information for an adequate control of fertilizing, plant protection and irrigation operations.

The chlorophyll content per unit area of the leaves alone cannot serve as an indicator of crop productivity potential because it ignores the leaf area that is available. Likewise, the leaf-area-index alone is not sufficient since it does not provide information about the chlorophyll that is involved. However, it seems reasonable to create the product of the leaf-area-index and the chlorophyll mass per unit leaf area. This product is the chlorophyll mass per unit of field area. It is commonly named chlorophyll index or **canopy chlorophyll** and can be regarded as a key criterion for estimating the productivity of crops.

This key position of the canopy chlorophyll for estimating the productivity of crops results from the fact that it controls the photosynthetic process. And following the original logic of Monteith (1972) it can be deduced that the **gross primary productivity (GPP)** of crops is linearly related to the amount of photosynthetically active radiation that is absorbed. It should be noted that this gross primary productivity (GPP) takes into account all plant parts plus roots as well as respiration energy. So it always overestimates the yield that farmers are interested in. However, this should not exclude the use of the GPP as a relative measure for the assessment of the site-specific situation regarding productivity.

The hypothesis is that a connection between reflectance sensing, absorbed radiation and finally the crop productivity exists. The relation between reflectance and absorbance is evident when regarding the fact that these two quantities together with the transmittance add up to 100, and that the transmittance in the visible range hardly counts within a well grown crop (Fig. 6.1, left).

**Fig. 6.9** Estimation of canopy chlorophyll for a maize- and a soybean crop by a red edge index (From Gitelson et al. 2005, altered)

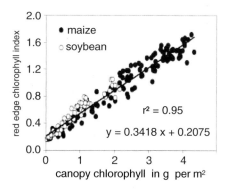

The logical consequence for recording the GPP is that at first information about the canopy chlorophyll is obtained via a suitable reflectance index. Secondly, by multiplying the canopy chlorophyll with the **photosynthetically active radiation (PAR)**, the gross primary productivity is estimated (Peng et al. 2010). The first step requires a suitable reflectance sensing technique, *e.g.* using the green- or the red edge chlorophyll index (Fig. 6.9). Among these, a red edge index (formula: $R_{NIR}/R_{720-730} - 1$) offers the perspective of a non-species specific use.

However, the use of the photosynthetically active radiation (PAR) as a **multiplicator** for the canopy chlorophyll inherently means that a temporally varying factor is involved. The canopy chlorophyll too changes during the growing season, but the PAR varies much more, namely diurnally and in addition with the weather. Frequent records and averaging would have to compensate for this.

A simpler approach is to use the canopy chlorophyll that is sensed by a green- or red edge index as a **direct proxy** of the gross primary productivity without multiplying it by the photosynthetically active radiation (Gitelson et al. 2008; Peng et al. 2010, 2011). Because the actual values of the chlorophyll indices already result from an interaction between the canopy chlorophyll and the radiation. So without any multiplication with the PAR, the effect of the radiation might be sufficiently included. Consequently, a simplified process of estimating the gross primary productivity seems viable. The path of its short sensing logic is:

*Chlorophyll index⇨Canopy chlorophyll⇨Gross prim. productivity*.

This procedure eliminates the short-term variations that might come from the multiplication with the PAR. However, the focus is on the results that can be obtained.

Sensed results were compared with data that were obtained via a so called "eddy covariance system" (Fig. 6.10). This method analyses the carbon dioxide fluxes around defined areas in the field and thus obtains information about the carbon fixation by the crop. The chlorophyll sensor was located on a vehicle, hence the results are based on proximal site-specific sensing. Yet similarly reliable information about the gross primary productivity can also be obtained with remote sensing techniques. Figure 6.11 shows the relation between a green chlorophyll index from satellite data and the gross primary productivity of maize derived from it. The carbon dioxide fluxes in the respective areas within the field again allowed the needed comparisons.

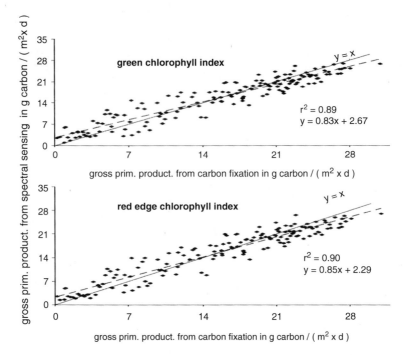

**Fig. 6.10** Comparing the daytime gross primary productivity of maize that was either proximal sensed by chlorophyll indices or measured via fixation of $CO_2$ fluxes in the field. For details to the chlorophyll indices see Fig. 6.8 and text to it. The results refer to 16 irrigated and rainfed maize fields within the years 2001–2008 in Nebraska, USA (From Peng et al. 2011)

**Fig. 6.11** Synoptic monitoring of gross primary productivity of maize using a green chlorophyll index based on Landsat data (From Gitelson et al. 2008, altered)

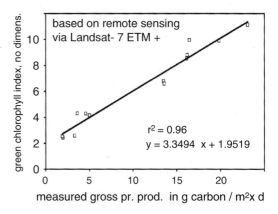

A prerequisite for productivity sensing from satellites is a **clear sky**. So depending on the respective climate, there may be temporal restrictions. And proximal sensing always facilitates high resolutions. However, the Landsat Thematic Mapper Plus satellite system – on which the comparison in Fig. 6.11 is based – also can provide a spatial resolution of 30×30 m. This probably suffices for most site-specific farm operations.

Another prerequisite for every method of crop canopy property recording via reflectance is that the amount of soil in the viewing area of the sensor is kept low enough. Because the reflectance from soil is completely different (Fig. 6.2). So closed canopies are needed. With closely spaced crops – *e.g.* small grains, colza, grass, clover and alfalfa – it is much easier to meet this premise than with widely spaced plants such as maize, beets and sunflower.

However, since the coefficients of determination ($r^2$) for sensing the gross primary productivity (GPP) are very high (Figs. 6.10 and 6.11), a small percentage of soil reflectance within the signals can be tolerated. Gitelson et al. (2008) even state that the effects of soil on the accuracy of GPP retrieval of maize are minimized once the canopy cover exceeds 60 %.

With proximal sensing and early growth stages of crops, **view directing** can help to avoid sensing errors that are caused by soil. This method aims at restricting the sensing view to small strips just above the plant rows and thus leaving out bare inter-row strips or at using an oblique view on canopies in order to avoid reflectance from soil. Yet with remote sensing, the use of such methods hardly seems possible. However an **elimination of soil errors** with row crops via special post-processing of signals might be feasible (Homayouni et al. 2008; Liu et al. 2008; Pacheco et al. 2008). Up to now, such post-processing to eliminate soil errors is not state of the art. It would lend itself for proximal as well as for remote sensing.

A point to consider is the frequency with which the gross primary productivity (GPP) should be monitored. High yielding crops often need several treatments with farm chemicals during the growing season. Accordingly, also several dates for recording the GPP might be reasonable. Whenever immediate processing of the signals and simultaneous use for the control of farming operations is feasible, proximal sensing during these operations would be desirable.

However, this probably would exclude manual inspection and correction of the results by the farmer prior to the control operation. These human interactions – that might take care of *e.g.* respective soil and water situations – could easily be implemented by recording and mapping the canopy productivity results in a separate first step. Processing and combining the results could then take place in a second and stationary step, thus preparing a final control map for the third step, the respective site-specific operation. This **multi-step procedure** would lend itself for **remote productivity sensing**. A definite point for remote sensing is that is can be repeated rather easily any time, provided no clouds obscure the view. With ground based proximal sensing, this is not feasible because of the labor that is involved.

How can a farmer use the signals from productivity sensing for defining the yield expectations? The information about the gross primary production (GPP) indicates the respective situation at intermediate stages within the growing season. This can at best help to get an estimation. If GPP maps were obtained at two different growing stages early in the season, the question is, how the site-specific signals should be combined into a single map.

A logical reasoning for this would be to expect the final yield to be proportional to averages or sums of the GPP from the sensing dates. However, the time span and the temperature between two sensing dates should be considered. The longer the time span and the higher the temperature within limits is, the more growth can be expected. Following this logic, Raun et al. (2001) concluded that the sum of the signals from two reflectance sensing dates should be divided by the cumulative **growing degree days** between the readings. For these calculations that were based on winter wheat in Oklahoma, USA, a growing degree day was defined as the sum of the daily maximum- and minimum temperatures minus 4.4 in °C. It is obvious that details of this procedure for estimating the yield potential must be adapted to respective crops and to local conditions. Raun et al. 2001 used the described method for site-specific sensing with the normalized difference vegetation index (NDVI) during the second half of the tillering time span and obtained an accuracy of 83 %. For high yielding crops, a red edge index instead of the NVDI is recommended (see Fig. 6.7 and text to it).

Though yield estimating via reflectance is not yet state of the art, it probably will become an important method for getting logic to site-specific control algorithms and thus for providing more efficiency to several farming operations.

## 6.4 Fluorescence Sensing

The denotation "fluorescence" indicates a flowing or a flux of radiation that is emitted. The emitter can be plants or also dead material. However, the latter case is left out here.

Contrary to reflectance, the radiation does not originate from irradiance that is simply thrown back from the canopy. Instead, fluorescent radiation can be traced back to photons that did enter an absorption process in plants. However, the photons that enter such processes in plants and induce fluorescence and those that leave the canopy as fluorescent light are different. The light that excites plants to fluoresce always has shorter wavelengths than the fluorescence that finally results from it. The development of fluorescent light implicates a **prolongation of wavelengths**.

Common ranges for plant fluorescence are either the **blue to green region** extending from about 400 to 600 nm or the **red to far-red region** from approximately 650 to 770 nm wavelength. The blue-green fluorescence is induced by ultraviolet light, thus by light that is not used by photosynthesis. It is assumed that the blue-green fluorescence develops within phenolic materials in the cell walls of plants (Buschmann and Lichtenthaler 1998). The red to far-red fluorescence can also result from ultraviolet radiation, yet in addition it can come from light that entered a photosynthetic process and hence chlorophyll molecules. Consequently, it is named **chlorophyll fluorescence**. The visible wavelengths that induce the chlorophyll fluorescence can range from the blue to the red region. But the exciting wavelengths always are below those of the final fluorescence.

The fact that the fluorescence always has higher wavelengths than the respective exciting light means that – in terms of energy per photon – the fluorescent light is less valuable (Fig. 3.1). The chlorophyll fluorescence is regarded as a **by-product of photosynthesis** because it is a means of getting rid of surplus energy. A question is, however, why are plants wasting energy?

# 6 Sensing of Crop Properties

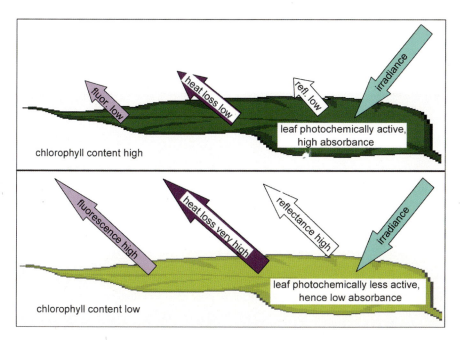

**Fig. 6.12** Two leaves with differences in the chlorophyll content and hence also in the photosynthetic activity, the absorbance and the reflectance of photochemically active radiation. But there are also principal differences in the fluorescence and in the heat loss

**Table 6.1** Use of absorbed light quanta[a]

| Energy used for … | Optimal photosynthesis (%) | Blocked photosynthesis (%) |
|---|---|---|
| Photosynthesis | 84 | 0 |
| Heat loss | 14 | 88 |
| Chlor. fluorescence | 2 | 12 |

[a]Compiled from data by Rosema et al. (1991)

Actually, plants are not wasting much of the solar energy that they receive if the growing conditions are good and the leaves are well equipped with chlorophyll. But the situation is quite different if photosynthesis is impeded or cannot take place at all though sufficient photosynthetic active radiation is available. For growing crops such situations can *e. g.* result from the lack or from a wrong supply of water or mineral nutrients. Diseases or pest infestations too can cause this. Hence plants need means to get rid of all or part of the solar energy if photosynthesis cannot take place. The dissipation of chlorophyll fluorescence and simply of heat to the environment serves this purpose (Fig. 6.12).

It seems reasonable to relate the thus dissipated energy to the energy of the absorbed light (Table 6.1). When the growing conditions are optimal, only 2 % of the absorbed energy is emitted as chlorophyll fluorescence. If on the other hand the photosynthetic process is completely blocked, the emitted chlorophyll fluorescence is six times as high. The heat dissipation as well goes up to the six-fold if no photosynthesis occurs. However, the absolute level of energy dissipation via heat is very much higher than via fluorescence. Under the same conditions, about seven times as much energy is lost via heat than via fluorescence (Table 6.1).

A prime objective in precision farming is obtaining site-specific information from the crop itself about the energetic efficiency with which the photosynthetic process is going on. Such information would supply essential knowledge for a logically controlled site-specific application of *e.g.* water and farm chemicals.

Theoretically, the dissipation either of **heat** or of chlorophyll **fluorescence** would be candidates for respective signals about the status of the photosynthetic process. From the amount of energy that is involved, it might be assumed that the heat dissipation could provide the best signals for this (Table 6.1). However, up to now, detecting the heat emission from plants in an accurate way is possible only in laboratories. For doing this with crop canopies in fields by remote or proximal sensing, no precise techniques exist. This has to do with the fact that the steadily changing weather affects the physical measuring conditions. Contrary to this for fluorescent radiation, remote sensing from aerial platforms and proximal sensing from farm machines in an online and on-the-go manner is state of the art.

### 6.4.1 Fluorescence Sensing in a Steady State Mode

The methods for fluorescence sensing operate either in a **steady state**- or in a **non-steady state** mode. This refers to the radiation that induces the fluorescence and which either is temporally constant (= steady) or for which a change is programmed during the time of sensing. Whenever passive sensing based on natural light occurs, a steady state mode exists. Because practically within the sensing time for a signal, the natural light is constant.

But in case active sensing takes place since the induction is based on artificial radiation (*e.g.* by laser light), this can be done either in a steady state- or also in a non-steady state mode. The latter aims at detecting how the fluorescence behaves when the illumination changes and therefore often in denoted as a **"kinetic" sensing method**. It is explained in detail in the next section. This section deals with steady state methods.

There are many factors that influence the course of the fluorescence spectra such as the exciting radiation and plant species. Yet an important feature are the high levels or peaks in the red (F680) and far red (F735) wavelength region. With a low chlorophyll concentration, the peak in the red range dominates. But with a high chlorophyll content, the peak shifts to the far-red region (Fig. 6.13). This phenomenon is explained by **re-absorption of photons** from the fluorescent light (Lichtenthaler 1996; Buschmann and Lichtenthaler 1998). If the wavelength is below 700 nm, the fluorescent photons coming from internal leaf cells can be re-absorbed when hitting adjacent chlorophyll molecules. With a wavelength above

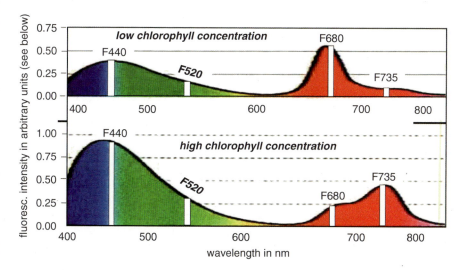

**Fig. 6.13** Steady state fluorescence spectra of maize leaves with low or high chlorophyll concentrations that were excited by ultraviolet light. The *white, thin* columns represent wavelength bands that are suitable for plant stress detection (Compiled from data by Guenther et al. 1999 and Lichtenthaler 1996). For details to the units on the ordinate see Schwartz et al. 2006

700 nm, re-absorption is not possible. And the higher the chlorophyll content within the leaves, the more re-absorption occurs. This explains why the fluorescence ratio F680/F735 can be regarded as a reliable indicator of the chlorophyll concentration in leaves.

The ratio of the blue to green fluorescence bands (Fig. 6.13) is less influenced by growth conditions. The blue band (F440) referenced to the far red band is used as an indicator for the sensing of fungi infections (Thiessen 2002).

The wavelengths of reflectance and fluorescence can be the same. Hence techniques that enable to differentiate between both types of radiation are needed. In case of active sensing with artificial light, **modulation** of the exciting radiation allows for this. This modulation can be obtained via pulses of the inducing laser light.

But in case of passive sensing, when natural light induces fluorescence, a modulation technique is not available. In addition, the chlorophyll fluorescence emitted by vegetation represents a small part of the radiation (Table 6.1). The sun – when it is used for inducing – poses a dilemma. It stimulates the formation of fluorescence. But simultaneously, it masks the fluorescence with reflectance. In order to prevent that the fluorescence is obscured by the reflected light, the sensing is done within narrow wavelength ranges where the incident natural light is either completely excluded or at least attenuated. The solar spectrum has several such very narrow ranges that are the result of light **absorption** by atmospheric molecules. These narrow spectral ranges are known as **Fraunhofer lines**, named after the physicist von Fraunhofer.

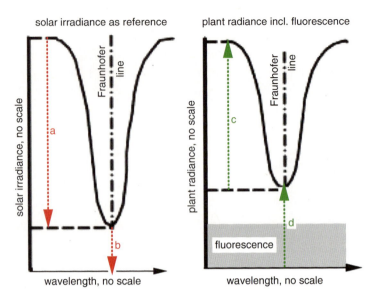

**Fig. 6.14** Basics for passive sensing of the chlorophyll fluorescence of vegetation using Fraunhofer lines. The radiance that is coming from the plants (*right*) is compared to the irradiance (*left*) that hits an adjacent reference panel situated in the same illumination conditions. The radiations denoted with the letters *a*, *b*, *c* and *d* are sensed. The components *a* and *c* are recorded adjacent to the wavelength of the Fraunhofer line, whereas *b* and *d* are measured precisely in this line. It should be noted that adding the components a and b or alternatively c and d does not make sense since the wavelengths are different

In order to be useful, such a Fraunhofer line must be in a region where fluorescence recording makes sense. Within the red and far red range, such Fraunhofer lines exist

- at 656.3 nm wavelength due to absorption by hydrogen
- at 687.0 nm wavelength due to absorption by oxygen and
- at 760.0 nm wavelength again due to absorption by oxygen.

The 687 nm oxygen wavelength almost coincides with the red peak of the chlorophyll fluorescence spectrum that is recommended for plant stress sensing (Fig. 6.13, top). However, at this wavelength as well as at the 656.3 nm hydrogen wavelength, the absorption of solar radiation by the atmosphere is much less than at the 760.0 nm oxygen wavelength. It is for this reason that generally the 760.0 nm oxygen band is preferred (Liu et al. 2005; Maier et al. 1999; Moya et al. 2004), which still is rather near to the recommended band in the far-red range.

The basic measurements needed to detect sunlight-induced fluorescence in Frauenhofer lines are outlined in a simplified way by letters in Fig. 6.14. The left inverted Gaussian curve indicates the band of **irradiance** that is directed towards the canopy. The intensity of this irradiance decreases on its way through the atmosphere as a result of absorption, *e.g.* by oxygen. The vertical Frauenhofer line represents the wavelength that is the center of the attenuation.

The right inverted Gaussian curve (Fig. 6.14) stands for the **radiance** band that the plants emit, hence for **reflected plus fluorescent radiation**. This radiation that is directed away from the canopy is compared to the irradiance (left).

The purpose of recording radiation lateral to the Fraunhofer line is to obtain an estimate of the reflectance R without any contribution of fluorescence. The quotient c/a provides this reflectance R, which can be regarded as a coefficient of reflection (see Sect. 3.1). Since the solar irradiance hits a reference panel that is non-fluorescent, its component b in the Fraunhofer line cannot contain energy that is converted to fluorescence. This means that the product of reflectance R and component b provides the absolute reflection part precisely within the Fraunhofer line. And because the plant radiance within the Fraunhofer line d contains reflection as well as fluorescence f in absolute values, subtracting the product R × b from it provides finally the fluorescence.

So fluorescence $f = d - R\,b = d - c\,b/a$.

Moya et al. (2004) and Liu et al. (2005) have shown that this method provides good results. However, this method inherently does not allow to sense vegetation stress via the F680/F735 ratio that can be used with active sensing.

### 6.4.2 *Fluorescence Sensing in a Non-Steady State Mode*

This method goes back to Kautsky and Hirsch (1931), who watched the fluorescence intensity of leaves that were held in the dark and then suddenly were illuminated. It showed up that starting from a very low level, the fluorescence intensity rose steeply to a maximum. While the illumination continued, the fluorescence then fell gradually. This descent of fluorescence took several seconds and sometimes included smaller intermediate maxima. Finally – with the illumination still on – the fluorescence got to a steady state level (Fig. 6.15). This reaction of fluorescence to varying illumination often is called the **Kautsky effect**.

The physiological background of this phenomenon is that the photochemical factory of a plant needs adjustments and time to get to the steady state operational mode. The **initial rise** of the fluorescence intensity with the start of the illumination is attributed to progressive saturation of the photochemical system. And the **slow decrease** of the fluorescence intensity after having attained the maximum is most likely due to protection mechanisms since the plant has to avoid adverse effects of an excess of light.

The significance of the Kautsky fluorescence curve is that it allows to get information about the most important crop property – its photochemical devices – under fairly controlled conditions. Because the light that enters the process is artificially programmed. And simultaneously, the waste energy that the crop sheds as fluorescence is recorded. Temporal optical indices that represent variations in the rise, in the maximum or in the decay of the Kautsky fluorescence have been developed and can assist in assessing the photochemical situation (Buschmann and Lichtenthaler 1998; Thiessen 2002).

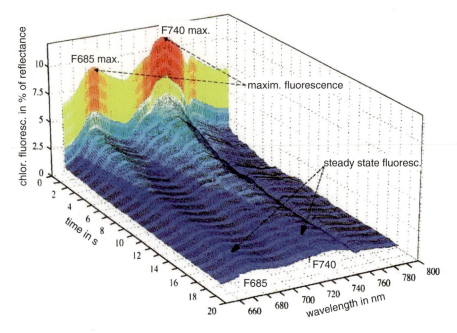

**Fig. 6.15** Spectra of chlorophyll-fluorescence that were acquired from maize leaves after 5 min of dark adaptation in the laboratory. The illumination with a 300 W xenon arc lamp began at time 0 and was to simulate constant solar radiation following the dark adaptation. The initial rise of the fluorescence according to the Kautsky effect cannot be seen. But the gradual decrease to the final steady state fluorescence is apparent. The graph also points out the maxima of the chlorophyll fluorescence – as explained in Fig. 6.13 – at 685 and 740 nm wavelength (From Entcheva Campbell et al. 2008, altered)

### 6.4.3 Fluorescence or Reflectance

Principally, both active as well as passive sensing of chlorophyll fluorescence can be done on a site-specific basis. However, passive sensing in Fraunhofer lines is not yet used in farming operations whereas active sensing of the chlorophyll content via the red/far red fluorescence ratio is state of the art. An advantage of fluorescence sensing – when compared to chlorophyll sensing via reflectance – is that radiation from the soil does not result in errors. Because the soil does not emit fluorescent radiation. So contrary to chlorophyll sensing by reflectance, a not **closed crop canopy** is not a problem.

But whereas reflectance allows to sense the chlorophyll content within the leaves as well as the leaf-area-index, this is not *per se* possible by fluorescence. Even if fluorescence sensing is done in combination with a device that scans the complete field surface, it is the canopy surface that supplies the signals and leaves below it are disregarded. Hence fluorescence cannot as well provide information about the chlorophyll content per unit of field area and thus about the **longterm yield** potential of crops – as suitable reflectance indices can (see Sect. 6.3). However, a combined use of fluorescence- and reflectance sensing may be reasonable.

The **Kautsky method** is not easy to apply in an online and on-the-go mode in field operations since the sensing of its effect needs time. However, a similar stimulating effect on fluorescence can be obtained with very short **laser light flashes** that hit the canopy and saturate the photochemical system for a tiny time period of 1 s or less. A subsequent sensing time span of 1–2 s can be realized while a tractor passes along crop rows if recording takes place successively from the front to the rear of the machine. Interesting approaches along this line have been dealt with by Thiessen (2002) and Hammes (2005).

The sensing via signals of reflectance on the one hand or from steady state- or from non steady state fluorescence on the other hand has different objectives within plants. With reflectance sensing, the main objective is the biomass – precisely the leaf-area-index – in combination with the chlorophyll content of the leaves. The product of the leaf-area-index and the chlorophyll content within leaves – thus the chlorophyll content per unit field area – is the classical objective. With **steady state fluorescence**, the objective mostly is just the chlorophyll content within leaves *per se* and sometimes – in case of blue/green fluorescence – the phenolic material in cell walls. Finally, with the non steady state **Kautsky fluorescence**, the prime objective is the rate of electron transport within the photosynthesis. In this case, the biomass and chlorophyll situation is taken as it is.

These differences in sensing objectives have temporal implications. The biomass of crops and the chlorophyll content per unit field area are rather long-term phenomena. Compared to these, the electron transport in the photochemical systems is a very fast matter. And the chlorophyll content within leaves is – on a temporal basis – in between of these extremes.

From this follows that reflectance sensing provides information about crop properties on the basis of rather long time spans, *e.g.* from several days to a few weeks. The steady state fluorescence methods can supply signals that are related to properties for a few days. Finally, the Kautsky fluorescence is able to indicate immediate problems in plant physiology.

A need for rapid detection exists in case of many fungal infections of crops. A fast detection and immediate crop protection action can provide the opportunity of preventing additional infection cycles and hence save fungicides. Because of its physiological background, the Kautsky fluorescence can supply such immediate control information about infections. By using indices that were derived from the non-steady state Kautsky fluorescence, it has been possible to indicate fungal infections on leaves several days earlier than the human eye was able to detect them (Buerling et al. 2009; Moshou et al. 2009; Thiessen 2002). So the perspectives for early detection of fungi are good. However, on-the go detection from farm machines is not yet state of the art. And since different crops, several fungi and various environmental conditions are involved, there are still problems that must be solved.

Crop properties will probably exist as short-term and long-term phenomena. So both reflectance- as well as fluorescence sensing might be reasonable. Yet since fluorescence signals are much weaker than those from reflectance, the preferred recording mode for it in the near future will be proximal- instead of remote sensing.

## 6.5 Sensing the Water Supply of Crops by Infrared Radiation

The perspectives of crop development depend not only on the chlorophyll situation but equally well on the water supply. Hence information about the water supply is needed too for the control of several farming operations such as irrigation or the application of farm chemicals. And getting this information directly from the crop instead of from the soil has distinct advantages. This automatically takes into account the site-specific influence of water-tension within the soil on the supply of the plants. Depending on the respective soil texture, the water tension varies considerably and thus the amount of water available. Furthermore, when sensing the supply of the crop instead of the soil, the influence of the root development on the situation is taken care of without any attention. And the root development varies during the growing season as well as on a site-specific basis.

Signals can be derived either from **reflected radiation** or from **emitted thermal radiation**.

### 6.5.1 Sensing Water by Near- and Shortwave-Infrared Reflectance

The visible wavelengths can be left out for water sensing from crop canopies. Instead, the reflected infrared domain can be considered as basis for signals. As with many other applications for spectral sensing, generally the best results are not obtained with wide bands that extend over long wavelength ranges within the spectrum, but instead from narrow bands or indices that use them. Such **hyperspectral bands** ranging from 1 to 10 nm width allow to select the most sensible region within the spectrum and to avoid the weakening or averaging effect that inevitably is associated with wide bands.

Soil- and canopy sensing differ not only in the region of the spectrum that is used (Fig. 6.2). With soil sensing via reflectance, a fundamental problem is the fact that it is based solely on the top-surface that is hit by the radiation. In order to get information from below the surface, sensing of vertical soil profiles is necessary.

This can only be done via proximal sensing from terrestrial vehicles or from farm machines that cut a slit into the soil. This limitation does not exist for sensing of crops by infrared radiation because this penetrates well into the vegetation and is reflected back from various layers below the canopy surface. Consequently, canopy sensing – contrary to soil sensing – lends itself for proximal sensing as well as for remote sensing from satellites or from aerial platforms.

Which wavelengths of narrow band reflectance can indicate the crop water supply? Table 6.2 shows results of experiments, for which the coefficients of determination ($r^2$) were above 0.60. The bandwidths of the wavelengths listed were between 1.5 and 10.0 nm with one exception, for which the range is indicated.

**Table 6.2** Spectroscopic sensing of water content of crop canopies (results listed by coefficients of determination $r^2$)

| Reference | Plant species, sensing methods | Reflect. indices R with wavelength in nm | Coeffic. of determinat. $r^2$ |
|---|---|---|---|
| Jones et al. (2004) | Maize | R1450 | 0.67 |
| | (*Zea mays*) | R2250 | 0.61 |
| | Sensing in the lab | | |
| | Spinach | R960 | 0.94 |
| | (*Spinacea oleracea*) | R1150 – 1260 | 0.93 |
| | Sensing in the lab | R1450 | 0.85 |
| | | R1950 | 0.80 |
| | | R2250 | 0.79 |
| Penuelas et al. (1993) | Gerbera, 3 varieties | R970/R900 | 0.79 |
| | (*Gerbera jamesonii*) | R970/R900 | 0.86 |
| | Sensing in the lab | R970/R900 | 0.84 |
| Sims and Gamon (2003) | Means of 23 species, | R960 | 0.75 |
| | Sensing in the field | R1180 | 0.81 |
| Sonnenschein et al. (2005) | Wheat | R900/R970 static[a] | 0.90 |
| | (*Triticum aestivum*) | R900/R930 – 990 var[b] | 0.90 |
| | Sensing in the field | | |
| | Wheat | R900/R981 | 0.90 |
| | (*Triticum aestivum*) | | |
| | Airborne sensing | | |
| Yang and Su (2000) | Rice | R697 | 0.81 |
| | (*Oryza sativa*) | R1508 | 0.90 |
| | Sensing in the field | R2113 | 0.96 |

[a]Denumerator and denominator static as listed
[b]Denumerator static, denominator was varied between R930 and R990

The wavelengths of the successful narrow band indices are located in a rather wide range extending from 697 to 2,250 nm, thus within the **near-infrared region** (700–1,300 nm) and the **short-wave infrared region** (1,300–2,500 nm). Within the whole spectrum, the near-infrared radiation has the highest foliar reflectance due to scattering in the canopy, and the short-wave infrared region has ranges of high liquid water absorption. But it is not clear whether these special properties of liquid water are important. High coefficients of determination can be seen in the near infrared- as well as in the short-wave infrared region (Table 6.2). However, sensing in the near-infrared is less expensive because these sensors are mass-produced and less sensitive to temperature.

Generally, it can be expected that *ceteris paribus* sensing in laboratories provides higher accuracies than proximal sensing in fields while remote sensing from satellites or aerial platforms supplies the least precise results. However, such an effect of the sensing location is not indicated by the coefficients of determination in Table 6.2. This probably is due to the fact that just comparing results from several investigations – among them many varying factors such as *e.g.* plant species, wavelength, illumination, sensing distance – does not supply knowledge about the basic effects of single parameters. This knowledge can be derived from investigations that allow a system analysis, in which the effect of varying factors can be controlled.

The graphs in Figs. 6.16 and 6.17 have been obtained by using modern canopy reflectance models (Clevers et al. 2010). These models offer the advantage that the results are independent of site and plant species and thus fairly universally valid. For details to these models see Jacquemoud and Baret (1990) as well as Verhoef (1984). The graphs have been supplemented with data to remote sensing preferences.

Across the whole near-infrared and shortwave-infrared region, a decrease of the water content causes a rise in the reflectance (Fig. 6.16). This is important, particularly when crops grow older. Small differences in the water content due to short term weather variations in early growth stages are more difficult to detect. However, the visual inspection of the course of the reflectance curves does not provide reliable information about the best wavelengths for sensing. This knowledge can be obtained from calculations of the respective correlations. Details to this cannot be dealt with here, yet important results are indicated in Figs 6.16 and 6.17.

Accordingly, good spectral information about canopy water content can be obtained from features centered at either **970** or **1,200 nm wavelength**. These wavelengths are approximately at the bottom of the two dips within the near-infrared spectrum. For best results, either first derivatives or simple **wavelength ratios** from the slopes of the spectrum just adjacent to these wavelength centers should be used. Recommended ranges for the derivatives or the simple ratios are either the left or alternatively the right slope from the 970 nm centre and also the left slope from the 1,200 nm center.

The green framed columns in Fig. 6.17, bottom, show the recommended slope ranges. Some of the simple ratios in Table 6.1 correspond closely to these ranges.

For **sensing from satellites**, absorption of radiation due to water vapor in the atmosphere within some wavelength ranges needs attention. Because sensing in these regions results in very noisy or inaccurate signals, they should be avoided. Within the short- wave infrared region (1,300–2,500 nm), there are two ranges that are affected. In Fig. 6.16, these ranges are indicated by red columns. It is not accidentally that these ranges are also close to dips within the spectral curves. Because the dips are the result of high spectral absorption by liquid water in the canopy. And the red columns indicate high absorption, yet by water vapor. Precisely seen, however, the centers of **water vapor** absorption are shifted approximately 50 nm to shorter wavelengths when compared to the dips in the curves for the **liquid water** in the canopy (Clevers et al. 2008, 2010).

In the near-infrared region, there are only two very thin bands of atmospheric absorption by water vapor. These narrow bands of absorption by water vapor – that too are radiation barriers – are located rather closely to the recommended slope ranges for canopy water sensing (see red vertical lines in Fig. 6.17). The centers of these bands in the near-infrared – which should be avoided – are at 940 and 1,140 nm (Gao and Goetz 1990). Thus "hyperspectral precision" in selecting the wavelengths for remote canopy water sensing is needed.

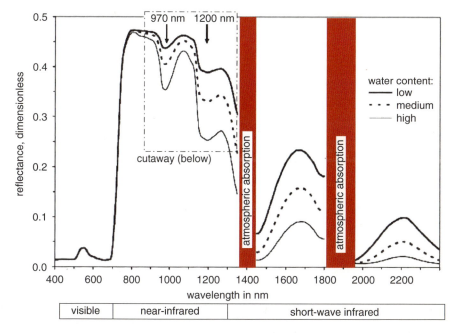

**Fig. 6.16** Canopy water content, reflectance and sensing barriers (*red columns*) due to atmospheric water vapor (From Clevers et al. 2010, altered and supplemented)

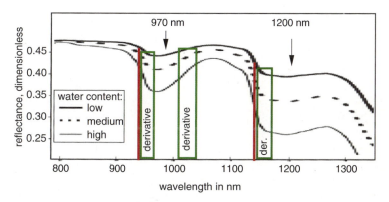

**Fig. 6.17** Enlarged cutaway from the figure on top with changed scales showing the locations of recommended bands for canopy water sensing and of barriers (in *red*) due to atmospheric water vapor. Either first derivatives or simple wavelength ratios based on narrow bands from within the *green* framed columns should be used (Drawn from data by Clevers et al. 2008 and 2010)

Summing up, it can be concluded that the perspectives for site-specific canopy water sensing look good. However, canopy water sensing is not yet state of the art. It could be practiced either in a proximal sensing mode simultaneously with other farming operations or in a remote mode, *e.g.* from satellites. The proximal sensing mode lends itself primarily for the direct control of farming operations, whereas remote sensing is especially suited for tactical inspections of large fields within a short time. The steadily increasing number of satellites improves the perspective to perform such remote tactical inspections frequently and hence to sense the occasionally fast changing water situations. This might be valuable for site-specific irrigation practices.

### 6.5.2 Sensing Water by Emitted Thermal Infrared Radiation

This method of sensing the water supply of crops relies on the **transpiration** that takes place at the surface of leaves. The sensing signals are based on the cooling effect of water transpiration on the canopy. The energy for converting the liquid water to the gaseous state causes a cooling of the canopy. Because of this, on a very hot day it can be more comfortable for humans to be inside a dense forest.

The more water is transpired, the more the canopy temperature is below the temperature of the surrounding air. The **crop water stress index (CWSI)**, developed by the U.S. Water Conservation Laboratory in Arizona (Jackson et al. 1981) depends on this.

The main criterion of the crop water stress index therefore is the **temperature difference** between the canopy leaves and the air. If a crop has water stress and therefore cannot transpire, there is hardly any difference between leaf- and air temperature. The red upper baseline in Fig. 6.18 stands for this situation. For the not water stressed crop, the transpiration depends on the relative humidity of the air. The lower the relative humidity is, the more the crop transpires. And the more the crop transpires, the lower the temperatures of the leaves. The green lower baseline (Fig. 6.18) represents the case of the fully transpiring, non water stressed crop. The vertical distances between the upper and lower baseline define the differences of the temperature span between leaves and air that occur when non transpiring crops on the one hand with fully transpiring plants on the other hand are compared. It should be realized that this comparison is about **differences of differences**.

The graphical interpretation of a specific case starts with a point of the actual situation, *e.g.* point B in Fig. 6.18. This point has a respective leaf- minus air temperature. In addition, it is located at a distinct distance to the lower baseline. This distance to the lower baseline represents the absolute situation. However, this absolute situation should be referenced or normalized. The **normalization** is done by dividing the absolute situation for point B by the total distance between the upper- and the lower baseline. This quotient is the crop water stress index (CWSI), which has differences of differences in the numerator as well as in the denominator (Fig. 6.18). A CWSI of 0 indicates that a crop is fully transpiring and hence is well supplied, whereas a value of 1 means maximal water stress. Generally, watering is recommended when the CWSI value for maize is above 0.22 (Irmak et al. 2000).

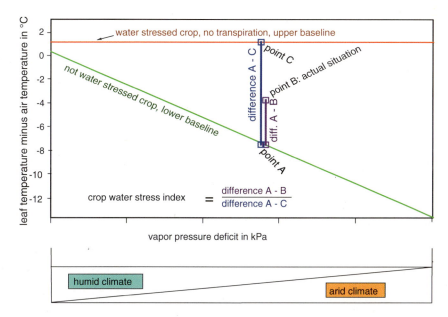

**Fig. 6.18** Graphical interpretation of the crop water stress index

Sensing by the crop water stress index has distinct advantages.

- The first advantage is that the sensing criterion – the transpiration – is a result of the plants own internal moisture control system. The transpiration reflects **response of the crop** to the existing water supply. This response is probably more pinpointed than other human estimation of the needed moisture.
- The second advantage has to do with the physical quantity that is involved. Recording temperatures is easily possible in a non-contact, site-specific mode, either by proximal on-the-go sensing or as well by remote recording (Sadler et al. 2002).

However, fact is also that sensing of the canopy- as well as of the air temperature plus the vapor pressure deficit alone does not suffice for the recording of the crop water stress index. The positions of the upper and lower baseline (Fig. 6.18) must be precisely defined. Theoretically, these positions depend on several meteorological factors – among others wind speed – and on crop properties. Rather complicated approaches have been proposed for determining the positions of the baselines (Jackson et al. 1988; Payero and Irmak 2006). A rather simple recommendation (Alchanatis et al. 2010; Moeller et al. 2007) that is empirically based is:

- to define the **upper baseline** by adding 5 °C to the sensed air temperature and subtracting this sum from the respective leaf temperature (see ordinate in Fig. 6.18).

- to define the **lower baseline** by using the temperature of an artificial wet reference surface – such as a small wet foam mat floating in water – instead of the leaf temperature. The air temperature is then subtracted from the temperature of this artificial wet reference surface. The latter provides for more stable temperatures than moist leaves do.

Provided the baselines are well defined, the crop water stress index can be regarded as a prime choice for estimating the water situation of crops (Meron et al. 2010). However, there are still limitations for its use in practice. An important limitation is that very erroneous signals result if the radiation hits soil or plant residues instead of active leaves. Hence closed canopies offer ideal conditions. But these might not exist in early growing seasons or with widely spaced crops. Correctives that can provide a remedy are either narrowing the sensors view to rows of crops in case of proximal sensing or alternatively sorting out soil data via post-processing by means of reflectance signals. For details to this see Sect. 6.3.

The crop water stress index or similar thermal indices are occasionally used for the control of large area irrigation in dry regions, yet presently barely ever employed in regions with humid climate. Of course, in humid regions there is less water stress of crops. But it is also more difficult there to get precise signals. This is because the lower the vapor pressure deficit is, the smaller the differences of the temperature spans are, which are the basis of the crop water stress index (Fig. 6.18). Advances in measurement techniques may help to get precise signals even under more humid conditions.

The water situation should be sensed under uniform conditions of illumination. In order to get signals from the time when the maximal expression of water stress exists, the sensing should be done around noon and possibly when the sky is clear. Remote sensing facilitates getting signals that are based precisely on the same time within the whole field, but the delivery of the results might delay the application. With proximal sensing the situation is *vice versa*.

A challenging perspective would be **site-specific irrigation** based on either the crop water stress index or on water sensing via reflectance. This could save water or improve yields in many cases. Real-time control should be aimed at since irrigation is a timely matter.

Provided the signals were available, the site-specific irrigation control for systems with center pivoting or linearly moving booms could be realized via nozzles thats either pulse or which have variable orifices. With **pulsing nozzles**, the "on" to "off" time ratio controls the water supply. In case of nozzles with **variable orifices**, moving a coned pin towards the direction of the orifice varies the flow rate. For actuating the site-specific irrigation by these devices, a sophisticated telemetric control network is feasible. Details to this have been dealt with by Camp et al. 2006; Pierce et al. 2006 and Sadler et al. 2000.

However, there still are obstacles to real time site-specific irrigation. At least this holds for site-specific irrigation that relies on sprinkler systems and is based on sensing the water supply via radiation. The path of the radiation that senses the moisture status should not be obstructed by water drops. These would cause large sensing errors. So any real-time control system for sprinklers will have to provide separate paths for the radiation that is used for sensing on the one hand and for the water drops on the other hand.

Guiding the water from the sprinkler boom down into the canopy via vertical flexible hoses would do away with this problem. This technique also reduces evaporation losses. However, it results in line watering instead of treating the whole area and is hardly used in large scale farming.

In short, site-specific irrigation in real-time based directly on the water supply of crops or soils for large scale farming is not yet state of the art (Evans and King 2010, 2012) though it is an urgent matter because of the increasing lack of water. Principally, site-specific irrigation could also rely on maps of **topography** and **soil texture**. Both factors can influence the water supply of crops. However, the direct control via the water stress of crops should be preferred.

## 6.6 Sensing Properties of Crops by Microwaves

Microwaves operate with the longest waves that are used for sensing – hence the name might be misleading (Fig. 3.1). The historical explanation for the name is attributed to the fact that microwaves are shorter than radiowaves. Compared to sensing by visible and infrared reflectance, observations in the microwave region are complementary and also more complex. And applications are still more limited, but they are growing.

A fundamental advantage is that microwaves can penetrate the atmosphere including clouds during day and night. Only heavy rain or snow can prevent the use. As far as the transmission through the atmosphere is concerned, microwaves are predestined for remote sensing. Yet the crop properties that can be detected are quite different from those sensed by visible and infrared radiation. Whereas the latter waves can sense chlorophyll, water and leaves, the main objectives of recording via microwaves up to now are canopy structure, vegetation type and biomass.

Satellites operate on artificially created microwaves, which are commonly called **radar-waves**. Some radar-wave configurations that are used on modern satellites are listed in Table 6.3.

**Table 6.3** Frequencies and wavelengths of radar bands within the 0.4–15 GHz range[a]

| Letter designation | Frequency in GHz | Wavelength in cm |
|---|---|---|
| P-band | 0.44 | 68 |
| L-band | 1.28 | 23 |
| S-band | 3.0 | 10 |
| C-band | 5.3 | 5.7 |
| X-band | 9.6 | 3.1 |

[a]Extracted from Ulaby et al. 1996 and altered. S-band wavelength corrected. Data refer to synthetic aperture radar (SAR). Additional bands can be available

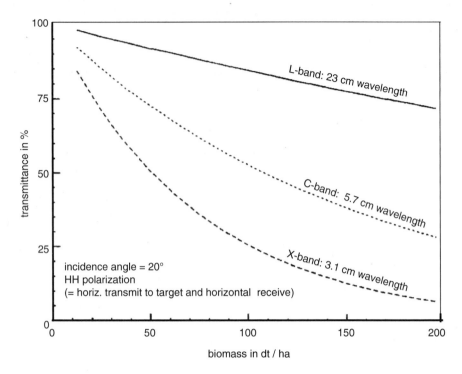

**Fig. 6.19** Transmittance of radar waves though a grass canopy (From Ulaby et al. 1996, altered)

Radarwaves might penetrate crop canopies deeper than visible radiation. A rule of thumb is that for radar waves the **penetration depth** in crops is about half the wavelength (Heinzel 2007), whereas with soils it is one fifth of it (Sect. 5.2.3.2). Hence the penetration in crops is approximately 2–3 times deeper than in soils. But this is rather a rough estimation. It is not the wavelength alone that determines the penetration. Among others, the biomass of a crop (Fig. 6.19) and its water content are important.

For microwaves, a plant canopy acts similar to a three-dimensional water-bearing structure. Within this structure, especially the leaves are the carriers of water. It is mainly the effect of the **water** on the reflection that allows to sense crop properties. And regarding this water effect, the basic functioning of microwaves within canopies is very similar to that within soils. Instead of the term "reflection", for microwaves generally the notation "**backscatter**" is used. It is the backscatter that can supply signals about crop properties via remote or proximal sensing. For details to the measuring of the backscatter see Fig. 5.18.

When a canopy is closed and the radar wavelength is approximately the same size as the leaves, not much of the radiation energy may reach the soil underneath. Yet depending on the crop and its canopy, long waves may get to the soil. So if the objective is to sense only the crop without any interference by the soil below, limitations in the length of the waves might be necessary. If the signals received are partly from the crop and to some extent are influenced by the soil below, it is very difficult to interpret them.

**Polarization** of microwaves has become an important feature when regarding the ability to sense crop properties. Whereas for non polarized microwaves the photons vibrate around the axis of propagation in all directions at random – though the wavelength is uniform – with polarized microwaves the vibrations are restricted to a common plane (Fig. 3.6). This plane might be horizontally (H) or vertically (V) oriented. Yet the waves that are sent to the canopy and those that are returned to the receiver might not be in the same plane. Only those microwaves, which get back to the recording receiver after a **single reflection** on the surface of the canopy, maintain their original polarization or their plane of vibration. These waves are recorded by the receiver as "**like-polarized**" and represent **surface scattering**. However, this type of scattering occurs mainly with short waves.

When very short waves are excluded, the majority of the waves are thrown back and forth within the crop before eventually being scattered towards the receiver. The result is that these waves – at least partly – get back to the receiver in a depolarized state. This means, the original polarization is lost. The depolarization does not alter the wavelength. And depolarization hardly occurs with waves that hit soil instead of a crop.

Hence the amount of depolarization can help to discriminate between soil and vegetation. A radar receiver can detect the amount of depolarization and thus record **volume scattering**. For this, the emitter of the satellite might just send polarized radiation within a vertical plane and receive in a horizontal plane, or vice versa. This would be a "**cross-polarized**" operational mode, which would indicate about volume scattering.

The main orientation of the plants can affect the backscatter. For many grain crops, the stems and its leaves are mostly oriented in a vertical direction, though this might depend on the growth stage If the incident radiation has a plane of polarization that is parallel instead of cross to the main orientation of the plants, the backscatter is lower. Hence using horizontally as well as vertically oriented polarizations can allow discriminating between crops according to their canopy structure or their habitus.

Another factor that affects the radar backscatter of crop canopies as well as of soils is the incidence angle. This is the angle with which the radiation coming from the satellite hits the earth.

Which crop properties can be sensed by means of radar waves when the present possibilities of varying frequencies or wavelengths, of different polarizations as well as incidence angles are exploited? Details to this question have been dealt with by Brisco and Brown 1998; Gherboudj et al. 2011; Jiao et al. 2010; Mattia et al. 2003; McNairn et al. 2009; Shimoni et al. 2009; Steingiesser and Kühbauch 1998. Summing up, the situation is:

- Chlorophyll content cannot be recorded.
- Indication of the crop water content is unstable. The results for water sensing via infrared radiation are better (Sect. 6.5).

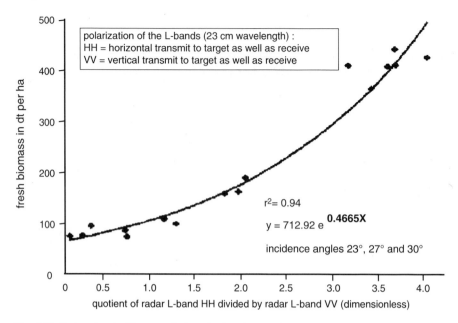

**Fig. 6.20** Estimating the biomass of winter barley via a quotient of like-polarized radar waves and remote sensing from satellite (From Kühbauch and Hawlitschka 2003, altered)

- Recording of the height and the biomass of vegetation is possible. Non-polarized radiation might just provide rough estimates, whereas fully exploiting the potentials of polarization allows precise indications (Fig. 6.20).
- Leaf-area-index can be indicated if the wavelength fits to the crop. With maize, the L-band (23 cm wavelength) provided good results whereas the X-band (3.1 cm wavelength) completely failed (Jiao et al. 2010).
- Crop classification for wide area monitoring and mapping is possible. This is the main present application. Supplementing the indications based on radar by estimations that rely on visible and infrared radiation is not necessary.

Modern radar satellites can provide the spatial resolution that is needed for site-specific farming. The resolutions that can be obtained go down to a few m. And regarding the temporal resolution, the prospects too are good because the number of satellites is increasing. This means that temporal limitations for receiving information about crop properties via remote microwave sensing probably will not exist any more in the future. The information can be supplied day and night irrespective of clouds and haze. Only heavy precipitation might prevent the use. Because of the almost unlimited temporal possibilities, proximal sensing of crop properties via radar does not seem attractive. At least not as long as proximal sensing needs the attention of a driver and therefore is difficult to repeat any time.

**Fig. 6.21** Georeferenced and mapped crops (*top left*) that were classified via polarized radar from satellite within a wide area (From Pottier and Ferro-Famil 2004, altered)

Up to now, the most frequent agricultural application of remote sensing via radar waves is **crop classification** (Fig. 6.21). It provides a fast record about the areas of crops that are growing within wide regions. This information is used by governmental departments, farm agencies and agribusiness institutions for planning purposes.

Monitoring and mapping the **condition of crops** – *e.g.* by recording the site-specific biomass or leaf-area-index several times during the season – could be a domain of application within single farms as well. It might be questioned whether it suffices to record the biomass or the leaf-area-index or whether in addition the chlorophyll within the leaves should be detected, as it is possible when using visible and infrared radiation. The answer to this question might – on the one hand – depend on crop type. Because precise yield estimates during the growing period for grain crops might require more information than those for forage crops. On the other hand, the operational possibilities must be considered. Because in maritime areas, long time spans can occur during which clouds prevent any remote sensing from satellites for visible and infrared radiation.

# References

Alchanatis V, Cohen Y, Cohen S, Moeller M, Sprinstin M, Meron M, Tsipris J, Saranga Y, Sela E (2010) Evaluation of different approaches for estimating and mapping crop water status in cotton with thermal imaging. Precis Agric 11:27–41

Brisco B, Brown RJ (1998) Agricultural applications with radar. In: Henderson FM, Lewis AJ (eds) Principles and applications of imaging radar. Manual of remote sensing, 3 edn, vol. 2. Wiley & Sons, New York

Buerling K, Hunsche M, Noga G (2009) Early detection of puccinia triticina infection in susceptible and resistant wheat cultivars by chlorophyll fluorescence imaging technique. In: van Henten EJ, Goense D, Lockhorst C (eds) Precision agriculture ´09. Wageningen Academic Publishers, Wageningen, pp 211–217

Buschmann C, Lichtenthaler HK (1998) Principles and characteristics of multi-colour fluorescence imaging of plants. J Plant Physiol 152:297–314

Camp CR, Sadler EJ, Evans RG (2006) Precision water management: current realities, possibilities and trends. In: Srinivasan A (ed) Handbook of precision agriculture. Principles and applications. Haworth Press, Binghamton

Ciganda VS, Gitelson AA, Schepers J (2012) How deep does a remote sensor sense? Expression of chlorophyll content in a maize canopy. Remote Sens Environ 126:240–247

Clevers JPC, Kooistra L, Schaepman ME (2008) Using spectral information from the NIR water absorption features for the retrieval of canopy water content. Int J Appl Earth Obs Geoinform 10:388–397

Clevers JPC, Kooistra L, Schaepman ME (2010) Estimating canopy water content using hyperspectral remote sensing data. Int J Appl Earth Obs Geoinform 12:119–125

Dash J, Curran PJ (2007) Evaluation of the MERIS terrestrial chlorophyll index (MTCI). Adv Space Res 39:100–104

Entcheva Campbell PK, Middleton EM, Corp LA, Kim MS (2008) Contribution of chlorophyll fluorescence to the apparent vegetation reflectance. Sci Total Environ 404:433–439

Evans RG, King BA (2010) Site-specific sprinkler irrigation in a limited water future. In: ASABE conference presentation, Phoenix, 5–8 Dec 2008, Paper no. IRR 10-8491

Evans RG, King BA (2012) Site-specific sprinkler irrigation in a water-limited future. Trans ASABE 55(2):493–504

Gao BC, Goetz AFH (1990) Column atmospheric water vapor and vegetation liquid water retrievals from airborne imaging spectrometer data. J Geophys Res 95(D4):3549–3564

Gherboudj I, Magagi R, Berg AA, Toth B (2011) Soil moisture retrieval over agricultural fields from multi-polarized and multi-angular RADARSAT-2 SAR data. Remote Sens Environ 115:33–43

Gitelson AA, Gritz Y, Merzlyak MN (2003) Relationships between leaf chlorophyll content and spectral reflectance and algorithms for non-destructive chlorophyll assessment in higher plant leaves. J Plant Physiol 160:271–282

Gitelson AA, Vina A, Ciganda V, Rundquist DC, Arkebauer TJ (2005) Remote estimation of canopy chlorophyll content in crops. Geophys Res Lett 32:1–4

Gitelson AA, Vina A, Masek JG, Verma SB, Suyker AE (2008) Synoptic monitoring of gross primary productivity of maize using Landsat data. IEEE Geosci Remote Sens Lett 5(2):133–137

Guenther KP, Dahn HG, Luedeker W (1999) Laser-induced-fluorescence: a new method for "precision farming". In: Bill R, Grenzdoerffer G, Schmidt F (eds) Sensorsysteme in precision farming, workshop, Universität Rostock, Rostock, 27/28 Sept 1999

Guyot G (1998) Physics of the environment and climate. Wiley, New York

Guyot G, Baret F, Major DJ (1988) High spectral resolution: determination of spectral shifts between the red and infrared. Int Arch Photogramm Remote Sens 11:750–760

Hammes E (2005) Biophysical research to effects of fungi infections in wheat on processes of photosynthesis and possibilities of infection recording from a moving tractor (in German). Doctoral dissertation, University of Kiel, Kiel

Heinzel V (2007) Retrieval of biophysical parameters from multi-sensoral remote sensing data assimilated into the crop growth model CERES-Wheat. Doctoral dissertation, University of Bonn, Bonn

Herrmann I, Pimstein A, Karnieli A, Cohen Y, Alchanatis V, Bonfil JD (2010) Utilizing the VENµS red-edge bands for assessing LAI in crops. In: ISPRS Archives, Vol XXXVIII, Part 4-8-2-W9, Core spatial databases-updating, maintenance and services – from theory to practice, Haifa

Homayouni S, Germain C, Lavialle O, Grenier G, Goutouly JP, Van Leeuwen C, Da Costa JP (2008) Abundance weighting for improved vegetation mapping in row crops: application to vineyard vigour monitoring. Can J Remote Sens 34(2):S228–S239

Irmak S, Haman DZ, Bastug R (2000) Determination of crop water stress index for irrigation timing and yield estimation of corn. Agron J 92:1221–1227

Jackson RD, Idso SB, Reginato RJ, Pinter PJ (1981) Canopy temperature as crop water stress indicator. Water Resour Res 17(4):1133–1138

Jackson RD, Kustas WP, Choudhury BJ (1988) A re-examination of the crop water stress index. Irrig Sci 9:309–317

Jacquemoud S, Baret F (1990) PROSPECT: a model of leaf optical properties spectra. Remote Sens Environ 34:75–91

Jensen JR (2007) Remote sensing of the environment, 2nd edn. Wiley, New York

Jiao X, McNairn H, Shang J, Liu J (2010) The sensitivity of multi-frequency (X, C and L-band) radar backscatter signatures to bio-physical variables (LAI) over corn and soybean fields. In: International archives photogrammetry remote sensing XXXVIII, Part 7B, symposium, Vienna, 2010, pp 318–321

Jones CL, Weckler PR, Maness NO, Stone ML, Jayasekara R (2004) Estimating water stress in plants using hyperspectral sensing. Paper no. 043065, ASAE, St. Joseph

Kautsky H, Hirsch A (1931) New experiments regarding carbon assimilation (in German). Naturwissenschaften 19:964

Kühbauch W, Hawlitschka S (2003) Remote sensing – a future technology in precision farming. In: POLINSAR. Workshop on applications of SAR polarimetry and polarimetric inferometry, ESA-ESRIN, Frascati, 14–16 Jan 2003

Lichtenthaler HK (1996) Vegetation stress: an introduction to the stress concept in plants. J Plant Physiol 148:4–14

Liu L, Zhang Y, Wang J, Zhao C (2005) Detecting solar induced chlorophyll fluorescence from field radiance spectra based on the Fraunhofer line principle. IEEE Trans Geosci Remote Sens 43(4):827–832

Liu J, Miller JR, Haboudane D, Pattey E, Hochheim K (2008) Crop fraction estimating from casi hyperspectral data using linear spectral unmixing and vegetation indices. Can J Remote Sens 34(1):S124–S138

Maier SW, Guenther KP, Luedeker W, Dahn HW (1999) A new method for remote sensing of vegetation stress. In: Proceedings of the ALPS 99, Meribel, 1999

Mattia F, Le Thoan T, Picard G, Posa FI, D'Alessio A, Notarnicola C, Gatti AM, Rinaldi M, Satalino G, Pasquariello G (2003) Multitemporal C-band radar measurements on wheat fields. IEEE Trans Geosci Remote Sens 41(7):1551–1560

McNairn H, Shang J, Jiao X, Champagne C (2009) The contribution of ALOS PALSAR multipolarisation and polarimetric data to crop classification. IEEE Trans Geosci Remote Sens 47(12):3981–3992

Meron M, Tsipris J, Orlov V, Alchanitis V, Cohen Y (2010) Crop water stress mapping for sitespecific irrigation by thermal imagery and artificial reference surfaces. Precis Agric 11:148–162

Mistele B, Gutser R, Schmidhalter U (2004). Validation of field-scaled spectral measurements of the nitrogen status in winter wheat. In: Precision Agriculture Center, University of Minnesota (ed) Remote sensing, seventh international conference on precision agriculture, Minneapolis, 25–28 July 2004

Moeller M, Alchanatis V, Cohen Y, Meron M, Tsipris J, Naor A, Ostrowsky V, Sprintsin M, Cohen S (2007) Use of thermal and visible imagery for estimating crop water status of irrigated grapevine. J Exp Bot 58(4):827–838

Monteith JL (1972) Solar radiation and productivity in tropical ecosystems. J Appl Ecol 9(3):744–766

Moshou D, Bravo C, Oberti R, Bodria L, Vougioukas S, Ramon H (2009) Intelligent autonomous system for the detection and treatment of fungal diseases in arable crops. In: van Henten EJ, Goense D, Lockhorst C (eds) Precision agriculture '09. Wageningen Academic Publishers, Wageningen, pp 265–272

Moya I, Camenen L, Evain S, Goulas Y, Cerovic ZG, Latouche G, Flexas L, Ounis A (2004) A new instrument for passive remote sensing. 1. Measurements of sunlight-induced chlorophyll fluorescence. Remote Sens Environ 91:186–197

Pacheco A, Bannari A, Staenz K, McNairn H (2008) Deriving percent crop cover over agriculture canopies using hyperspectral remote sensing. Can J Remote Sens 34(1):S110–S123

Payero JO, Irmak S (2006) Variable upper and lower crop water stress index baselines for corn and soybeans. Irrig Sci 25:21–32

Peng Y, Gitelson AA, Keydan G, Rundquist DC, Leavitt B, Verma SB, Suyker AE (2010) Remote estimation of gross primary production in maize. In: Proceedings of the 10th conference on precision agriculture, Denver, 18–21 July 2010

Peng Y, Gitelson AA, Keydan G, Rundquist DC, Moses W (2011) Remote estimation of gross primary production in maize and support for a new paradigm based on total crop chlorophyll content. Remote Sens Environ 115:978–989

Penuelas J, Filella I, Biel C, Serrano L, Save R (1993) The reflectance at the 950–970 nm region as an indicator of plant water status. Int J Remote Sens 14:1887–1905

Pierce FJ, Chavez JL, Elliott TV, Matthews GR, Evans RG, Kim Y (2006) A remote-real-time continuous move irrigation control and monitoring system. In: Paper no. 062 162, 2006 ASABE annual international meeting, Portland, 2006

Pottier E, Ferro-Famil L (2004) Analyse temps-frequence de milieux anisotropes a partir de donnees SAR polarimetrique. In: Journee "Imagerie Polarimetrique" GDR ISIS, GDR Ondes, Paris, 16–17 Mar 2004

Raun WR, Solie JB, Johnson GV, Stone ML, Lukins EV, Thomason WE, Schepers JS (2001) In-season prediction of potential grain yield in winter wheat using canopy reflectance. Agron J 93:131–138

Reusch S (1997) Development of an optical reflectance sensor for recording the nitrogen supply of agricultural field crops (in German). Doctoral dissertation, Department of Agricultural Systems Engineering, University of Kiel, Kiel. Forschungsbericht Agrartechnik, VDI-MEG 303

Rosema A, Verhoef W, Schroote J, Snel JFH (1991) Simulating fluorescence light- canopy interaction in support of laser–induced fluorescence measurements. Remote Sens Environ 37:117–130

Sadler EJ, Evans R, Buchleiter G, King B, Camp C (2000) Venturing into precision agriculture. Irrig J 50:15–17

Sadler EJ, Camp CR, Evans DE, Millen JA (2002) Corn canopy temperature measured with a moving infrared thermometer array. Trans Am Soc Agric Eng 45(3):581–591

Schwartz JW, Piston D, DeFelice LJ (2006) Molecular microfluorometry: converting arbitrary fluorescence units into absolute molecular concentrations. Handb Exp Pharmacol 175:23–57

Serrano L, Fillela I, Penuelas J (2000) Remote sensing of biomass and yield of winter wheat under different nitrogen supplies. Crop Sci 40:723–731

Shimoni B, Borghys D, Heremans R, Perneel C, Acheroy M (2009) Fusion of PolSAR and PolInSAR data for land cover classification. Int J Appl Earth Obs Geoinform 11:169–180

Sims DA, Gamon JA (2003) Estimation of vegetation water content and photosynthetic tissue area from spectral reflectance. Remote Sens Environ 84:526–537

Sonnenschein R, Jarmer T, Vohland M, Werner W (2005) Spectral determination of plant water content of wheat canopies. In: Zagajewski B et al. (eds) Proceedings of the 4th EARSel. Workshop of imaging spectroscopy. New quality in environmental studies. Warsaw University, Warsaw, pp 727–737

Steingiesser R, Kühbauch W (1998) Recording of fresh and dry biomass and estimation of fresh biomass from winter barley in various regions of Europe by an airborne radar-sensor (in German). J Agron Crop Sci 181(3):145–152

Sticksel E, Huber G, Liebler J, Schächtl J (2004). The effect of diurnal variations of canopy reflectance on the assessment of biomass formation in wheat. In: Precision Agriculture Center, University of Minnesota (ed) Remote sensing, seventh international conference on precision agriculture, Minneapolis, 25–28 July 2004

Thiessen E (2002) Optical sensing techniques for site-specific application of farm chemicals (in German). Doctoral dissertation, University of Kiel, Kiel. Forschungsbericht Agrartechnik, VDI-MEG 399

Ulaby FT, Dubois PC, van Zyl J (1996) Radar mapping of surface soil moisture. J Hydrol 184:57–84

Verhoef W (1984) Light scattering by leaves with application to canopy reflectance modelling. Remote Sens Environ 16:125–141

Yang CM, Su MR (2000) Analysis of spectral characteristics of rice canopy under water deficiency. In: Proceedings of the Asian conference on remote sensing 2000. Session agriculture and soil, Taipei, 4–8 December 2000

# Chapter 7
# Site-Specific Soil Cultivation

Hermann J. Heege

**Abstract** Site-specific soil cultivation has several objectives. In primary cultivation, the main objective is the control of the working depth. Signals for this control can be derived from the clay content, the organic matter content, the hydromorphic properties and the slope of the soil. An algorithm can combine these signals to control the working depth. The soil resistance to penetration is a suitable control signal for sensing hardpans below the topsoil, but not for the working depth within the topsoil, since it depends mostly on the water content.

In secondary cultivation, clod size reduction is an important objective. The site-specific control signals for this can be obtained from the forces acting on a sensing tine of a cultivator. The standard deviations of the forces can provide for suitable control parameters.

In stubble- or fallow cultivations either fast- or slow decomposition of the residues should be aimed at depending on rotations, climate and risk of soil erosion.

The introduction of controlled traffic farming or of unmanned farm machinery will promote crop production without tillage well beyond the present use.

**Keywords** Controlled-traffic • Depth-control • Fractionated seedbed • Impact-sensing • Moisture-line • Tilth-control

## 7.1 Basic Needs

The need for soil cultivation mainly results from crop- or plant successions in modern agriculture. In case of perennial vegetation without any abrupt crop- or plant succession as *e.g.* with permanent grassland, no one thinks about cultivation. Human

---

H.J. Heege (✉)
Department of Agricultural Systems Engineering, University of Kiel,
24098 Kiel, Germany
e-mail: hheege@ilv.uni-kiel.de

needs for specific and uniform plant products necessitate annual monoplant fields and thus soil cultivation, either as **primary cultivation** (ca.10–30 cm deep), **secondary cultivation** or **seedbed preparation** (less than 10 cm deep) and in some cases also postharvest **stubble- or fallow cultivation**.

Precision in soil cultivation should be focussed at

- creating the bulk density needed for efficient plant growth
- providing the soil aggregate sizes that guarantee a high emergence of seeds
- weed control
- crop residue management
- water management.

All these efforts can be queried. In fact, some farmers plant crops without any previous cultivation. A soil, which is high in organic matter content may – without any cultivation – provide the bulk density as well as the soil aggregates needed. Weed control can be taken care of completely by herbicides. And if crop residues just are left on the soil surface, they can efficiently reduce soil erosion. So why worry to get them buried?

Yet again, there are also serious limits to these options. Not all soils have a sufficiently high organic matter content. Very often soils get compacted by heavy harvesting machinery as well as by post-harvest rain, hence necessitating cultivation. Weed control solely by herbicides can be costly. And last but not least, crop residues in the seedbed can seriously impair the emergence of seeds. This applies especially to narrowly spaced crops such as the widely used small grains in high yielding areas and when short time spans between harvesting and sowing prevail.

In short, despite the advances in zero- and minimum tillage practices, the majority of the farmers still has to cultivate. For these farmers, cultivation might take between 30 and 50 % of the energy needed for all field operations. Yet more precision in cultivation is urgently needed and might help to reduce the present controversies about soil cultivation techniques.

## 7.2 Primary Cultivation

Two effects deserve attention:

- the soil inversion and
- the depth of cultivation.

It is generally known that only the **mouldboard plough** can provide for an effective soil inversion and thus for a rather complete burial of weeds as well as of crop residues. All other implements just mix the soil and therefore leave some weeds and residues on the surface. The result is that the plough is the most effective tool for mechanical weed control and does away with all problems arising from residues in sowing operations.

Yet on the other hand, the plough is a very energy- plus labour consuming tool. And the bare surface left by the plough induces soil erosion. This holds true especially for sloped fields with silty soils in continental climate.

The steady advances in **chemical weed control** induce farmers more and more to rely on herbicides instead of the plough. From an environmental point of view, the herbicides as well as the plough can be questioned. Yet the future looks brighter for weed control by herbicides than for weed control by the plough. The advances which have been made for efficient and environmentally safe herbicides as well as for crops made resistant to herbicides by genetic progress are remarkable. It will hardly be possible to offset these advances by better ploughs.

There still may be negative effects of residues from the previous crop on the emergence of seeds. They mainly result from a less accurate seed placement in the soil by the openers of the sowing machines because of the residues. Yet this obstacle to cultivation without soil inversion by a plough too is losing its impact. Several means of eliminating negative effects of **crop residues** on or in the seedbed on the emergence have been developed. Examples are openers, which hardly are affected in their seed placement by residues or raking devices which operate ahead of the coulters and move the residues in the area between the rows (row cleaners). At least with widely spaced crops such as maize, soybeans and beets, these row cleaners can be used successfully. Modern harvesting machines too substantially can alleviate the problem by leaving finely chopped- and uniformly distributed residues on the field. In short, crop residue management also less and less requires ploughing.

The subsequent text therefore deals primarily with problems and solutions for precision cultivation with tined implements.

## 7.2.1 Factors for the Depth of Primary Cultivation

The bulk density needed for efficient plant growth is mainly regulated by the depth of the primary cultivation. The deeper the primary cultivation tools operate, the more the soil is elevated above the original surface, thus the more the bulk density is lowered. Methods for recording the bulk density in the laboratory are state of the art. However, sensing techniques capable of recording the bulk density in a continuous manner online on-the-go are not available. But a feasible approach is to adjust and to control the depth of the primary cultivation according to surrogates such as

- water supply
- texture
- organic matter
- slope
- resistance to penetration.

### 7.2.1.1 Water Supply

Precipitation increases bulk density of soils. This holds especially for bare soils. But when it comes to adapting cultural practices to precipitation, it is helpful to differentiate between climate and weather. **Climate** acts on large, contiguous areas and is

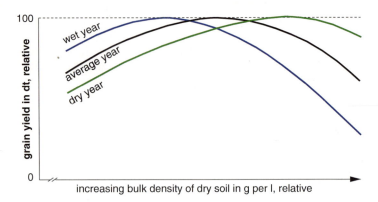

**Fig. 7.1** Principal relation between soil bulk density and grain yield (From Hakansson et al. 1974, altered)

defined on a longterm basis, *i.e.* for decades and centuries. **Weather** is a matter of smaller areas and shorter time spans. Very important is the forecasting. The climate of an area is fairly constant. Forecasting therefore hardly is a problem. Yet forecasting the weather for more than a few days still is a problem in large parts of the world.

Therefore, adapting the soil cultivation to the respective climate has been state of the art for a long time. A global traveller attentive to farming operations will notice that in humid regions the primary cultivation is done deeper than in regions with low rainfall. This is, because in humid areas the water supply of the crop seldom is a problem, but the soil aeration often is. In areas with low rainfall, the situation is *vice versa*, consequently here the cultivation is either very shallow or even left out completely.

But what about the adaptation of the **cultivation depth** to short term weather expectations? Ideally the farmers would have to provide for a high bulk density in a dry growing season and *vice versa* in a wet year (Fig. 7.1).

Despite the fact that many weather satellites now operate in combination with modern computers programs, the **weather forecasts** still are not reliable beyond a few days. Much better adapted cultivation could be realized, if in the future the forecasts were accurate for at least some weeks.

An alternative approach would be adapting the depth of cultivation not to the precipitation expected in the next days but instead to the respective **water content** in the soil. This would be a response to the rainfall and the evaporation of some days of the past. Sensing techniques for recording the water content in the soil are available (Chap. 5). The problem is that the depth of primary cultivation is expected to have its effect over the whole growing season of the crop. Hence the incidental water content just at the time of cultivation may not always be a reliable indicator for the best depth of cultivation. Increasing the depth of cultivation with the water content of the soil might even be completely wrong, if the tools do not break up any more but instead of this deform and compact the wet plastic soil. Cultivating deep in regions with a humid climate still might be reasonable, but this should preferably be done during dry spells in order to prevent damage to the soil.

# 7 Site-Specific Soil Cultivation

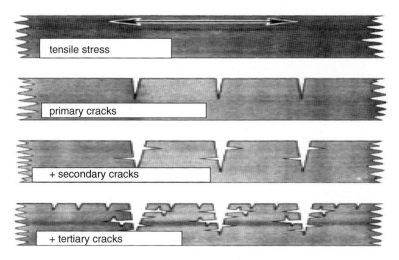

**Fig. 7.2** Gradual formation of cracks in a drying clay soil (From Dexter 1988, altered)

Dry periods are especially important when it comes to **hydromorphic soils**. These soils are associated with temporal water saturation in the lower part of the topsoil, *e. g.* in most parts of the world during the winter period. Most commonly this waterlogging occurs due to a concentrated water flow into topographic depressions that are not well drained. Thus in rolling areas there may be just some spots of hydromorphic soil within a field. This hydromorphic soil can easily be recognized when breaking it up with a spade by its blue-grey colour. When the field is prepared under dry conditions, these areas need deep cultivation. Further details are dealt with below.

### 7.2.1.2 Texture

It is well known that the sequence freezing-thawing can substantially reduce the strength of soils and thereby promote **break-up** of clods. A similar effect can be obtained free of charge by weather sequences that cause wetting followed by drying. However, this effect depends very much on soil texture and its interaction with water. The question is, how wetting or drying of a soil affects its volume. With sandy soils, the water content hardly changes the volume. On the other hand, clay soils swell during wetting and conversely shrink while drying. These processes induce tensile stresses within the soil, and this in turn causes cracking of the dry soil (Fig. 7.2).

Therefore, if the texture within a field is not uniform, it can be reasonable to adjust the site-specific depth of cultivation for a promotion of soil break-up to this. Below it is shown how such a control can be realized (Sect. 7.2.2.1).

### 7.2.1.3 Organic Matter

Organic matter in a soil generally counteracts any compaction. Therefore, soils that are low in organic matter content need more cultivation. The primary cultivation tools must go deeper. The problem is that with the intensity of cultivation the **decomposition** of the organic matter too increases. Hence farmers might get into a vicious circle: a low organic matter content induces to cultivate deeper, but this in turn and in the long run might reduce the organic matter content even more.

The way out of this dilemma is not necessarily less cultivation, if this means lower yields. Because lower yields might in turn mean less incorporation of organic residues. The best solution probably is precise adaptation of cultivation needs to the respective situation, thus preventing any superfluous cultivation without any sacrifice in yield combined with care for organic matter in the soil by suitable residue management. In essence, this means precise soil and crop management. Since many fields do not have a uniform organic matter content and sensing the respective organic matter situation is possible, the control of the depth of primary cultivation via this soil property seems reasonable.

### 7.2.1.4 Slope

On slopes, the runoff of water deserves attention. The result of this runoff and its effect on soil erosion is accumulation of finely textured soil with much organic matter in depressions to the disadvantage of hills. Hence from the outset, sloped fields do not have uniform soils and need site-specific attention.

Site specific cultivation can

- either provide for improved adaptation to results of erosion that occurred in the past or
- be targeted at reducing erosion in the future.

The improved adaptation to erosion from the past is possible by controlling the depth of cultivation according to the respective texture and organic matter content on hills, slopes and in depressions of the fields.

And reducing erosion in the future can be obtained by improving the infiltration of water and thus decreasing the runoff. Several practices can contribute to this such as keeping the soil surface vegetated as much as possible, leaving crop-residues at the soil surface and finally – if bare soil cannot be avoided – increasing the depth of cultivation. So whenever feasible, vegetation or its residues should be used to improve infiltration and to prevent erosion. Deep cultivation of bare soil as a means for this should be regarded as an exceptional matter.

There are cases where improving the infiltration still cannot prevent that a substantial portion of the precipitation runs off. This inevitable runoff should occur without much soil erosion. For this, the water should be slowly channelled downwards via broad, saucer shaped, grassed waterways into ditches. These grassed waterways can act as perennial runoff strips.

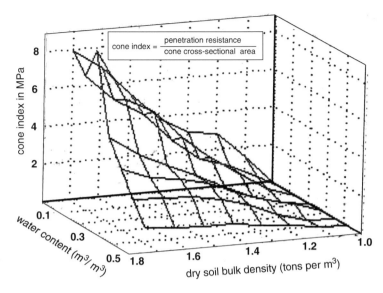

**Fig. 7.3** The effect of bulk density and water content on the cone index (From Sun et al. 2003, altered)

### 7.2.1.5 Resistance to Penetration

Soil resistance to penetration has for a long time been an object of investigations, *e.g.* in search of a **substitute** for bulk density. Traditionally, probes with a sensing cone on the tip are pushed **vertically** into the soil. Yet this mode of operation hardly is suitable for on-the-go sensing. Therefore, the interest has shifted to probes, which move in a **horizontal** direction through the soil and can easily be drawn by a tractor (Lüth 1993). So continuous- instead of punctual signals are recorded. Adamchuk et al. (2008) placed several horizontally oriented sensing cones that were staggered along a vertical blade for simultaneous recording of the horizontal resistance at several depths below the soil surface.

The signals obtained from penetrometers are strongly influenced by the **water content** of the soil regardless of the sensing direction. The results shown in Fig. 7.3 are based on a cone index, which is penetration resistance divided by cone cross sectional area. The effect of the water content of the soil on the cone index is several times higher than the influence of the bulk density.

This explains why it is attempted to sense resistance to penetration and water content simultaneously. The objective in doing this is to correct the cone index data for the influence of the water content (Sun et al. 2003; Hartung and Drücker 2009). Without such corrections, the resistance to penetration cannot provide reliable signals to control the cultivation depth, since the water content of a soil can vary considerably on a temporal- as well as on a spatial basis.

Notwithstanding the influence of the water content, penetrometers are well suitable to detect **hardpans** in soils, *e.g.* those that might exist below the topsoil as a

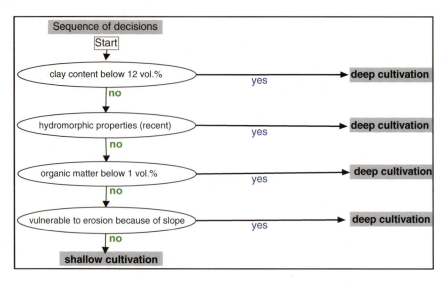

**Fig. 7.4** Sequence of decisions for the control of the primary cultivation depth (From Sommer and Vosshenrich 2004, altered)

result of geological deposits or might develop from longterm plowing at the same depth (Gorucu et al. 2006). The density of usual hardpans differs very substantially from that within the topsoil. Any differences within the topsoil are smaller, hence more difficult to sense.

### 7.2.2 Site-Specific Control of the Primary Cultivation Depth

#### 7.2.2.1 Algorithm for the Control

Sommer and Vosshenrich (2004) developed a control system for the cultivation depth. It is based on several soil properties and a control algorithm (Fig. 7.4).

Whenever the **clay content** is below 12 % by volume, **hydromorphic properties** exist, the **organic matter content** is below 1 % by volume or **slope induced erosion** might occur, deep cultivation is practiced. It should be noted that with this control system every one of the four indications alone – or in a combination – brings about deep operation. Whenever none of these situations exists, the control system goes to shallow cultivation. All of these indications are not short timed and therefore lend themselves for control via maps and their overlay (Fig. 7.5). Particularly regarding the clay- and the organic matter content, the maps prepared for this overlay control system might be used for many consecutive years, since these soil properties are rather constant over a very long time.

The hydromorphic situation might change when the drainage has been improved. And the precautions necessary to prevent erosion depend not only on the respective

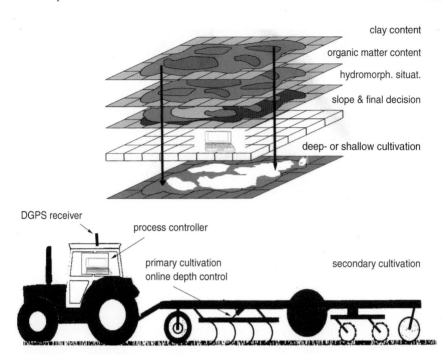

**Fig. 7.5** Map overlay for a site-specific depth-control of primary cultivation (Vosshenrich and Sommer 2005a, altered and supplemented)

slope, but on crop residues left near the soil surface as well. So for these two factors the situation might vary over time.

There are fields where neither hydromorphic soil nor prevention of erosion need attention, and only a varying clay- and organic matter content remain as control factors. Both factors can be recorded online and on-the-go. So principally a control in real-time would be possible (Vosshenrich and Sommer 2005b). However, for most cases it seems reasonable to control by **georeferenced maps** instead of employing a real-time system. Because a dual-map overlay control system – once created – can be used for many years.

But what means deep- or shallow cultivation in absolute terms? Vosshenrich and Sommer (2005a) defined 20 cm for deep- and 10 cm for shallow cultivation. These absolute depths refer to Northern German conditions. As outlined above, the respective climate should be considered when the local adjustment for deep- and shallow operation is defined in absolute terms.

An **interaction** of the factors clay- and organic matter content concerning the needed depth of cultivation can be expected. The higher the organic matter content is, the lower the clay content probably can be, and *vice versa*. This is, because both factors inherently promote soil break-up without any cultivation. Vosshenrich and Sommer (2005a) state that with an organic matter content of 1.5 % or more a clay content of 8–10 % would be sufficient for shallow cultivation. However, until now this interaction is not included in a control algorithm that is state of the art.

It is generally recommended to use this site-specific cultivation system only with soils that do not have compacted zones. The advice therefore is to start with this system after a first deep cultivation of the whole field under dry conditions.

In principle, such a system is feasible with many different cultivation tools. Differentiating between primary- and secondary cultivation in this respect might make sense in humid climates. Yet this distinction hardly helps under dry conditions, where the depth of cultivation might not substantially go below the level of the seedbed.

### 7.2.2.2 Economics

Provided the soil is not uniform within a field, which is the rule, the savings from site-specific cultivation in a humid climate can justify the additional investment needed for the depth control. This is the result of a study by Hartung and Druecker (2009) as well as by Isensee and Reckleben (2009) undertaken in Northern Germany on rolling fields and soils of glacial origin.

The site-specific cultivation was done with the technology and the equipment as explained in Figs. 7.4 and 7.5. The depth of the primary cultivation was mostly either 20 or 10 cm, since for a more gradual control system no information existed. The result of the depth control system was that on the average approximately half the area was cultivated shallow.

The **benefits** realized from the site-specific primary cultivation were 6 € per ha in fuel costs and 2 € per ha from the increased work-rate, thus a sum of 8 € per ha. At first sight this sum of benefits may seem trivial. However, the total charge including labor cost for primary cultivation by tined implements according to contractor rates is about 25 € per ha. Hence the benefits in fuel costs and from the increased work-rate amount to about 32 % of the usual contractor expenses. In the long term, these benefits can easily cover the additional costs for the control needed.

There were no significant effects of this site-specific primary cultivation on the **yields** of small grains (Sommer and Vosshenrich 2004; Isensee and Reckleben 2009).

It can be expected that this cultivation method reduces the decomposition of the soil organic matter because of less aeration at locations where deep work is not needed. However, this benefit of site-specific operation is difficult to evaluate.

## 7.3 Secondary Cultivation

Except for sandy fields or for soils high in organic matter content, usually some **clod break-up** by secondary cultivation is needed for a good crop emergence. In a cloddy soil, the seeds do not get close contact with the soil water, hence the germination is reduced. Figure 7.6 shows the emergence of small cereals depending on the mean

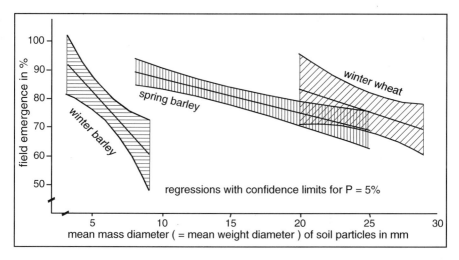

**Fig. 7.6** Soil clod size reduction and field emergence (Heege and Vosshenrich 2000)

mass diameter of soil clods from experiments in Germany. This parameter of soil tilth is often denoted as mean weight diameter (MWD). It is determined by a sieving process and by weighing the fractions.

The wide span from 3 to 20 mm mean diameter that can provide for a high emergence may be surprising. Yet the balance between the water- and air supply in the seedbed is important. In Germany, winter barley is sown substantially earlier in autumn than winter wheat at a time, when the soil still is rather dry. Therefore, for a good balance of water to air, winter barley needs a finer seedbed than winter wheat, which gets enough water anyway because of the rather wet soil in late autumn or early winter.

There are exceptions from the general rule that the field emergence increases with the soil break-up. These exceptions apply to soils with a high silt content in a climate, where heavy showers instead of drizzling rains show up. Under these conditions, the surface of a fine seedbed can get puddled by showers and in subsequent dry weather it can become crusted. The result is a cut off in aeration and thus poor emergence. However, these exceptions do not invalidate the general rule.

### 7.3.1 Sensing Soil Tilth

Since the soils in many fields are not uniform in texture or organic matter content and compaction by the harvesting machines is irregular too, the tilth differs within a field. The knowledge that the farmer has about the site-specific clod sizes in the seedbed depends on his visual impression and therefore is fragmental. There is a need for real-time sensing of the site-specific clod sizes produced during seedbed preparation.

**Table 7.1** Approaches to sense soil tilth

| Sensing techniques | Authors |
|---|---|
| **Non contact sensing techniques** | |
| Gamma-ray attenuation | Oliveira et al. (1998) |
| Image analysis and -processing | Bogrekci and Godwin (2007a) |
| | Stafford and Ambler (1990) |
| Laser relief metering | Bertuzzi and Stengel (1988) |
| | Destain and Verbrugge (1987) |
| | Harral and Cowe (1982) |
| Reflectance of visible-and/or infrared radiation | Bowers and Hanks (1965) |
| | Orlov (1966) |
| | Zuo et al. (2000) |
| Ultrasonic relief metering | Scarlett et al. (1997) |
| **Contact sensing techniques** | |
| Horizontal mini penetrometer | Olsen (1992) |
| Strain gage on spring tine | Bogrecki and Godwin (2007b) |

This would allow to adjust the **cultivating intensity** on-the-go or alternatively to adapt the site-specific **seed-rate** in order to compensate for coarser seedbeds.

The traditional method to examine the clod size distribution is mechanical sieving of soil samples. This method is laborious and probably completely unsuited for on-the-go sensing in real-time. However, it is still used as the standard reference method in research for the assessment of tilth. Many attempts to realize a sensing of the clod size distribution in a more convenient way have been made (Table 7.1).

The methods can be classified into non contact- and contact techniques, depending on the physical principles employed. Up to now, none of these methods is used in practical farming. And on-the-go sensing in real-time has been realized only with either relief sensing by ultrasonics or with a strain gage on a spring tine of a cultivator. The latter principle is dealt with below in detail.

### 7.3.1.1 Principle of Tilth Sensing via Impact Forces Acting on a Tine

The basis of this method are the **impact forces** that act on a spring type tine during secondary cultivation. These forces result from the collision of the tine with soil clods. They can be recorded by measuring the distance, for which the tine is deflected because of the soil's resistance.

A rather simple method to sense the deflection of the tine is the use of a strain gage. The deformation of the tine is measured by bonding a small electrical conductor to this machine part. Since this electrical conductor is stretched, distorted or compressed together with the tine, its current output changes. From this – after careful calibration – the forces can be obtained.

There is a fundamental difference in the objectives of sensing the soil forces that act on a tine or a cone depending on whether the control of either primary- or secondary cultivation is aimed at.

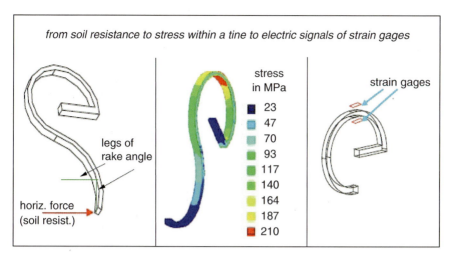

**Fig. 7.7** Sensing instrumentation for recording the soil break-up in a seedbed (From Bogrekci and Godwin 2007b, redrawn and altered)

With primary cultivation the main question is the **depth of operation**. The concept is that the higher the soil resistance to a tine or a cone moving in the soil is, the more the soil density should be reduced and the deeper the primary cultivator should operate. Yet up to now the realization of this idea has not been successful. This might be explained by the fact that for most soils the resistance to a tine or a cone is predominantly influenced by the water content and much less by the soil density (Fig. 7.3). And since the water content of a soil can change fast in a spatial- as well as in a temporal mode, the sensing objectives for the soil density hardly are met.

With secondary cultivation the situation is different: the main objective is not the density, it is the **tilth** or break-up of the soil. The emergence of crops mainly depends on the soil break-up within the seedbed as shown in Fig. 7.6 whenever zero tillage methods are excluded and tillage is needed.

Bogrekci and Godwin (2007b) have shown that a **strain gage** located at a suitable location on a spring tine of a cultivator can provide a reliable and an inexpensive source of signals about the soil break-up in a seedbed. Figure 7.7 shows, at which location of the spring tine the strain gage should be bonded. It is the highest point of the tine and the position, where the soil resistance causes the maximum load or stress on the tine.

### 7.3.1.2 Results with Tilth Sensing via Impact Forces

The general situation under *ceteris paribus* conditions is that the finer the soil structure is, the lower the mean horizontal force on the tine is. This results from the fact that for the finer soil particles it is easier to flow backwards around the tine. Therefore they exert less horizontal resistance. But despite this it must be asked, whether the **mean** horizontal force is the most suitable parameter for sensing the break-up of the soil.

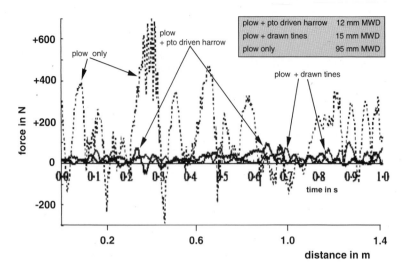

**Fig. 7.8** Signals on the sensing tine for three different tilths of a clay soil (From Bogrekci and Godwin 2007b, supplemented and altered). *MWD* mean mass diameter of soil clods

In principle, soil clod sensing is **impact sensing**. Almost every clod in a seedbed is a unique case in terms of size, mass and shape. Therefore, the impacts delivered by the clods on the recording tine must vary constantly. The sensitivity of the sensing instruments including the cultivator tine must allow for indicating this. An investigation of several spring tine types in this respect showed that the tine type as shown in Fig. 7.7 gave the best results.

An important factor of cultivator tines is the **rake angle**, which is the angle between the tine part operating in the soil and the undisturbed soil surface, seen perpendicular to the direction of travel. The lower the rake angle is, the smaller the soil resistance is in most cases. However, the larger the rake angle is, the more the soil is broken up. This is, because with a larger rake angle it is more difficult for the clods to pass the tine, hence the soil resistance as well as the break-up are higher.

The tine in Fig. 7.7 left has a rake angle of approximately 90°, which is a common angle in implements for secondary cultivation. Bogrekci and Godwin (2007b) have shown that using a 90° rake angle enables the soil break-up to be distinguished more effectively than when using smaller angles. This as well probably is due to the fact, that the clods cannot go around the tine as easy as with smaller angles, and this improves their recording.

In Fig. 7.8, the forces recorded by the sensing tine are shown for a soil, which was either only ploughed or secondary cultivated as well. The secondary cultivation was either by a drawn tined harrow or by a harrow that was operated via the power-take-off (pto). All curves show a **high resolution** on a time basis. They hold for a travel speed of 5 km/h. Since the duration is 1 s, the length of travel represented by the abscissa in Fig. 7.8 is about 1.4 m. Within this distance, all curves of the three different methods of soil cultivation show more than ten amplitudes. Therefore, the impact of each clod on the sensing tine seems to be well recorded.

Along all curves there are short instances, within which the forces become negative. This is the result of vibrations of the spring tine. The differences in the forces before and after secondary cultivation are evident; however, those between the seedbed preparation by the drawn harrow tines on the one hand and the power- take-off driven harrow on the other hand are not. Some evidence of a finer seedbed after secondary cultivation by means of the power-take-off drive is given by the comparison of the respective mean weight diameters (see insert of Fig. 7.8).

However, when the erratic course of the curves is taken into account, it cannot be expected that the mean force is the best benchmark for the site-specific tilth sensing. A mean criterion can hide many details of the data from which it was obtained. A close observation of the course of the curves suggests that concentrating the analysis on the respective amplitudes, frequencies or standard deviations of the forces might be reasonable. In a statistical analysis by Bogrekci and Godwin (2007b), which was focussed on an indicator for the soil tilth, the best results were supplied by the **standard deviation of the forces**.

It should be noted, that this analysis had the mean weight diameter (MWD) of the soil aggregates as an indicator of the soil tilth. There are several factors that might influence the signals of the soil break-up that are obtained via the standard deviation of the forces. Such factors can be the driving speed, the depth of cultivation, the soil type and the direction of cultivation compared to the direction used in the previous primary cultivation.

The influence of the **driving speed** as well as of the **soil type** seem to be rather small. As to soil type this might be expected, since – provided the clod sizes do not differ – its effect on the mass of soil aggregates is rather small. At least this holds, when peat soils are excluded. As to the driving speed, this might be surprising, since the driving speed must have a considerable influence on the impact forces. But with all implements that use drawn tines for soil cultivation, the driving speed also significantly affects soil break-up. Higher driving speeds increase the break-up as a result of the higher impact forces. So if the tilth after cultivation is compared, the effects of the driving speed on the soil break-up as well as on the forces exerted on the sensing tine go in the same direction and therefore probably even out.

The results shown in Fig. 7.9 are based on pooled data for driving speeds ranging from 5 to 15 km/h and for two soil types. Despite this, the root mean square error is only 3.5 mm for the mean weight diameter of the soil aggregates.

But the **sensing-depth** of the tine is important. In most cases, forces on tines increase more than proportional with the depth of operation in the soil. So if the cultivation implement does not provide a constant depth, a special depth control device for the sensing tine is necessary.

It seems reasonable to compare the soil tilth that this tine sensor indicates with results obtained by the standard sieving procedure. A correlation based on the respective mean weight diameters that were recorded after careful calibration is shown in Fig. 7.10. The result is a rather simple linear relation, as might be expected

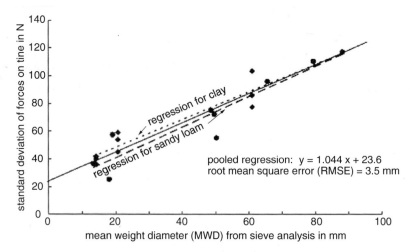

**Fig. 7.9** Sensing of soil tilth by the standard deviation of the forces acting on a sensing tine (From Bogrekci and Godwin 2007b, altered)

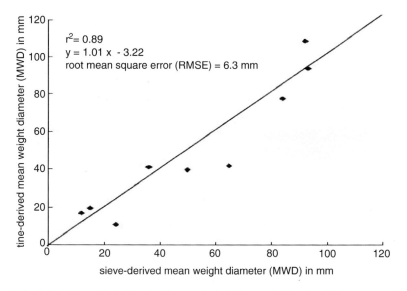

**Fig. 7.10** Soil tilth recorded either by the standard sieving method or by the tine sensor (From Bogrekci and Godwin 2007b, altered)

from the data in Fig. 7.9. The root mean square error of recording the soil break-up via the tine sensor instead of the standard sieving method is 6.3 mm of the mean weight diameter. Taking into account the reactions of the crops on the mean weight diameter (Fig. 7.6), this error seems acceptable.

### 7.3.1.3 Implications for Soil Tilth Sensing

The important prospect is that a tine sensor instead of a sieving method can deliver signals about soil tilth online and on-the-go. Thus it is possible to react on the respective site-specific situation in real-time. The object of control would be either the soil **break-up** by the secondary cultivation or the **seed-rate**, assuming that a higher seed-rate might be needed with a coarser seedbed and *vice versa*. However, the concept of adjusting the seed-rate to the respective soil break-up is like repairing a symptom and leaving its cause unaltered. The better approach for most crops probably is to avoid local deficiencies in the relative emergence as much as possible. Hence, controlling the soil break-up on the basis of the tilth sensed is the more logical procedure.

Where should the sensing-tine take the signals during the cultivation procedures? It would be possible to take the signals after primary cultivation by placing the sensing-tine **ahead** of the secondary cultivation implement. With suitable implements, this would allow to set up a control system for the secondary cultivation on the basis of the soil break-up by the primary tool. However, in order to get the desired seedbed, it still would be necessary to know the final soil break-up by the secondary implement in advance. This knowledge – on a soil specific basis – is hardly available. A simpler and more logical control system ensues if the sensing tine is placed **behind** the secondary cultivation implement. Thus the effect of the respective primary- and secondary implement is included in the final control. However, this position of the sensing-tine would need very fast control results in order to prevent too late adjustments since the sensing occurs afterwards. A compromise between these positions ahead or behind would be to place the sensing tine within the secondary cultivation tools, *e.g.* within the last row of tines.

The emergence of the seeds depends largely on the water transferred to them from the soil, therefore on the clod break-up as well as on the water content in the soil. But knowing the water content at the time of secondary cultivation and sowing alone is not sufficient. The short-term water supply in the first days **after** sowing is the most important criterion for emergence. Because this is so, there is still a huge lack of certainty when the question of the soil break-up needed for high emergence comes up. With much rain in the first days after sowing a coarse seedbed suffices and might be even beneficial, and with dry days after sowing it is *vice versa*.

But the short-term weather forecasts still are not reliable in most parts of the world. The expectations for reliable rain-forecasts, which were raised since the introduction of **weather satellites** and of new meteorological techniques, have not been met. The situation is that the existing uncertainties about short-term weather prospects bring about one of the biggest problems with many precision farming operations in rainfed areas. This applies especially to secondary cultivation and sowing operations. The objectives with the use of a soil tilth tine sensor could be defined much more precisely, if the rainfall in the next few days were known, because then the soil break-up needed could be defined more accurately.

Mapping the results of a tine-sensor would be possible. However, this would be reasonable only if additional soil properties, which were sensed too, were important. Such a property could be the site-specific water content of the soil instead of the rain expected. The lower the water content of the soil, the finer the seedbed should be. So if recent data about the water content were mapped, both the water situation as well as the soil break-up data could be combined in a new control map for subsequent cultivation- or sowing operations. Yet experience with such a concept up to now is not available.

The ideal concept of site-specific secondary cultivation would be a system of real-time on-the-go control for soil tilth, possibly obtained in one operation based on a tine sensor. Up to now such a concept can only be realized with power-take-off driven implements. These allow adjusting the soil break-up by altering the speed of the tines as well as their cuts per unit of travel distance on-the-go. Modern drives for the **power-take-off** and for the travel speeds of tractors provide the means for such on-the-go adjustments.

However, **drawn tools** on secondary cultivation implements are dominating in most parts of the world. And with solely drawn tools it hardly is possible to adjust the soil break-up on-the-go during one pass.

In short, sensing the soil break-up during cultivation in real time is feasible, but the possibilities of using the signals are limited. The restrictions in use on the one hand come from present deficiencies in the precision of the weather forecasts, and on the other hand are caused by present limits in the adjustment of drawn secondary cultivation implements. But these conditions can change. Hopefully, in the future the meteorological advances will provide for more precise weather forecasts, and more cultivation implements will be at hand, which allow for adjusting the soil break-up while operating. So the long-term outlook for site-specific secondary cultivation still might be good.

With primary cultivation, the situation is completely different. The main objective of the control is not precise clod break-up, but the decrease of soil bulk density via depth of operation. The latter can rather easily be adjusted on almost all implements. And furthermore, the short-term weather in the next days hardly is important with primary cultivation. Instead of this, the depth of the primary cultivation must be adjusted to long-term precipitation effects and to soil inherent properties. The knowledge about these factors is available or can rather easily be obtained. Therefore, as compared to secondary cultivation, the control of site-specific primary cultivation is a matter of present day realization.

### 7.3.2 *Precision in the Vertical Direction Within the Seedbed*

A uniform soil tilth in the vertical direction within the seedbed might be adequate with soils that do not tend to surface crusting after heavy rain and subsequent dry weather as well as with crops that emerge rather easily. But soils with a high silt- and low organic matter content can get crusted under the weather conditions described above. And the seeds of some crops – *e.g.* sugar beets, spring sown small cereals and some oil crops – emerge better if the seedbed created by secondary

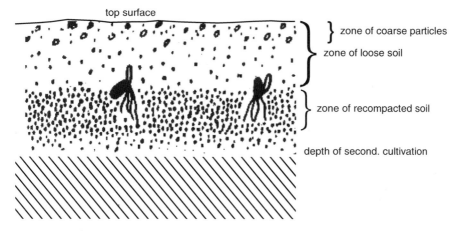

**Fig. 7.11** Fractionated seedbed for the prevention of crust formation on the top surface by some coarse soil particles (vertical cross-section)

cultivation is not uniform in the vertical direction. So for these cases, soil size reduction within the seedbed should be controlled not only in the horizontal direction, but in the vertical direction as well.

The soil particles that have direct contact with the seeds always should be fine in order to ensure water transfer. Possibly immediately underneath the seeds, the soil should be denser than above them. This helps to draw some water from below in a dry spell. Loose soil above the seeds provides for the aeration needed. And if surface crusting might develop, there should be coarser soil particles on or near the surface to prevent this. So a **fractionated seedbed** (Fig. 7.11) would be needed (Heege and Vosshenrich 1998, 2000; Heinonen 1985; Hakansson et al. 2002; Satkus and Velykis 2008).

Up to now, no farm machines are available that allow precisely to adjust the degree of fractionation. But at least the seedbed could tend to be structured as outlined above. To a small extent, the more dense soil immediately underneath the seeds is created by the pressure of the seed openers. And some vertical fractionation of soil particles can be obtained by segregating small soil particles from larger ones during secondary cultivation. While the soil is stirred, small particles have a better chance than the larger ones to sift downwards in voids, *e.g.* behind tines. As a result, more large aggregates remain near the surface.

This **segregation by sifting** needs rather slow moving tools, such as the tines of drawn implements. With fast moving tines, such as those of power-take-off driven implements, no segregation occurred (Heege and Vosshenrich 2000). Probably this can be explained by the time that is needed for small particles to sift downwards in voids behind tines. With fast moving tines, this time is too short for a remarkable segregation to take place.

The fractionation depends on the **rake angle** of the tines (Fig. 7.7, left). A low rake angle and therefore tines pointed forward in the direction of travel improve the segregation. This is, because these tines tend to lift more coarse clods to the surface and hence to create larger voids for the small particles to sift downwards in.

Many secondary cultivation implements have tools that operate with rather low speeds. Examples for these tools – in addition to drawn cultivator tines – are drawn discs and rollers or power-take-off driven rod weeders (Fig. 7.12). These tools hence can leave a partially fractionated seedbed (Winkelblech and Johnson 1964; Maze and Redel 2006).

Recording the fractionation of the seedbed may be possible on the basis of the soil resistance or of the impact forces exerted on horizontal sensing tips that are mounted at suitable depths of a vertical tine (Adamchuk et al. 2008). As with sensing of the general soil break-up (Fig. 7.7), the best results probably will be obtained by using the standard deviations of the impact forces (Figs. 7.9 and 7.10). Clods on top of a fractionated seedbed can be expected to deliver higher standard deviations than the fine particles in the zone provided for seed placement.

Yet the problem is not the sensing method. It is the lack of cultivation tools, which allow for precisely adjusting the **intensity of fractionation**. The development of cultivation tools suitable for such an adjustment is an important prerequisite for more precision in the vertical direction of the seedbed. There is an urgent need for advances in this direction.

## 7.4 Stubble- and Fallow Cultivation

Stubble- and fallow cultivations are "interim" operations between two successive crops. When a crop is sown a few days after harvesting of the preceding crop, they may be left out completely. Often planting occurs some weeks or a few months after harvesting. Then only one operation may be necessary, which after small grains is called stubble cultivation. But when a complete fallow year (summer-fallow) is practised before another crop is established, several passes of fallow cultivation may be needed. So the length of the fallow period after harvesting is an important criterion.

Generally, humid, high yielding areas have short fallow periods, since the water supply for the crops hardly is a problem. In dry-land regions the situation can be *vice versa*. Here there may be plenty of land, but not enough water. Therefore, a long fallow time can be reasonable in order to accumulate and store precipitation in the soil, which then later can be useful to get a better crop yield. The rationale of summer-fallow is to use the precipitation of 2 years for a crop growing season of 1 year.

The objectives of stubble- or fallow cultivation are **weed control, residue management**, sometimes **erosion control** and **conservation of soil moisture**.

**Weed control** as an objective of cultivation has lost in importance as a result of modern possibilities with efficient herbicides (chemical fallow), which can easily be realized in a site-specific mode (see Sect. 6.2).

The objectives in **residue management** depend on the climate. In humid, high yielding areas with short fallow periods, the main objective is to get fast decomposition of the large amounts of residues in order to avoid interference with seed openers. The decomposition of the residues depends on its contact with the soil, therefore on its incorporation by cultivation. A rule for small cereals in Germany is a depth of incorporation or a soil layer of 1–2 cm per every 10 dt of straw per ha (Taeger-Farny 2003). Since high yielding small cereals can leave about 90 dt of straw per ha, this would mean a depth between 9 and 18 cm.

The grain yield within a field can vary considerably, and the same holds true for the residues. Consequently, the depth of incorporation should not be uniform, instead it should be adapted to the site-specific amount of residues. Pforte and Hensel (2006) as well as (2010) propose a control system for the **depth of incorporation** that is based on percent residue cover. Their online approach relies on vehicle based reflectance spectroscopy and wavelengths in the near-infrared range from about 800 to 1,400 nm. Within the visible- and infrared spectrum, this is the wavelength range that has the maximum difference in reflectance between straw on the one side and bare soil on the other side. This means that this near-infrared range can provide signals for an online on-the-go control of the incorporation-depth based on percent residue cover.

A varying site-specific residue load can also result from an uneven straw distribution by combines. This applies especially to high capacity combines with wide cutter heads. However, an even distribution and short chopping of the straw are technically possible and can hopefully be taken for granted in the future. It would not be reasonable to adjust the depth of straw incorporation to deficiencies of combines that can be corrected.

It might be possible to sense the site-specific residue load in other ways, namely either by converting from site-specific crop yields to amounts of residues or by remote sensing from satellites (Zheng et al. 2012). However, none of these methods has been field-tested for site-specific operations.

The objectives of residue management are quite different in regions with dry-land farming and long fallow periods. Here fast decomposition is not a topic; it might be even a disadvantage. Instead, the prevention of soil erosion – mainly via wind – is a much more important point (Schillinger and Papendick 2008). Residues left on or near the soil surface are very effective in reducing erosion by wind as well as by water. Therefore, when summer-fallowing of erosion prone soils is practiced, incorporation of the residues should be mostly avoided. The objective is having at least partly "**anchored crop residues**" on the field. These residues extend well above the soil surface, yet still are attached to the soil, *e.g.* by their roots.

**Conservation of soil moisture** as an objective of stubble- or fallow cultivation too is highly dependent on the respective climate. In a humid climate this objective hardly counts. Yet in dry-land regions, success in farming strongly depends on the ability to manage water.

How should in this respect fallow fields be managed? The important point here is to minimize evaporation by weeds as well as from bare soil. Based on the same area, plants always evaporate more than bare soil. The rationale of summer-fallowing rests on this. Plants with their elaborate root system just are very effective in

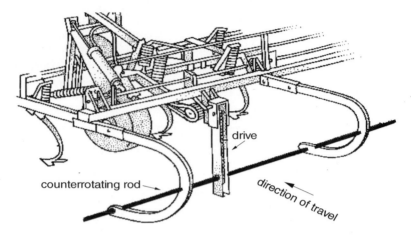

**Fig. 7.12** Rod-weeder (From Gist 2002, altered)

sucking water to the surface. Therefore, weeds must be completely kept off the field, either by a site-specific application of herbicides or by cultivation. But capillaries of soils as well can suck water to the surface. This applies particularly to non-disrupted, continuous capillaries. Hence a bare, undisturbed, settled soil evaporates more water than the same field with a shallow, cultivated top layer (Wuest and Schillinger 2011). The problem is that this top layer can promote erosion. So this top layer either needs a special protection against erosion, or the disruption of the soils capillaries must occur below the surface with targeted precision. Technical solutions for both alternatives are available.

The classical instrument for shallow cultivation with at least some inherent protection against erosion is the **rod-weeder** (Fig. 7.12). Its main tool is a horizontal square rod with vertical cross-sectional side lengths of about 25 mm. This rod is oriented perpendicular to the direction of travel and is rotated against the direction of travel. Its depth of operation is between 5 and 10 cm. This implement creates stratification or fractionation within its operating depth. Due to the rotations of the rod against the direction of travel, coarse clods, weeds and residues mainly pass by overhead. Small soil particles instead are not carried upwards and therefore squeezed downwards. The result comes close to the ideal stratification and fractionation as pointed out in Fig. 7.11. This implement up to now mainly is used for dry-land farming in Northwestern areas of Canada and the USA. Its prospects in other parts of the world as a means of preventing surface crusting and erosion in combination with modern sowing techniques deserve attention.

When it comes to the disruption of the soil capillaries with targeted precision in order to prevent evaporation, farmers in dry-land areas refer to a process of "**setting a moisture line**" in the field. The rod-weeder does this quite well because of its rather shallow operation. And in addition it accumulates weeds, straw and clods on or near the surface to assist in erosion control. However, it does not leave "anchored crop residues" as well as "anchored killed weeds" on the field. The residues and weeds just are lifted and mainly deposited on the surface, they are not anchored to the soil any more.

**Fig. 7.13** Setting a moisture line in a fallow field by an under-cutter. The *insert* shows the main tool. The soil surface is covered with anchored crop residues (Photo from Schillinger, altered)

Yet leaving an anchored protection is possible, if the soil is not tilled at all at the surface and weed control completely left to herbicides. Setting a moisture line still would be necessary to reduce evaporation from the fallow land (Zaikin et al. 2007). This can be achieved by an implement, which just cuts the soil well below the surface in a horizontal direction, but hardly breaks it up vertically. Such an implement is the **under-cutter** (Fig. 7.13). Its main tools are wide, horizontal blades, which cut through the soil about 10 cm below the surface. The soil surface with anchored protection against erosion is almost left undisturbed. Petrie (2009) has shown that for winter wheat after summer-fallow the combination of once undercutting in the spring followed by herbicide applications for weed control was as effective as several times rod weeding during the fallow period in terms of water conservation and grain yields.

So this combination of undercutting and chemical, site-specific fallowing deserves attention in dry-land regions. It protects the soil in two ways. Firstly, it prevents erosion effectively. Secondly, because this combination widely eliminates cultivation, less soil organic matter is decomposed. This weighs heavily in dry-land areas, since the effect of cultivation in reducing the soils organic matter is much more pronounced in dry and warm areas than in humid regions with lower temperatures.

## 7.5 No-Tillage: Prerequisites, Consequences and Prospects

The prerequisites as well as the prospects for no-tillage practices depend largely on the precision exerted in farming. An important trigger for the interest in no-till has been and still is the development of efficient and competitive herbicides. Without these, the present use of no-till practices – *e.g.* on soils with a high organic matter content – would not be possible.

But there is also an important development of the past decades, which constrains no-till. It is the increasing use of heavy farm machinery. Especially on wet soil, this machinery induces soil compaction. The use of wide, low-pressure tires alleviates the problem only partly. The situation often is that cultivation has to correct the damage that was exerted on wet soil in previous operations, *e.g.* during harvesting.

It is possible to inform the farmer via an online sensor whether his machinery is compacting the soil in an undue way. A **field-trafficability-sensor** indicates the depth of sinkage of the tires and thus can inform the operator on-the-go about the site-specific situation (Brunotte 2007). However, what are the options when a farmer realizes that his machine is damaging the soil? It is very unlikely that he will stop a harvesting operation: the soil might be even wetter after having waited a few days. The farmer is better informed about the situation, yet his options are rather limited. One option – however – generally is reasonable. In case the farmer is operating the machine with an inflation pressure of the tires, which is too high for field use and should have been adjusted before starting the work, he can correct this.

A challenging approach to solve the compaction problem with heavy machinery is the "**Controlled Traffic Farming**" concept (Chamen 1998; Chamen et al. 2003). It is based on the experience that normally plants grow better on non-trafficked soil and wheels run better in tracks. Consequently, cropping areas in the field are strictly separated from wheel lanes. This means that the same wheel tracks are used for every sowing-, fertilizing-, spraying- as well as harvesting operation, and this year after year. Since compaction by wheels in the cropping area does not occur any more, tillage can be omitted and no-till practices can be used. Despite the area lost because of the tracks, higher yields can be obtained than with conventional farming (Chamen et al. 2003; Kelly et al. 2004; Tullberg et al. 2007). Precise maneuvering is essential, yet this is easy to provide by GPS guidance systems.

The agronomical advantages of this concept are striking. However, the consequences for the machinery and its management can create problems. The running gears of all tractors and machines should be able to use the same tracks. Only small deviations in the track widths are possible as a result of differences in tire widths (Fig. 7.14, top). In case wagons or trucks are loaded on-the-go from the harvesting machines and not at special places on the headlands, their track width too must be adapted.

A precise track use is not possible with round trips in the field. Instead driving up and down is needed – preferably along the longest side of the field. This is no problem in flat areas. But this driving pattern does not fit well to rolling areas and especially not to contour farming practices. Furthermore, the system does not allow random turns within the field. So any loading vehicles irrespective of the filling level should not leave the lane before arriving at the headlands.

Adapting all implements to the same track width is not easy because of the **wide tracks** of present day high capacity grain combines. Their track width exceeds that of modern heavy tractors by about one third. Technically, enlarging the track widths of tractors and trailers is possible. Yet problems arise in densely populated countries from public road-traffic regulations for tractors and trailers on such wide tracks. It is mainly for this reason that controlled traffic farming with tractors, trailers and combines

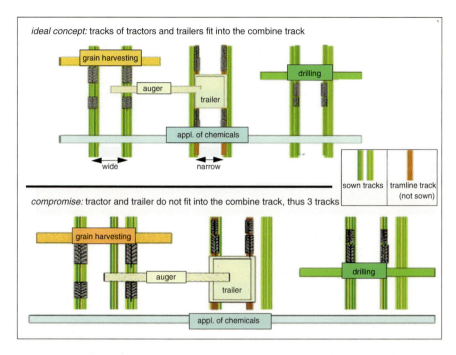

**Fig. 7.14** Arranging the tracks of various farming operations for controlled traffic farming in order to avoid soil compaction in the growing area. The objective is the separation of growing area from trafficked soil (From CTF Controlled Traffic Farming 2009, altered)

running in the same tracks seldom is practised in densely populated countries. There are alternatives, *e.g.* employing two standard track widths and to let the narrower running gears use only one track of the combine (Fig. 7.14, bottom). Yet this increases the tracked area of the land.

Controlled traffic farming has gained considerably in Australia, where problems with wider tracks on tractors and trailers on public roads are less common. This makes it possible to operate there with one **standard track width** for all machines (Fig. 7.14, top). In Europe – where the concept was initiated (Chamen et al. 2003) – it is mainly used for rotations of small cereals and rape (colza). Concepts for rotations with forage- or root crops should be developed. The prospects of getting rid of technically induced soil compaction and thus to pave the way for a widespread application of no-till practices deserve attention. It should be borne in mind that all standards of present day track widths are man-made and therefore can in principle be altered.

Another approach of reducing soil compaction and by this to eliminate cultivation needs is the introduction of **unmanned farm machinery** or **robots**. The present large and heavy farm machinery results from its ability to provide for a high work-rate per time unit for the driver. But modern automatic guidance systems such as RTK- GPS and other control systems can eliminate the need for a permanent

**Fig. 7.15** Unmanned tractor (Courtesy John Deere Co.)

driver. And as soon as no permanent driver or operator is needed on the machine any more, the incentive for using large and heavy equipment is gone. Having several small machines instead of a large one then does not imply higher labour costs any more. Robotic implements can operate around the clock. They can be used in a more flexible way with small and irregularly shaped fields. Because of the higher number of implements, the default risk is smaller than that of large conventional machines. But above all, the smaller robots might reduce soil compaction and therefore also help to introduce no-till practices.

However, presently robots are available only for rather simple operations such as mowing or rice transplanting. Small robots, which move up and down in fields for **scouting purposes** such as detecting of weeds, of diseases, of insects or for sampling soil are in the focus of research (Blackmore 2007). A medium sized prototype farm robot has been presented by the farm machinery industry (Fig. 7.15).

Yet there still remain many farming operations, for which until now robotic solutions do not exist. There is still much to do before soil compaction by giant machines is eliminated by swarms of small robots that move through fields and make it possible to farm generally without any tilling of the soil.

## References

Adamchuk VI, Ingram TJ, Sudduth KA, Chung SO (2008) On-the-go mapping of soil mechanical resistance using a linear depth effect model. T ASABE 51:1885–1894

Bertuzzi P, Stengel P (1988) Measuring effects of tillage implements on soil surface geometry with a laser relief meter. In: Proceedings of the 11th international conference of the International Soil Tillage Research Organization (ISTRO), Edinburgh, 1988

Blackmore BS (2007) A systems view of agricultural robots. In: Stafford V (ed) Precision agriculture '07. Wageningen Academic Publishers, Wageningen, pp 23–31

Bogrekci I, Godwin RJ (2007a) Development of an image-processing technique for soil tilth sensing. Biosyst Eng 97:323–331

Bogrekci I, Godwin RJ (2007b) Development of a mechanical transducer for real-time soil tilth sensing. Biosyst Eng 98:127–137

Bowers SA, Hanks RJ (1965) Reflection of radiant energy from soils. Soil Sci 100:130–138

Brunotte J (2007) Measuring the actual trafficability by online sensors – a practical solution? (in German). In: Strategien zum Bodenschutz – Sachstand und Handlungsbedarf. Gesellschaft für konservierende Bodenbearbeitung e.V., Neuenhagen

Chamen WTC (1998) Soil cultivation: new methods and technologies. In: Club of Bologna, Proceedings of the 9th meeting, Bologna, 15–16 Nov 1998 (Editione Unacoma Service srl), pp 73–85

Chamen T, Alakukku L, Pires S, Sommer C, Spoor G, Tijink F, Weisskopf P (2003) Prevention strategies for field traffic induced compaction: a review. Soil Till Res 73:161–174

CTF Controlled Traffic Farming (2009) What is controlled traffic farming? http://www.controlledtrafficfarming.com/downloads/DL%20fliers.PDF

Destain MF, Verbrugge JG (1987) Measurement of soil surface profiles with an optical displacement transducer. In: Proceedings of the 9th international conference of the international society of terrain vehicle systems, Barcelona, 1987

Dexter AR (1988) Advances in characterization of soil structure. Soil Till Res 11:199–238

Gist J (2002) Steel in the field. In: Bowman G (ed) A farmers guide to weed management tools. Sustainable agriculture networks, Handbook series, Book 2, USDA, Beltsville

Gorucu S, Khalilian A, Han YJ, Dodd RB, Smith BR (2006) An algorithm to determine the optimum tillage depth from soil penetrometer data in coastal plain soils. Appl Eng Agric 22(5):625–631

Hakansson I, Eriksson J, Danfors B (1974) Soil compaction – field structure – crops (in Swedish) Jordbrukstekniska institutet, Meddelande nr. 354, Uppsala: 33

Hakansson I, Myrbeck A, Etana A (2002) A review of research on seedbed preparation for small grains in Sweden. Soil Till Res 64(1–2):23–40

Harral BB, Cowe CA (1982) Development of an optical displacement transducer for the measurement of soil surface profiles. J Agric Eng Res 27:421–429

Hartung E, Druecker H (2009) Promotion of mulch-seeding by precision in cultivation (in German). In: Schriftenreihe der Agrar– und Ernährungswissenschaftlichen Fakultät der Universität Kiel 114, pp 163–173

Heege HJ, Vosshenrich HH (1998) Soil cultivation: new methods and new technologies. In: Club of Bologna, Proceedings of the 9th meeting, Bologna, 15–16 Nov 1998 (Editione Unacoma Service srl), pp 53–64

Heege HJ, Vosshenrich HH (2000) Interactions between soil cultivation and climate. In: International Soil Tillage Research Organisation. Proceedings of the 15th ISTRO conference, Fort Worth, July 2000, Paper No. 73

Heinonen R (1985) Soil management and crop water supply, 4th edn. Swedish University of Agricultural Sciences, Uppsala

Isensee E, Reckleben Y (2009) Effects of site-specific soil cultivation by a tined implement (in German). Rationalisierungs-Kuratorium für Landwirtschaft, 4.1.1.3, Rendsburg (http://www.rkl-info.de)

Kelly R, Jensen T, Radford B (2004) Precision farming in the northern grains region. Soil compaction and controlled traffic farming. Queensland Government, Department of Employment, Economic Development and Innovation, Brisbane. http://www2.dpi.qld.gov.au/fieldcrops/3166.html

Lüth HG (1993) Development of a horizontal penetrograph as a sensing method for soil compaction (in German). Doctoral thesis, Department of Agricultural Systems Engineering, University of Kiel, Kiel. Forschungsbericht Agrartechnik der Max-Eyth-Gesellschaft Agrartechnik im VDI (VDI-MEG) 235

Maze RC, Redel BD (2006) Discussion paper on seedbed finishing. Government of Alberta. Agriculture and rural development. Publication on the web on 16 Feb 2006: 7 http://www1.agric.gov.ab.ca/$department/deptdocs.nsf/all/eng8078/$file/Discussion_Paper_on_Seedbed_Finishing.pdf?OpenElement

Oliveira JCM, Appolini CR, Coimbra MM, Reichardt K, Bacchi OOS, Ferraz E, Silva SC, Galvao Filho W (1998) Soil structure evaluated by gamma ray attenuation. Soil Till Res 48:127–133

Olsen HJ (1992) Sensing of aggregate size by means of a horizontal mini-penetrometer. Soil Till Res 24:79–94

Orlov DS (1966) Quantitative patterns of light reflection by soils. Influence of particle size on reflectivity. Soviet Soil Sci 13:1495–1498

Petrie S (2009) Using the undercutter sweep in a reduced tillage summer fallow system. Oregon State University, Agricultural Experiment Station. 2009 Dryland agricultural research annual report. Special report 1091. http://extension.orgonstate.edu/catalog/html/sr/sr1091-e/sr1091.pdf

Pforte F, Hensel O (2006) Online-measurement of percent residue cover for implement control in conservation tillage. In: XVI CIGR world congress: agricultural engineering for a better world, Bonn, 3–7 Sept 2006. Book of Abstracts, pp 175–176

Pforte F, Hensel O (2010) Development of an algorithm for online measurement of percent residue cover. Biosyst Eng 106:260–267

Satkus A, Velykis A (2008) Modeling of seedbed creation for spring cereals in clayey soils. Agronomy Res 6 (special issue): 329–339

Scarlett AJ, Lowe JC, Semple DA (1997) Precision tillage: in field real time control of seedbed quality. In: Stafford JS (ed) Precision Agriculture 1997. Proceedings of the first European conference on precision agriculture, Warwick, Sept 9–10 1997. Bios Scientific Publishers, Oxford, pp 503–511

Schillinger WF, Papendick RI (2008) Then and now: 125 years of dryland wheat farming in the Inland Pacific Northwest. Agron J 100:166–188

Sommer C, Vosshenrich HH (2004) Soil cultivation and sowing. Management system for site–specific crop production (in German). Verbundprojekt Preagro. Endbericht, Kapitel 4. Herausgegeben vom KTB (CD-ROM): 101–120. http://www.preagro.de/Veroeff/Liste.php3

Stafford JV, Ambler B (1990) Computer vision as a sensing system for soil cultivator control. In: Proceedings of the Institution of Mechanical Engineers (ImechE) C419/041, IME Conference on Mechatronics, Cambridge, UK, Sept. 1990, pp 123–129

Sun Y, Schulze Lammers P, Damerow L (2003) A dual sensor for simultaneous investigation of soil cone index and moisture content. Agric Eng Res 9:12

Taeger-Farny W (2003) Working depth as needed. Experiences with site-specific soil cultivation (in German). Neue Landwirtsch, Sonderheft, pp 66

Tullberg JN, Yule DF, McGarry D (2007) Controlled traffic farming. From research to adoption in Australia. Soil Till Res 97:272–281

Vosshenrich HH, Sommer C (2005a) Not too much and not too little. How much break up with site-specific soil cultivation (in German). Neue Landwirtsch, Berlin, Heft 11: 54

Vosshenrich HH, Sommer C (2005b) Is online controlled site-specific soil cultivation possible? (in German). Neue Landwirtsch, Berlin, Heft 12: 41

Winkelblech CS, Johnson WH (1964) Soil aggregate separation characteristics of secondary tillage tools components. T ASAE 7(1):29–33

Wuest SB, Schillinger WF (2011) Evaporation from high residue no-till versus tilled fallow in a dry summer climate. Soil Sci Soc Am J 75(4):1513–1519

Zaikin AA, Young DL, Schillinger WF (2007) Economic comparison of undercutter and traditional tillage systems for winter wheat-summer fallow. Farm business management report. Washington State University, Pullman

Zheng R, Campbell JB, de Beurs KM (2012) Remote sensing of crop residue cover using multi-temporal Landsat imagery. Remote Sens Environ 117:177–183

Zuo Y, Erbach DC, Marley SJ (2000) Soil structure evaluation by using a fiber-optic sensor. Trans ASAE 43:1317–1322

# Chapter 8
# Site-Specific Sowing

**Hermann J. Heege**

**Abstract** Site-specific control of the seed-density can rely on maps of soil texture. The seed-density should rise from sand to silt and loam and fall again towards clay. In this way either the yields can be increased or seeds can be saved.

For the sowing depth, the site-specific control should be based either on texture or on water content of the soil. In regions with maritime climate and consequently frequent rain, the control via soil texture seems reasonable. This could be realized by using texture maps and adjusting the depth of openers on-the-go by means of ultrasonic distance sensing.

In areas with continental climate and thus longer dry spells, a control based on the water content of the soil is a good choice. Here a soil moisture seeking control system that adjusts the sowing depth on-the-go to the drying front in the soil via infrared reflectance- or electrical resistance seems reasonable.

A special challenge is the increasing conflict between no-till with crop residues on the surface and sowing techniques. The trend to smaller row widths in order to realize yield increases adds to this conflict. But there are concepts available that can cope with this conflict. These are dealt with.

**Keywords** Crop residues • Inter-row sowing • Row cleaners • Seed density • Seeding depth • Soil moisture • Ultrasonics

---

H.J. Heege (✉)
Department of Agricultural Systems Engineering, University of Kiel,
24098 Kiel, Germany
e-mail: hheege@ilv.uni-kiel.de

## 8.1 Seed-Rate or Seed-Density

Site-specific sowing has several objectives. A dominant objective is the control of the **seed-rate** or instead of it the **seed-density**. The former is defined in kg per ha, the latter in the number of seeds per unit area. Important is also the spatial placement of the seeds in the field. This has to do with two objectives: the **seed distribution over the area** and the **sowing depth**. All these objectives depend heavily not only on the respective crops, but on the seeding methods employed as well.

Different crops, soils and climates necessitate a very wide range of seed-densities (Fig. 8.1). The seeding methods depend largely on the respective seed density. Precision drilling – targeted placing of every individual seed – can only be realized for crops with rather low seed densities because of the costs. For crops with high seed-densities – such as small grains, grasses, clover and alfalfa – it still is indispensable to practice bulk-drilling.

However, adjusting machines for bulk-drilling via the seed-rate or seed-mass in kg per ha should be replaced by a control that is oriented directly at the required number of seeds per unit area, the seed density. Because it is this number and the field emergence that define the **plant density per unit area**. Therefore, the aim should be online and on-the-go control based on sensing the number of seeds that pass through the seed tubes.

The traditional adjustment of the seed-mass in kg per ha with bulk-drilling can cause substantial **deviations** from the number of seeds per unit area. These deviations can arise from varying slip of the driving wheel within a field due to firm- or soft ground as well as dry- or moist soil. Deficiencies can be due to the varying bulk density of the seeds, since these drills are designed for bulk- or volume metering. And the bulk density changes with the species, the variety, the provenance, and because of vibrations in the hopper, which influence the settling of the seeds. Finally, rather large deficiencies can arise from the varying average mass per seed. With European wheat varieties, the mean mass per seed fluctuates between 40 and 55 mg. All these deficiencies can be overcome by a reliable method of closed loop control of the seed numbers per unit area. For this, sensing the number of seeds that are passing through the seed tube is necessary.

Such a control of the seed density via optical sensors – which act as seed counters – is state of the art with precision drilling. With these drills, sensing the seeds is facilitated because the distances between the falling seeds are rather even. Usually, the **coefficient of variation** for these distances with precision drills is only about 10–20 %.

With bulk-drilling, the corresponding distances of the seeds at the end of the seed tube are very uneven because of more gaps as well as of seed clusters. The mean coefficient of variation for the seed distances is 100 %. Sensing the number of seeds online and on- the-go therefore is more difficult. But it is possible, at least for small grains. It is necessary to adapt the seed sensing technique to the irregular seed sequences. This adaptation can be done by either compensating for the seeds that were not recorded as a result of clusters by a **computer program** (Feldhaus 1997; Heege and Feldhaus 2002) or alternatively by more elaborate **optical sensors** for

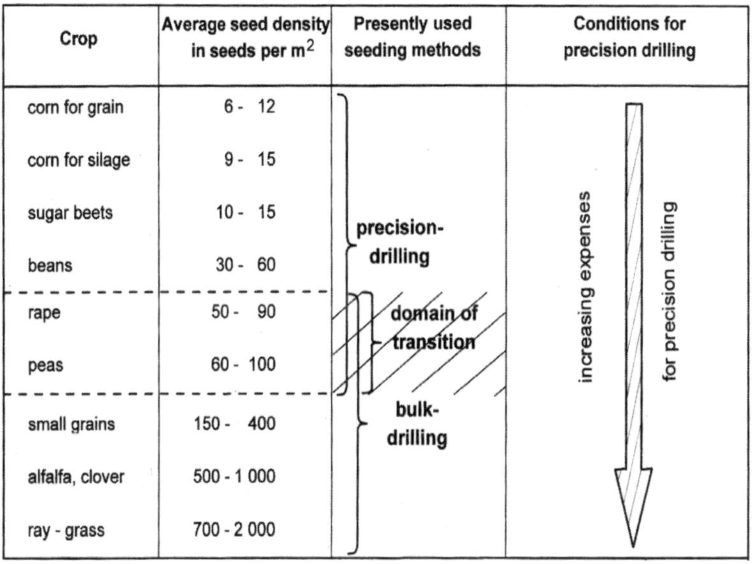

| Crop | Average seed density in seeds per m$^2$ | Presently used seeding methods | Conditions for precision drilling |
|---|---|---|---|
| corn for grain | 6 - 12 | | |
| corn for silage | 9 - 15 | | |
| sugar beets | 10 - 15 | precision-drilling | |
| beans | 30 - 60 | | |
| rape | 50 - 90 | domain of transition | increasing expenses / for precision drilling |
| peas | 60 - 100 | | |
| small grains | 150 - 400 | bulk-drilling | |
| alfalfa, clover | 500 - 1 000 | | |
| ray - grass | 700 - 2 000 | | |

**Fig. 8.1** Conditions for seeding methods (Heege and Billot 1999)

the seed counting (Müller et al. 1994, 1997). For an online and on-the-go control of the seed-density it is necessary to supply the control computer with data about the site-specific implement speed as well (Fig. 8.2).

## 8.1.1 Site-Specific Control of Seed-Density

The seed-densities for crops as listed in Fig. 8.1 vary within a wide span. The range indicated for *e.g.* small grains goes from 150 up to 400 seeds per m$^2$. Several parameters bring about this wide range, such as species, variety, sowing time, water supply of the crop and soil texture. In rain-fed areas, it is only soil texture, which lends itself to site-specific control within individual fields.

All other parameters mentioned above usually do vary from field to field within a farm and therefore need attention on this basis. But they can be regarded as being constant within individual fields with present day farming techniques. On the basis of these parameters, therefore, site-specific control within individual fields does not make sense

The influence of soil texture and precipitation on the seed-rate needed for a high yield of winter wheat is shown in Fig. 8.3 via a factor, by which the respective seed-density is multiplied. With a low annual precipitation, a smaller seed-density suffices. Because the water supply does not support a dense and lush crop. However, the

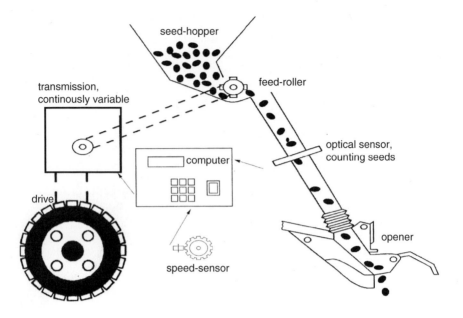

**Fig. 8.2** Closed loop control of the seed density of small grains by means of seed counting (Heege and Feldhaus 2002)

seed-density needed rises up only to an annual precipitation of 700–800 mm. Beyond a precipitation of 1,000–1,100 mm it drops again. This may be due to the fact that wet soils provide a higher emergence and thus might save seeds.

But as shown in Fig. 8.3, the influence of **soil texture** is more distinct than the effect of precipitation. Silty- or loamy soils always should be provided with higher seed-densities than sandy- or clay soils. This too probably can be explained by the fact that these soils supply plants more evenly during the growing season with water than sandy- or clay soils do and therefore are better able to support a dense crop.

Site-specific control of the seed-density on the basis of the annual variation of precipitation does not make sense, since knowledge about the supply of parts within the field in rain-fed areas does not exist *a priori*. But data about the site-specific soil electrical conductivity as a **substitute of texture** can rather easily be recorded, georeferenced and mapped with present day sensor technology (Sect. 5.2.1). Whether as an alternative also site-specific sensing of texture based on infrared reflectance will become feasible, is not clear (Sect. 5.3). Whatever method will be preferred, because practically these soil properties do not vary temporally, any maps generated can be used for several decades as a basis for on-the-go control of seed-density. Since fields with varying soil texture are common in many parts of the world, a substantial gain in sowing precision could be obtained if differences in the soil texture were taken care of in this way.

**Crop plasticity** is often mentioned as an argument against a need for site-specific control of seed-density. This feature is based on the ability of individual plants of a

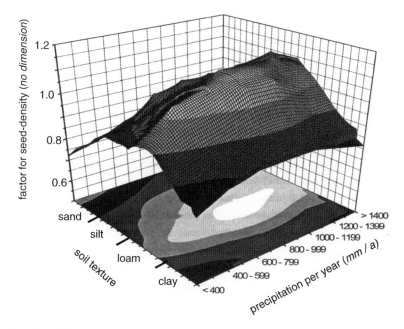

**Fig. 8.3** Factor for seed-density of winter-wheat depending on soil texture and precipitation for flat fields and a mean annual temperature of 8 °C (From Wiesehoff 2005, altered)

crop to act on inaccuracies in the local seed-rate as well as in the seed distribution over the area by either compensatory- or *vice versa* restricted growth. Within limits, the effect is such that the spatial growth of the plants automatically adjusts itself to the space available. Especially with small grains, the plasticity of a crop can be remarkable due to a matched development of lateral plant branches by tillering.

The result of a high crop plasticity is that the relation between seed-density and yield presents itself as a rather flat curve. This means that the effect of a more precise seed-density on the **yield** is rather small. However, at the same time this implies that more targeted sowing can result in substantial savings of **seed costs**. So the benefit of site-specific seed-density control just shifts from yields to seed costs. With crops that have a low plasticity – *e.g.* beans, sugar beets or potatoes – it is *vice versa* (Fig. 8.4)

Having in mind that besides crop plasticity the actual differences in soil texture within fields will influence the benefits achieved from site-specific control of the seed-density, it must be expected that results from field experiments will vary greatly. Schneider and Wagner (2005) presented a survey of results based on experiments with small cereals and maize from four different places in Germany. In all cases there were benefits from site-specific sowing, partly in savings of seed costs, but partly also from yield increases. The monetary value of these agronomic benefits – site-specific control costs not subtracted – was between 13 and 92 Euros per ha. In case an average agronomic monetary benefit of 50 Euros is assumed, it should easily be possible to cover the site-specific control costs from this.

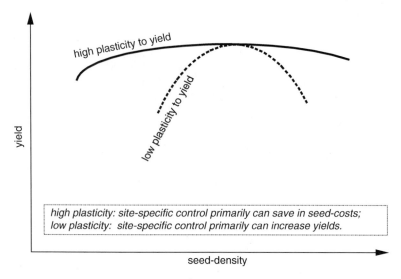

**Fig. 8.4** Influence of seed-density on yield depends on crop plasticity (schematic curves)

## 8.2  Seed Distribution over the Area

Precision farming also implies precision in the seed distribution over the area. The area per seed is the product of row spacing and mean seed distance within the row. With all crops, presently row spacings are much wider than seed distances within the rows. For wheat – the most common crop on earth – a row spacing of about 12 cm and a mean seed distance within the row of only 2–3 cm are frequently used. Thus the plant distances perpendicular to the row directions are about five times the distances within the rows. The situation is similar with most other crops.

There are no indications that plants grow better when the distances to its nearest neighbours in the row direction are smaller than in the direction perpendicular to the rows. On the contrary, research at many places has shown, that *ceteris paribus* decreasing the row spacing and accordingly increasing the seed distances within the row improves growth of plants (Cox and Cherney 2011; Heege 1993; Heege and Billot 1999; Lambert and Lowenberg-DeBoer 2003; Neumann et al. 2006). Probably plant distances should be optimized irrespective of spacial direction. This means that for **precision sown crops**, ideally the plants should be arranged in the pattern of squares (row spacing equals mean seed distance within rows) or even better theoretically in the pattern of equilateral triangles (Heege 1967).

For **bulk drilled crops**, reducing the row spacing to the mean distance of seeds within the rows too would result in a very substantial improvement of the arrangement of plants. However, because of the irregular seed distances within rows, in this case the theoretical optimum in plant arrangement still would be not be obtained with a row spacing, which equals the **mean seed distance** within rows. The theoretical optimum for bulk drilling would be a row spacing that approaches zero, which is broadcasting (Heege 1967, 1970, 1993). Therefore, as far as the spatial seed arrangement with bulk sowing is concerned, **broadcasting** must be regarded as the best method. The problem is that not all sowing methods for broadcasting provide for a sufficiently even depth control of the seeds. And good depth control can be even more important than improvements in the seed distribution over the area. Yet both objectives – precision in the seed distribution over the area as well as an even depth – can be obtained by using wing-coulters. The seeds are broadcast within the empty space below the wings and the flattened soil underneath it.

Provided an adequate depth control is obtained, **improvements in the seed distribution** over the area can under *ceteris paribus* conditions (*e.g.* same seedrate per ha or seed density) result in

- small increases in emergence of seeds
- less weed growth
- small increases in yield.

An effect of improvements in the seed distribution over the area on the **emergence** of seeds has been observed with closely spaced crops such as *e.g.* small grains (Mülle 1979; Mülle and Heege 1981). This probably can be explained by wider distances to nearest neighbouring seeds. For closely spaced crops, this reduces competition for growth factors that promote emergence, *e.g.* water.

Less **weed growth** with improved plant distribution over the area is the result of quicker canopy closing and hence faster shading of the soil.

The small increases in **yield** finally can be explained by a more even allocation of several growth factors such as light, water, air and mineral nutrients to plants. There is less light interception within the canopy, which promotes photosynthesis.

Yield responses of small grains, rape (colza) and beans to the seed distribution over the area are shown in Fig. 8.5. Similar responses to improvements in the seed distribution over the area have been reported for maize and soybeans (Lambert and Lowenberg-DeBoer 2003; Cox and Cherney 2011) as well as for cotton (Reddy et al. 2009).

However, there are reasons why in practice any changes in seed spacing towards more equidistant arrangements between neighbouring seeds irrespective of the direction are realized rather slowly.

For **bulk drilled crops** with a high seed density such as *e.g.* small grains, a decrease in opener spacing for drilling in narrower rows can present problems. Especially on fields that were not ploughed, clogging of crop residues or clods between adjacent openers can increase. This can impair the precision of the seeding depth and hence the emergence (Heege 1993). Hence again in this respect too both precise seed distribution over the area and accurate seeding depth must be observed. Furthermore, the expenses for sowing machines increase. And finally, with widely-spaced crops that are harvested by individual row units such as maize and soybeans, the expenses for harvesting machines too are higher.

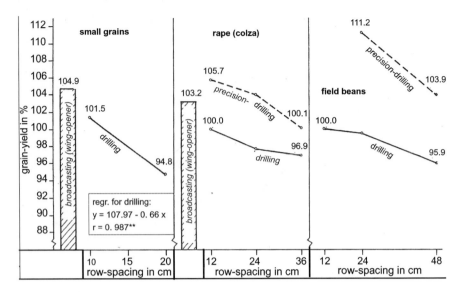

**Fig. 8.5** Seed distribution over the area and yield. In all cases, the respective yield for drilling with 12 cm row-spacing is set to 100. Broadcast seeding is done by wing-coulters, which allow a similar precision in depth control as conventional openers for drilling. The seed-rates per ha are constant. For details see Heege 1993 or Heege and Billot 1999

## 8.3 Seeding Depth

A precise control of the seeding depth is important for a uniform emergence and development of all plants. This is in turn essential for treating all plants of a crop at the same growth stage with agrochemicals and thus for high yields.

Prerequisites for the germination of seeds are an adequate temperature, enough water and sufficient oxygen for respiration. The farmer can influence the temperature only by selecting the best time of the year. Control of water- and oxygen supply is possible via the seeding depth, which is a means to provide these factors in the best proportion.

An increase of the seeding depth generally improves the water supply for germination, since in most cases the water content of a soil rises with the distance from the surface. But the higher the water content, the lower is the oxygen supply, since the soil pores are either filled with water or air. A very high seeding depth hence can put the oxygen supply of seeds at risk.

The energy for the growth of seedlings before assimilation starts can only be provided by the seeds themselves. Hence, even after germination took place, too deep planting can result in either in no emergence or in weak plants arriving at the soil surface.

Thus low or uneven emergence can be the result of either a **lack of water**, a **lack of oxygen** or a **lack of energy**. Generally, due to the influence of the weather, controlling for an adequate supply of water is the main problem.

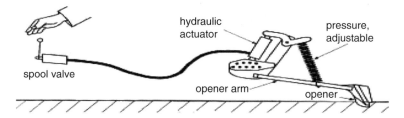

**Fig. 8.6** Manual pressure control for a general-purpose drill (From Scarlett 2001, altered)

## 8.3.1 Control of the Seeding Depth

Commonly, the seeding depth depends either on the **pressure** that the free swinging drill coulters exert on the soil or alternatively on the height of **spacers** such as skids or wheels that are attached to the openers of heavier sowing units. The adjusting by means of the pressure is used only for some bulk-metering machines. An interesting strategy would be to adjust the depth of the coulters such that **moist soil** is attained.

#### 8.3.1.1 Control via the Pressure Exerted on the Openers

With adjusting the **pressure** (Fig. 8.6), the seeding depth depends on the resistance of the soil. The higher the resistance of the soil, the lower the seeding depth is and *vice versa*. And since the resistance goes up with the speed, the seeding depth decreases with it. Hence the travel speed should be kept rather constant. However, controlling the pressure can also have advantages. On fields with uniform soil moisture but with varying soil resistance, the openers deposit the seeds deeper in loose soil than in more dense soil. This can be beneficial since the water transfer from the soil to the seeds is worse in loose soil. And for getting a sufficiently recompacted soil zone directly underneath the seeds (Fig. 7.11), deeper operation of the openers in loose soil is helpful too – as long as the energy of the seeds suffices for emergence.

But the situation is opposite in case the moisture content varies within a field. Because the lower the moisture content, the higher the soil resistance is (Fig. 7.3). And consequently, the less deep the seeds are placed. This is contrary to the respective needs. So in summary, constant pressure can improve or impair the emergence.

#### 8.3.1.2 Control via the Depth from the Soil Surface by Spacers

Controlling the depth of the openers from the soil surface by means of adjustable **spacers** – in most cases gauge wheels – does away with these influences. This method is state of the art with precision drills and is increasingly used with bulk-metering drills as well. For a precise depth control, the best location of the gauge wheels is as close as possible to the openers and lateral of them (Karayel and Özmerzi 2008; Morrison and Gerik 1985). The problem with depth control by

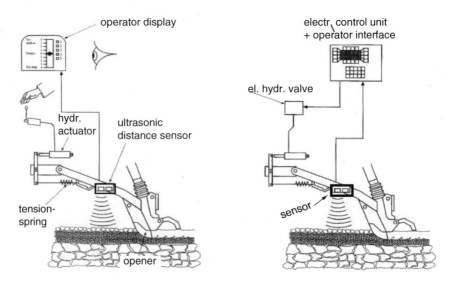

**Fig. 8.7** Use of ultrasonic distance sensors for either manual (*left*) or for computer assisted control (*right*) of the sowing depth (From Auernhammer 1989 and Scarlett 2001, altered)

gauge wheels is that their adjustment must be done manually on every row unit. This excludes site-specific depth control on-the-go, which is possible with pressure control.

A way out of this situation is a combination of pressure- and depth control, but without gauge wheels. Instead of using gauge wheels, the penetrating depth of the openers is sensed by **ultrasonic sound waves**, which are emitted by small sensors that are fixed to the arms of the openers. The time that the vertically oriented sound wave echos need to return to the sensors defines the distance of the opener arms above the ground and thus the actual seeding depth (Fig. 8.7).

This sensing technique for indicating and controlling the seeding depth on-the-go was developed by Dyck et al. (1985) and commercialized in Australia. Openers of air seeders with or without these ultrasonic sensors were tested at the Alberta Farm Machinery Research Centre (1994). The ultrasonic sensing resulted in much smaller differences between the intended and the finally measured seed depths. However, a continuous use of such seeding depth control by ultrasonic waves in commercial farm machinery did not occur. High prices for ultrasonic sensors in the past may have been responsible for this. But in the meantime these prices have dropped substantially as a result of mass production for backup-distance sensors used in automobiles as driver aids for parking. So it may be time to reconsider ultrasonic sensing of the seeding depth.

Kiani et al. (2010) used an experimental control device for automatic and online adjusting of seeding depths based on ultrasonic sensing in a similar manner as outlined in Fig. 8.7. The targeted seeding depths were attained with very high precision

($r^2 = 0.94$ or even higher). However, this alone does not mean that ultrasonic control of seeding depths is ready for applications.

Precise sensing of the sowing depth with this technique implies an unimpeded access of the ultrasonic waves to a resting soil surface. Therefore, the waves should hit the soil well in advance of the opener front in order to prevent inaccuracies arising from soil in motion due to mechanical stirring. Bulky crop residues must be removed from the row area (see Sect. 8.4.1). Interference by clods must and can be taken care of by averaging of signals and by adjusting the response time. And finally, under very dry conditions, dust may influence the results.

But there are important advantages too. Control adjustments are possible on-the-go, which is a prerequisite for site-specific operations. And since the control devices need less space in the lateral direction than gauge wheels, realizing of seeding in narrow rows is facilitated.

The tests at the Alberta Agricultural Research Centre (1994) and by Kiani et al. (2010) were about the precise adjustment of sowing depth to preset **constant** numerical values for the field. But when it comes to site-specific control, the question is, on which criterion a **varying** seeding depth within a field should be based. Such a control could be oriented *e.g.* at the temperature, the texture, the organic matter or the water content of a soil. Sensing techniques for site-specific recording and mapping of these soil properties are available. However, what is reasonable?

It is well known that a specific temperature range is needed in the soil for every plant species for germination as well as for emergence. However, except for selecting the proper season of the year, predicting the soil temperature around the seeds for several days is hardly possible. Therefore, sensing the site-specific temperatures at the tip of an opener does not make sense. This would provide for the situation just at the time of seeding, not more. Varying weather can change the temperature within a short time.

Contrary to this, the **texture** as well as the **organic matter content** of a soil are constant on a time basis. In most cases, these factors are interrelated. Organic matter contents of soils tend to increase with the clay content. This means that a site-specific control oriented on the clay content or texture of a soil simultaneously includes differences in the organic matter content, at least to some extent. All site- specific recommendations that are based on clay content or texture probably intrinsically will do this.

Generally, it is recommended to increase the crop-specific seeding depth with decreasing clay- and hence with rising sand content of a soil. This holds especially for regions in continental climate. Yet the widely used control systems that keep the positions of the openers constant to the soil surface prevent a site-specific adaptation of the seeding depth in fields with varying soil texture. Ultrasonic distance sensors would allow to adjust the seeding depth on-the-go to the respective texture. The logical procedure for realizing such a control system would be starting with maps about the site-specific texture that was estimated from sensing electrical conductivities or perhaps infrared reflectances. These maps could then be used for many years as a base for a site- as well as a crop-specific seeding depth control via ultrasonic distance sensing.

However, the most important factor for an adequate control of the seeding depth probably is soil moisture. This is dealt with in the next section.

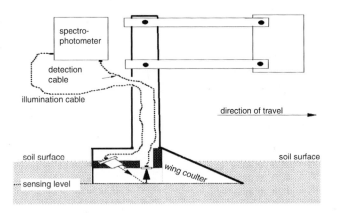

**Fig. 8.8** Principle of moisture sensing at various depth levels below the soil surface by infrared reflectance (not to scale)

#### 8.3.1.3 Control via Soil Moisture

In many cases, the moisture content rises with the vertical distance from the soil surface. The logical consequence of this would be to increase the seeding depth until the soil moisture needed is arrived at. This leads to a site-specific control of the **depth** of the openers that is based on sensing the soil moisture. The moisture levels, which are needed for various seed species for germination and emergence, are well known.

Such a depth control calls for a precise sensing of the water content in successive horizontal soil layers. Up to now, no sensing technique that is simply directed at the soil **surface** can do this. Yet scratching the soil surface by an opener so that the planes of different soil layers are accessible allows to sense the situation (Fig. 8.8).

The sensing principles used for this rely either on the reflectance of near-infrared radiation (Price and Gaultney 1993; Mouazen et al. 2005) or on electrical resistance or -voltage (Carter and Chesson 1993; Bowers et al. 2006). Details on water sensing principles are dealt with in Chap. 5.

However, sensing the moisture at a predetermined level below the soil surface alone does not suffice. A **moisture seeking control algorithm** is needed, which automatically searches for the water content necessary by sensing at various depth levels while preventing unneeded deep placement of the seeds. Price and Gaultney (1993) developed such a control algorithm for maize.

The need for such a control system in rainfed regions probably depends very much on the climate. Its usefulness can be questioned for regions with a distinct **maritime climate**. Rather frequent drizzling rains here are the rule. Even if at the time of sowing the soil moisture at the normal seeding depth is not quite sufficient, the probability of some rain within a short time span is rather high. This also means that the concept of rising moisture with increasing depth below the soil surface – which relies on rather long time periods of evaporation into the atmosphere without rain – quite often is not valid here.

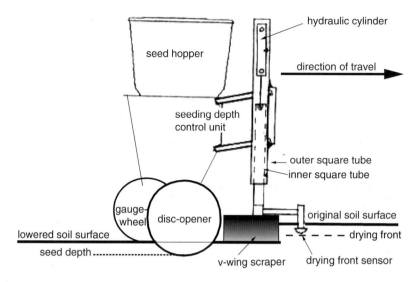

**Fig. 8.9** Schematic of experimental precision seeding unit with depth control based on the position of the drying front combined with a reduced seed cover. The final press wheel for closing the seed furrow is not shown. Details are no to scale (From Bowers et al. 2006, simplified and altered)

In regions with **continental climate**, the rain is less frequent, but instead when it occurs, it is heavier. And the periods without rain are longer. As a consequence, here the concept of rising moisture with increasing depth is more reliable. Hence probably it is not incidentally that the concept of "moisture seeking" started in regions with continental climate.

In areas where long dry spells frequently occur, the difference between **germination** on the one hand and **emergence** in the field on the other hand can get important. Seeds placed into a moist soil zone that is rather deep below the soil surface might germinate, but not generate an emerged seedling because of a long distance to the soil surface. Under these circumstances, seeding to moisture alone is not sufficient. It must be supplemented by seeding close enough to the surface. This obvious conflict of objectives can be sorted out by scraping aside some soil from the top along the seed row and thus creating a lower soil surface for emergence.

Figure 8.9 shows how – based on **sensing of the drying front** in a field – on the one hand seeding into moist soil and on the other hand a lowered soil surface for safe emergence can be obtained. The drying front within the vertical soil profile is where the water evaporates. Therefore, it is just beneath the top layer of dry soil, in which the water content is in balance with the relative humidity of the air. The vertical position of the drying front is sensed via the electrical resistance or voltage that exists between an insulated electrode at the tip of the drying front sensor (Fig. 8.9) and the seeding unit. Locating the drying front then allows to set the scraper in such a way that the air dry soil and in addition a thin layer of subjacent moist soil are swept into the inter-row area. The seeds thus are put into moist soil.

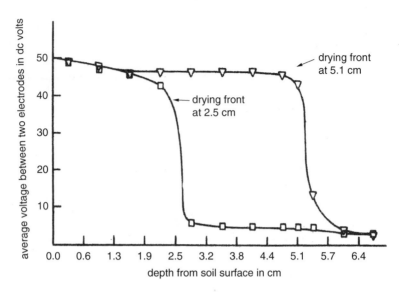

**Fig. 8.10** Average voltages between two uninsulated electrodes that move horizontally in a sandy loam. The depths of the electrodes were varied so that drying fronts of the soil were crossed (From Bowers and Bowen 1975)

Site-specific control for such seeding according to a moisture seeking program and for a limited seeding depth is possible (Weatherly and Bowers 1997). Locating the drying front via electrical resistance or voltage can be realized with a high precision (Bowers et al. 2006; Carter and Chesson 1993). An adaequate sensor control results in deviations on either side of the drying front of not more than about 3 mm (Fig. 8.10). The air-dry soil is above the point, where the voltage decreases rapidly with depth.

This concept of seeding depth control via sensing of the drying front and adapting the soil cover above the seeds seems to be a logical approach for regions with continental climate. But commercial use of this concept up to now is lacking. The same holds for comprehensive knowledge about scraping- and seeding depths that should be searched for by the control system. It should be borne in mind that **contact sensing** of resistance or voltage within a soil implies wear by abrasion, corrosion and collisions with stones for the tool.

**Non contact sensing** based on infrared radiation is less subjected to these wear factors. However, it is not known whether this method of sensing the moisture of successive soil layers allows for a similar precision.

## 8.4  Less Tillage, Crop Residues and Sowing Methods

There is a distinct conflict of objectives between precision of seed placement on the one hand and less cultivation on the other hand. Improvements in the seed distribution over the area as well as a high emergence are more difficult to obtain, when as a result of less cultivation, the residues of the previous crop remain on or near the soil surface.

Yet there are important points in favour of reduced- or even **no-tillage**:

- substantial savings in the expenditure of energy and labour
- less decomposition of soil organic matter
- less or no moist soil is moved upwards to the surface, therefore less- or no artificial clod production occurs
- crop residues on or near the soil surface can substantially reduce erosion, especially in areas with continental climate and on slopes.

However, some **consequences** of less or no tillage must also be dealt with:

- hitherto mechanical weed control via soil cultivation must be compensated for by more herbicides. This holds especially for regions with maritime climate.
- since less fungi on crop residues are buried, more fungicides must be used. This consequence too is more important in maritime regions.
- the seeding techniques have to cope with crop residues on or near the soil surface.

As a result of more efficient herbicides and fungicides, counteracting to the consequences for weeds and fungi becomes easier and easier. Yet the conflict between seeding techniques and crop residues remains and must be taken care of. With conventional sowing techniques, the residues can seriously reduce the emergence. This can be the result either of less precise seeding depths or of residues that are placed close to seeds and hence impair the water transfer from the soil to the seeds. Several strategies to cope with these residue problems are available.

### 8.4.1 *Vertical Discs, Cleaned Rows or Inter-Row Sowing*

Openers that employ heavily loaded **vertical discs** often can solve the residue problem in dry regions with small grains or oil crops, when the soil is rather hard and the straw is brittle. Under these conditions, the discs just cut through the residues. In humid areas, this method often fails, since here the soil and the straw are more flexible. Therefore, the straw is not cut by the disc openers, but instead it is pushed into the seed furrow. This can seriously impair the emergence.

In maritime climates, the moist straw also tends more to hair-pinning around seeding devices and hence to clogging.

Sowing into cleaned rows relies on a **local segregation** of the respective areas for sowing and residues. **Row cleaners** accomplish this by moving the residues ahead of the openers into the inter-row areas. The implements used for this usually are vertical, ground-driven, fingered wheels, which are slightly slanted in the direction of travel as well as perpendicular to it (Fig. 8.11, left). The rotating fingers act like rotary rakes and push the residues sidewards. Depending on the depth of operation, also some strip cultivation (strip tillage) just ahead of the openers can result (Fig. 8.11, right). Site-specific depth control of the row-cleaner units via data from field maps (*e.g.* about soil texture or residues) is possible.

**Fig. 8.11** Field sown with maize after maize without any previous cultivation by using row cleaners in front of the seeding units (*right*). The *insert* (*left*) shows a row cleaner for site-specific depth control in front of a seeding unit (Courtesy of Yetter Mfg. Co. Inc., rearranged and altered)

**Fig. 8.12** Inter-row sowing of wheat after wheat. The protection against soil erosion is evident (Courtesy of gps-Ag, Kangoroo Flat, Australia)

Seeding into strips that are less loaded with residues can also be achieved by **inter-row sowing**. This method relies on placing the seeds in furrows that are located precisely between the stubble rows of the previous crop (Fig. 8.12). A prerequisite for this is that subsequent crops either have the same- or multiple row widths. Precise seed placement is obtained by using machinery guidance via a global positioning system with real-time kinematic corrections (RTK-GPS). The positioning control is arranged in such a way that year after year the rows of the previous crop can be located again at seeding time. An auto-steering function is set for seed placement precisely between the rows of the previous crop.

In case the field in not flat, it can be necessary to use a control system that is not solely based on the position of the tractor, because on side-hills or contours the implement can drift sidewards. To prevent inaccuracies resulting from this, the positioning control system has to auto-steer the implement as well. This is state of the art. The same seeding machine and the same track direction should be used year after year. Because this ensures that small inaccuracies in the lateral opener spacings occur subsequently in all crops and tracks in the same pattern and hence hardly affect the result. It is obvious that the best results with this method are obtained if the straw is harvested as well as the grain.

Inter-row sowing is used successfully in Australia for wheat, canola (rape) and lentils in no-till rotations with row spacings of 22 cm or more (McCallum 2007). A special advantage of this method is the lower incidence of the take-all fungal disease (*Gaeumannomyces graminis*) in cereal rotations. This probably can be explained by some **spatial separation** of the seedlings from infections on roots of the previous crop. And due to less straw being next to the emerged seedlings than with other no-till sowing methods, the efficacy of soil applied herbicides near the seedlings might be improved. This holds for the use of row-cleaners as well.

Yet inter-row sowing and row cleaning differ in their response to **long stubbles**. Inter-row sowing between long stubbles is no problem as long as the vertical clearance of the machinery allows for this. In contrast to this, row cleaners operate better in residues that are not attached to the soil any more. With long, bulky residues the best solution for work without interference can be using both inter-row sowing and row-cleaning (Fig. 8.11, right).

### 8.4.2 Seeding into Cover Crops

In humid areas, cover crops can provide for the preservation of soil fertility and for the suppression of weeds between the growing periods of main crops. Yet no-till sowing into fields with cover crops presents special problems. At seeding time, the cover crop might be still growing and might have a dense canopy. Therefore, inter-row sowing is not possible. And cover crops also complicate the use of row-cleaners, because the **standing, dense residues** easily get caught in the moving parts of the machines and thus cause clogging.

Torbert et al. (2007) developed an attachment to the front of row-cleaners of seeder units that pushes the standing, killed residues of cover crops in a forward and sideward direction. In this way they prevented clogging.

The use of **roller-crimpers** for coping with the clogging problem for no-till sowing into cover crops originated in Brazil, Argentina and Paraguay (Derpsch 1998; Kornecki et al. 2009). This method for sowing into cover crops relies on the concept that the position of the residue plants – that still are anchored to the soil – should be in the **direction of travel** of the seeding implements. No hair-pinning of residues on parts of the seeding implements can occur if the position of the stems

**Fig. 8.13** Roller-crimper in front of a tractor for no-till seeding into a cover crop. The metal strips on the roller drum crimp the stems. They should not cut or chop the stems (Courtesy of I & J Manufacturing, Gap, Pennsylvania)

and leaves is and remains exactly in the direction of travel. This condition is achieved by flattening out the cover crop on the soil via rolling and crimping precisely in the **seeding direction** (Fig. 8.13).

Hence no residue parts are present, which are oriented at random and because of this can cause blocking. Since the plants still are fixed to the soil by their roots, by and large they remain in their place and are not taken along with the machinery. With tall cover crops, it is recommended to use row-cleaners in front of the seeding-units as well.

Since precise orientation of the flattened plants is very important, rolling and crimping cannot solve the problems for no-till seeding, if a cover crop lodges in several directions. However, with well managed cover crops this hardly occurs. So the roller-crimper technique provides new perspectives for no-till methods after cover crops.

However, the agronomic prerequisites of cover crops still remain. These are firstly the extra water supply for a crop instead of a fallow period within the rotation, and secondly the need to fit this crop into the seasonal schedules of the main crops. In areas that are short in water supply, the first prerequisite cannot be come up to. And if the second prerequisite – maintaining the seasonal schedules of the main crops within the rotation – cannot be met, yield losses with these can result. So seeding into cover crops still needs to be well adapted to the existing conditions.

## 8.4.3  Seeding and Loose Residue Sizes

Loose residues from the previous crop can affect the emergence in two ways

- by deteriorating the precision of the sowing depth
- by impeding the water transfer from the soil to the seeds.

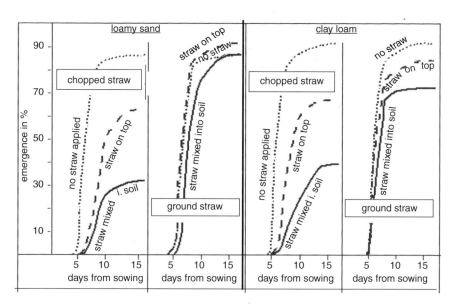

**Fig. 8.14** Emergence of rape (colza) depending on the size reduction and the location of barley straw within soils from experiments in a laboratory (From Heege and Vosshenrich 2000)

The effect of both factors depends on sizes of loose residues.

It is well known that substantially reducing the sizes of loose crop residues can do away with clogging problems of sowing machines and hence provide for a precise sowing depth. Yet how is the water transfer from the soil to the seeds – and thus the germination and emergence – affected by a reduction of residue sizes and the location of residues within the soil? Figure 8.14 shows results to these questions for the interactions between rape (colza) seed and barley straw from laboratory experiments with two soils. The straw was either chopped, ground or left out completely.

The order of increasing emergence – with one exception – was straw mixed into the soil, straw on the top of the soil, no straw applied. The exception occurred with the loamy sandy soil and ground straw; for this combination the alternatives in straw placement hardly did matter. In all cases, grinding the straw instead of chopping it clearly improved the emergence. Similar results as with rape were obtained for small grains with the exception that the differences between the treatments were somewhat smaller.

In short, a distinct reduction of straw sizes during harvesting can help to solve residue problems. Utilizing this effect would necessitate to have grinders instead of choppers on combines and to increase their engine-power for doing this. An appealing prospect of this option would be that it allows to realize a good seed distribution over the area via small row widths with no-till of small grains and rape (colza) and thus to have the yield advantages associated with this (Fig. 8.5).

However, it should be pointed out that significantly smaller straw sizes would decrease the erosion control by residues. This effect would probably weigh heavily in regions with continental climate and susceptible soils, whereas in areas with maritime climate it probably can be neglected.

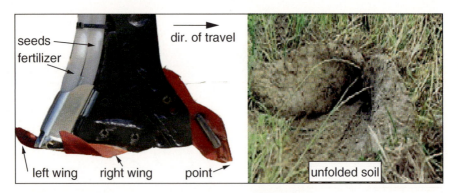

**Fig. 8.15** Broadcasting seeds underneath a "Collovati" undercutter wing opener. The unfolded soil (*right*) falls back into place automatically. Fertilizer application in a line underneath the undercutter plane is optional (Courtesy of Tonutti Spa, Remanzacco, Italy)

## *8.4.4 Seeding Underneath Undercutters*

An alternative no-till sowing method that makes it possible to obtain a good seed distribution over the area for crops with a high seed density is sowing underneath the wings of undercutters (Collovati 2008). The operating principle is shown in Fig. 8.15.

The seeds can be broadcast over the whole area provided the lateral distance between the openers is set accordingly. As a result, the seed distribution over the area is even better than with narrow row drilling (see Sect. 8.2). Concerning the precision of the seeding depth, there is principally no difference between either wing-openers, disc-openers or shoe-openers (Heege and Billot 1999).

It should be noted that both ends of the undercutter wings are curved slightly upwards (Fig. 8.15). Because of this, the soil movement sidewards is restricted, instead the soil falls back in its original position (Vamerali et al. 2006). The front chisel and the wings do some mixing of loose residues on the surface with soil above the undercutter plane. Preceding cultivation does not improve the results with this opener; in case it incorporates residues into the soil, it might even cause blockages. In short, seeding underneath these undercutters is a no-till method.

## References

Alberta Farm Machinery Research Centre (1994) Evaluation report 713. Kee Ultrasonic depth control system. Prairie Agricultural Machinery Institute, Humboldt, pp 1–10

Auernhammer H (1989) Electronics in tractors and machines (in German). BLV Verlagsgesellschaft, Munich

Bowers CG, Bowen HD (1975) Drying front sensing and signal evaluation for planters. Trans Am Soc Agric Eng 18(6):1051–1058

Bowers CG, Collins CA, Harris EP (2006) Low soil moisture planting of cotton for optimum emergence. Appl Eng Agric 22(6):801–808

Carter LM, Chesson JH (1993) A soil moisture seeking planter control. ASAE Winter Meeting, Chicago, Paper no. 931553

Collovati M (2008) Frustration led to new no-till opener design. No-Till Farmer, April 2008, pp 14–15. www.no-tillfarmer.com

Cox WJ, Cherney JH (2011) Growth and yield responses of soybean to row spacing and seeding rate. Agron J 103(1):123–128

Derpsch R (1998) Historical review of no-tillage cultivation of crops. In: Proceedings of the first JIRCAS seminar on soybean research, Iguassy Falls. Working report no. 13, pp 1–18

Dyck FB, Wu WK, Lesko R (1985) Automatic depth control for cultivators and air seeders developed under the AERD program. In: Proceedings of the Agri-Mation I. Conference and exhibition, 25–28 Feb 1985, St. Joseph, pp 265–277

Feldhaus B (1997) Seed counting and closed loop control for drills with volume-metering (in German). Doctoral thesis, University of Kiel, Kiel. Forschungsbericht Agrartechnik des Arbeitskreises Forschung und Lehre der Max Eyth Gesellschaft Agrartechnik im VDI, Nr. 302

Heege HJ (1967) Equidistant spacing-, drillling- and broadcast sowing of small cereals with spezial reference to the seed distribution over the area (in German). KTL-Berichte über Landtechnik 112, Helmut-Neureuter Verlag, München-Wolfratshausen

Heege HJ (1970) Seed distribution over the soil surface with with drilled and broadcast cereals (in German). Grundlagen der Landtechnik 20: 45 (in English: Translation No. 529 of the National Institute of Agricultural Engineering, Wrest Park, Silsoe, 1985)

Heege HJ (1993) Seeding methods performance for cereals, rape and beans. Trans Am Soc Agric Eng 36(3):653–661

Heege HJ, Billot JF (1999) Seeders and planters. In: Stout BA, Cheze B (eds) CIGR handbook of agricultural engineering, vol III. American Society of Agricultural Engineers, St. Joseph, pp 217–239

Heege HJ, Feldhaus B (2002) Site specific control of seed-numbers per unit area for grain drills. Agric Eng Int CIGR J. IV, PM 01 012. http://www.cigrjournal.org/index.php/Ejournal

Heege HJ, Vosshenrich HH (2000) Interactions between soil cultivation and climate. In: ISTRO 2000. 15th conference of the international soil tillage research organization, Fort Worth, 2–7 July 2000, Paper no. 73, pp 1–10

Karayel D, Özmerzi A (2008) Evaluation of three depth-control components on seed placement accuracy and emergence for a precision planter. Appl Eng Agric 24(3):271–276

Kiani S, Kamgar S, Racufat M (2010) Automatic online depth control of seeding units using a non-contacting ultrasonic sensor. In: XVIIth world congress of the international commision of agricultural and biosystems engineering, Quebec City, 13–17 June 2010

Kornecki TS, Raper RI, Arriaga FJ, Schwab EB, Bergtold JS (2009) Impact of rye rolling direction and different no-till row cleaners on cotton emergence and yield. Trans ASABE 52(2):383–391

Lambert DM, Lowenberg-DeBoer J (2003) Economic analysis of row spacing for corn and soybean. Agron J 95:564–573

McCallum M (2007) Inter-row sowing using 2 cm auto-steer. Research update. Australian Government. Grains Research and Development Corporation (GRDC). http://www.grdc.com.au/director/events/researchupdates?item_id=AC90061EB8AA0DAB16BE3008184EDB96&pageNumber=1

Morrison JE, Gerik TJ (1985) Planters depth control. Part I and part II. Trans ASAE 28(5): 1415–1418 and 28(6):1745–1748

Mouazen AM, De Baerdemaker J, Ramon H (2005) Towards development of on-line soil moisture sensor using a fibre-type NIR spectrophotometer. Soil Till Res 80:171–183

Mülle G (1979) Research on precision seeding of small grains (in German). Doctoral dissertation, University of Bonn, Bonn. Forschungsbericht Agrartechnik der Max Eyth Gesellschaft 32:21

Mülle G, Heege HJ (1981) Seed spacing over the area and yield of grain (in German). J Agron Crop Sci 150:97–112

Müller J, Rodrigues G, Köller K (1994) Optoelectronic measurement system for evaluation of seed spacing. In: AGENG meeting, Milano, Report No. 94-D-053

Müller J, Kleinknecht C, Köller K (1997) Optosensor-recording seed distances with grain drills (in German and English). Landtechnik 52:76–77

Neumann H, Loges R, Taube F (2006) The wide row system – innovation for organic winter wheat cultivation? (in German). Ber Landwirtsch 84(3):404–423

Price RR, Gaultney LD (1993) Soil moisture sensor for predicting seed-planting depth. Trans Am Soc Agric Eng 36(6):1703–1719

Reddy KN, Burke IC, Boykin JC, Williford JR (2009) Narrow-row cotton production under irrigated and non-irrigated environment: plant population and lint yield. J Cotton Sci 13:48–55

Scarlett AJ (2001) Integrated control of agricultural tractors and implements: a review of potential opportunities relating to cultivation and crop establishment machinery. Comput Electron Agric 30:167–191

Schneider M, Wagner P (2005) Economic viability of precision farming with a whole farm approach (in German). In: Verbundprojekt preagro II. Zwischenbericht 2005: 278 http://www.preagro.de/

Torbert HA, Ingram JT, Prior SA (2007) Planter aid for heavy residue conservation tillage systems. Agron J 99:478–480

Vamerali T, Bertocco M, Sartori L (2006) Effects of a new wide-sweep opener for no-till planter on seed zone properties and root establishment in maize (Zea mays, L.): a comparison with double-disc opener. Soil Till Res 89:196–209

Weatherly ET, Bowers CG (1997) Automatic depth control of a seed planter based on soil drying front sensing. Trans Am Soc Agric Eng 40(2):295–305

Wiesehoff M (2005) Site-specific sowing of winter wheat (in German). Doctoral dissertation, University of Hohenheim, Hohenheim. Forschungsbericht Agrartechnik des Arbeitskreises Forschung und Lehre der Max Eyth-Gesellschaft im VDI 430:62

# Chapter 9
# Site-Specific Fertilizing

**Hermann J. Heege**

**Abstract** Fertilisation aims at providing soils with nutrients for high crop yields without adversely affecting the environment. Since in most cases the properties of soils as well as of crops vary within individual fields, site-specific fertilization is needed. The challenge is to find **sensing methods** that provide suitable signals for the site-specific control of fertilizer application. Feasible approaches to meet this challenge are based on

- recording the yield of previous crops and the nutrient removal derived from it
- electrochemical indication of nutrients in soils by ion-selective electrodes
- sensing the nutrients either in soils or in crops via optical reflectance.

The best choice depends on a variety of factors such as *e.g.* nutrient type, properties of soils, properties of crops and climate. The last listed method – sensing via optical reflectance – can be used in a proximal mode from farm machines or also in a remote mode from satellites provided clouds do not obstruct the radiation. Its use for in-season nitrogen application with proximal sensing from farm machines is becoming a leading technology.

**Keywords** Bending resistance • Benefits • Control algorithm • Crop-reflectance • Fluorescence • In-season crop properties • Ion-selective electrodes • Nutrients removed • N-sensor • Reflectance of soils

H.J. Heege (✉)
Department of Agricultural Systems Engineering, University of Kiel,
24098 Kiel, Germany
e-mail: hheege@ilv.uni-kiel.de

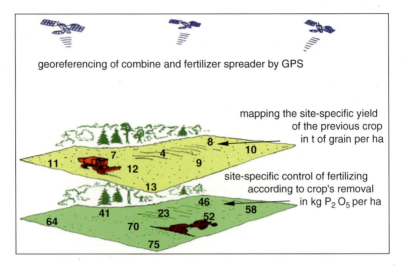

**Fig. 9.1** Concept of site-specific fertilizing according to nutrients removed in the previous harvest. The *inserted numerals* are based on the harvest of wheat in tons of dry matter – straw not considered – as well as on a removal of 5.8 kg of $P_2O_5$ per ton (From Swedish Institute of Agricultural Engineering 1988/89, supplemented and altered)

## 9.1 Fertilizing Based on Nutrient Removal by Previous Crops

Harvesting crops always results in a removal of mineral nutrients from fields. The concept that these nutrients should be replaced in order to avoid depletion of soils has been accepted in the agricultural sciences since long. However, in the past, removal and replacement respectively referred to whole fields. The yields as well as the fertilizing had a whole field basis.

Modern georeferencing, fertilizing and harvesting techniques (see Chap. 12) have changed the situation. If during harvesting, the yields in units of mass are recorded and georeferenced with a high spatial resolution for *e.g.* grid areas of 100 m², the **removal of nutrients** for these small areas can be determined. Because knowledge about the gravimetric nutrient content of the harvested products is available. Consequently, a georeferenced distribution of the fertilizer for the next crop can be controlled in such a way that precisely this amount of nutrients is applied in a site-specific way (Fig. 9.1). The result is a **site-specific maintenance application**, since the nutrients removed by the harvest are replaced.

This concept for detecting the site-specific nutrient need is reasonable if the supply of the previous crops was adequate, hence the soil not deficient in nutrients. A second prerequisite is that the removal by the plants is the only factor or at least the dominating criterion for the nutrient depletion of the soil.

The first prerequisite mentioned above can be met approximately by starting the system with careful conventional manual sampling, traditionally analysing the supply in a laboratory and providing soils with nutrients in such a way that a rather uniform level of supply is obtained. It might require some efforts and costs to do this preparatory first step in a site-specific way. The needs can be somewhat eased if a small average oversupply – at least for the major nutrients phosphate and potash – is accepted.

The second prerequisite – the removal by the harvest being the main factor of changes in the nutrient level – implies that losses of nutrients by leaching, by runoff of water or by erosion do not occur. For well managed fields, losses via runoff or by erosion will be on the average rather small and will represent mainly exceptional cases that take place in situations of adverse weather. But this does not hold for **losses by leaching** that result from natural rain or irrigation. In many agricultural areas, these losses by leaching into the not-rooted subsoil and into the groundwater horizon can be substantial as well as rather unpredictable. Whether the losses via leaching still allow to rely on nutrient removal alone for a site-specific control of fertilizing, depends on the ions that are involved. The main point is whether the nutrients are absorbed by soil particles or bound within compounds that are not soluble in water and whether hence their leaching downwards is prevented.

The **nitrate** and **sulphate ions** hardly are absorbed by soil particles, they are easily dissolved in water and hence are disposed to leaching into the non-rooted subsoil and into the groundwater. The leaching of these ions occurs especially on sandy soils and somewhat less on soils with a higher clay content, however, soil texture alone cannot eliminate the problem. Another factor that considerably can reduce the leaching is nutrient uptake by growing crops in combination with precise fertilizing that is in line with the seasonal needs of plants. This is simply because ions that are taken up by the roots of crops cannot leach any more. Yet this fact can only help to diminish the leaching problem somewhat, it cannot do away with this problem in geographical regions where during winter there is no nutrient uptake. In short, for a site-specific control of nitrogen- and sulphur fertilizing, in most areas it cannot be expected that the concept of nutrient sensing via removal supplies reliable information.

The situation for fertilizing of **phosphate** is completely different. The phosphate ions are very immobile within soils. Leaching of these ions does not occur. This is mainly due to fixation in calcium-, aluminium- and iron phosphates that are not soluble in water. Hence sensing the removal of phosphate via yield recording can be regarded as a good choice for mapping the site-specific removal. In a subsequent operation, the map can be used to control the spreading. The information about the removal is obtained by simply applying a factor to the site-specific yield.

**Potash** ions again present another situation. They do not leach anything as fast as nitrate- or sulphate ions do. Yet they do not possess the immobility of phosphate ions either. Potash that is not taken up by the crop can be held in the soil by clay minerals or by organic matter. However, in order to prevent leaching in humid areas and when no crops are growing, a clay content of 5 % or more is needed (Potash Development Association 2006). And for a site-specific concept, this minimum in clay content is needed for about all cells within a field. So for sensing the potash supply via removal by crops, restrictions in the soil texture must be considered.

For **calcium** and **magnesium**, depletion of reserves within soils due to removal by crops seldom is a problem. Thus sensing the removal by crops hardly is

**Table 9.1** Removal of phosphate and potash per ton of dry matter of harvested crop parts[a]

| Crop | Phosphate in kg $P_2O_5$ | Potash in kg $K_2O$ |
|---|---|---|
| Winter wheat, only grain | 5.8 | 7.6 |
| Winter wheat, straw plus grain | 4.0 | 18.6 |
| Winter barley, only grain | 6.5 | 8.8 |
| Winter barley, straw plus grain | 4.6 | 19.7 |
| Maize, only grain | 6.8 | 4.8 |
| Soybeans, only grain | 14.0 | 22.0 |
| Rape or canola, only grain | 18.4 | 9.2 |
| Grass, leys, parts aboveground | 8.0 | 30.0 |

[a]Extracted from International Fertilizer Use Manual (2007) and Soil Science Society of America (2008), data partly converted

necessary. But large amounts of fertilizers that contain these chemical elements might be needed to maintain a specific acidity level of soils (pH value) that promotes growth. Site-specific sensing of this acidity level is dealt with is Sect. 9.2.2.

In principle, also the removal of **micronutrients** such as copper, zinc, manganese, boron and molybdenum by the harvest can be recorded via yield sensing. For doing this, the information about the micronutrient content within crops and parts thereof is available (International Fertilizer Association 2007; Osmond and Kang 2008). Especially for micronutrients, recording the site-specific situation via removal by yield sensing appears highly attractive, since no other methods are available. Because the laborious traditional soil sampling plus analysing in laboratories is by far too expensive on a site-specific basis. However, with micronutrients – as with macronutrients – sensing the situation via removal might not be the only factor that defines the need of a crop. Leaching as well as the release of micronutrients by soils must be considered. Both might depend on the respective ion and the pH of the soil. Up to now, recording the removal of micronutrients via site-specific yield sensing is not state of the art. Research directed to this topic is urgently needed.

Summing up, phosphorus as well as of potash – with exceptions in the latter case (see above) – are candidates of application. Removal rates of these nutrients are listed in Table 9.1. They depend on crop species and on the harvested crop parts. The more the harvested crop parts are composed of vegetative- instead of reproductive plant matter, the higher the ratio of potash units to phosphate units is.

It should be noted that the **removal rates** in Table 9.1 are per ton of dry matter. Removal rates per ton of harvested product that are based on moist plant material and not on its dry matter can be very misleading since the water content can vary widely with most crops. As an extreme example: when forage in harvested from grasslands or leys, the water content of fresh material can be 85 %, but with dry hay it may be 15 %. With all crops, the water content of the harvested product changes steadily due to varying weather. And the water does not contain any nutrients.

Hence for exact records, the respective **site-specific dry matter** content at harvest time must be taken care of. For several crop species, yield sensing instruments in harvesting machines that record the respective dry matter situation and simultaneously supply site-specific maps about this are state of the art.

Hence the logical procedure for defining the nutrient removal is:

- harvested wet crop mass times dry matter content = harvested dry crop mass
- harvested dry crop mass times removal per dry mass unit = nutrient removal.

As for the units:

In Table 9.1 the nutrient removal is expressed per ton of dry matter of the harvested crop parts. A harvested wet crop mass per site-specific cell in t would have to be multiplied by the dry matter content on the wet basis in decimal fractions and not in %. The final nutrient removal would then be defined in kg of $P_2O_5$ or $K_2O$ per site-specific cell in the field.

## 9.2 Fertilizing Based on Soil Sensing by Ion-Selective Electrodes

### 9.2.1 Basics

This method of detecting the site-specific nutrient supply of the soil relies upon the **electrochemical series of potential** (voltage) that holds when chemically different conductors of electricity get into contact. As a result, the difference in potential can cause a flow of electrons in case of metals or alternatively of ions in case of liquids or slurries. The function of galvanic cells is based on this phenomenon.

However, with ion-selective electrodes the objective is not to produce electricity, but just to use the indicated voltage for detecting specific ions. And in order to get information about specific ions – such as those of the nutrients – the respective electrodes contact the soil or the soil: water mixture via a **membrane** (Fig. 9.2). The function of this membrane is to transmit only the respective ions that are to be sensed. So in case of soil sensing, membranes are selected that just transmit either

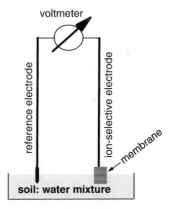

**Fig. 9.2** Operating principle of electrochemical sensing by ion-selective electrodes. Both electrodes can be within one probe (simplified and not to scale)

H⁺ or $NO_3^-$ or K⁺ *etc.* ions. The main objective in the development of ion-selective electrodes has been to find membranes that effectively prevent the passage of unwanted ions and just let pass specific ions of mineral nutrients or of water for the recording of soil pH.

Presently, ion-selective electrodes are used in numerous laboratories for analysing soils, foods, water and for clinical chemistry. They are widely employed for measuring the pH of soils either via portable handheld instruments or in a stationary mode in laboratories. Despite this, the ability to precisely just let pass ions of the nutrient that is to be sensed and to retain all other ions with other chemical formulas still is a topic of efforts and research. The question is how well membranes – which consist of various materials – are able to separate ions that have different formulas so that interferences are avoided.

The **voltage output** of ion-selective electrodes generally is proportional to the logarithm of the measured specific ion activity. This relation must be included in the data processing program if it is targeted at the active ions.

As for site-specific fertilizing, the challenge is online and on-the-go sensing by ion-selective electrodes. Up to now, these instruments cannot record in a continuous mode in a similar way as implements that sense soil properties via electrical conductivity do. The present procedure is to use on-the go operations, during which periodically small samples of soil or mixtures of soil and water are prepared and brought into contact with ion-selective electrodes. In order to avoid incorrect measurements, the electrodes must be cleaned by rinsing with water following each contact. The cleaning process and sometimes also the sample preparation necessitate an **intermittent sensing operation**.

Selecting the best membrane for the respective ion is a crucial point. Due to the many applications in various laboratories, ion-selective electrodes with a variety of different membranes are commercially available. Some recommendations are to use (Adamchuk et al. 2005; Kim et al. 2007a, b; Lund et al. 2005):

- electrodes with glass- or antimony membranes for sensing pH
- electrodes with polyvinyl-chloride membranes for potassium and nitrate
- electrodes consisting of cobalt rods for phosphates.

The polyvinyl-chloride membranes may be treated with special chemicals in order to improve the performance. Electrode aging and mechanical wear due to abrasion by soil still are problems, especially with membranes based on glass or polyvinyl-chloride. For sensing of pH, the wear problem can be reduced by using antimony instead of glass for the membranes (Adamchuk 2008).

### 9.2.2 Sensing pH and Nutrients in Naturally Moist Soils

An important point is the preparation of the soil samples. The traditional procedure of stationary sensing via ion-selective electrodes in laboratories has been to do this based on **solutions** or **slurries** of soil samples. On-the-go operation from farm vehicles is facilitated if **naturally moist soil** samples can directly be sensed. However,

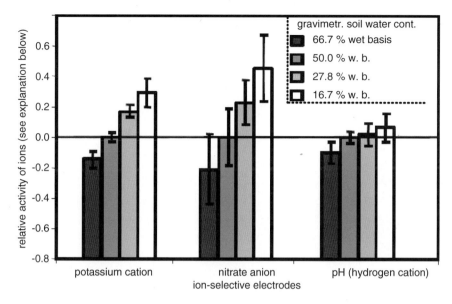

**Fig. 9.3** Relative activity of ions and the water content of the soil samples. The ion-selective electrodes that were used correspond to the recommendations mentioned in the previous section. The results have been standardized by subtracting the respective average records for 27.8 % water content from each measurement. The *bars on top* of the columns indicate standard deviations (From Sethuramasamyraja et al. 2007, altered)

sensing the soil as it presents itself in the field implies that the moisture varies to some extent. This holds between different fields and times as well as within a field when site-specific sensing takes place.

Experimental results for sensing of soil samples with varying water content did reveal that the ions react on this differently (Fig. 9.3). If hydrogen ions are sensed when mapping the pH of soils, there hardly is an effect of varying soil water content. Contrary to this, the results for sensing of nitrate- as well as of potash ions do depend on the soil: water ratio. Especially with nitrate ions, the effect is obvious (Fig. 9.3).

The technique as shown in Fig. 9.4 provides **pH readings** from naturally moist soil samples in cycles of about 10 s. This means that with a travel speed of 2 m/s (= 7.2 km/h) and a transect distance of 20 m, a reading is obtained for successive cells of 400 m$^2$. So compared to the conventional method of taking one sample per ha for analyzing in special laboratories, the **spatial resolution** is 25 times better.

In non-uniform fields, this is a substantial improvement. Because in most cases, it is not possible to solve the site-specific problem that is associated with the conventional sampling resolution by **interpolation** between the mapped points. Any interpolation method can only help to alleviate the problem if the adjacent points are spatially related. Yet with a sampling cell size of 1 ha, the distance between neighboring locations is about 100 m. For the pH situation of most fields, this is well beyond the range of semivariance (Mulla and McBratney 2000). This means that any interpolation is useless (see Sect. 2.3).

**Fig. 9.4** Implement for simultaneous sensing of pH and electrical conductivity of soils. For pH sensing, a horizontally oriented sampling shoe (*insert top right*) is periodically lowered into the soil and when lifted pushes a naturally moist soil sample against ion-selective electrodes. After this, the electrodes are rinsed with water. Crop residues – if present – are swept aside by a row cleaner. Calibrating of the electrodes is essential for precise sensing (From Veris Technologies, Salina, USA, altered)

On the other hand, solving the problem with a substantially better resolution while using the conventional method of analyzing in laboratories is economically not possible. Because for the traditional laboratory method, the costs rise almost proportional to the number of samples. Whereas with modern proximal or remote sensing methods, the number of samples hardly affects the costs.

It is important to differentiate between **soil water pH** on the one hand and **soil buffer pH** on the other hand. Ion-selective electrodes that sense hydrogen cations indicate via their voltage output only the soil water pH, which reflects what the plants experience. But the soil water pH explains nothing about the ion absorbing capacity of the soil particles. This **ion absorbing capacity** results in reserve- or exchange acidity that is indicated by the soil buffer pH. When conventional soil laboratories analyze soil in order to present recommendations for liming, both soil water pH as well as soil buffer pH are measured. The soil buffer pH is not an *a priori* existing feature of the soil, it is generated in the laboratory. Basically, soil buffer pH is the resulting pH that develops after in the laboratory there was added a defined amount of liming material. In the laboratory, the difference between the original soil pH and the final pH that is created by the liming material, a buffering solution, is recorded. This difference depends heavily on soil properties. It indicates how much lime is needed to raise the soil pH to the level that is aimed at.

So for recording the **liming requirement** in a site-specific manner by on-the-go sensing, the "buffering capacities" of the soils need to be taken care of in addition to the soil water pH. This means either a procedure for sensing pH buffer in a similar way as in conventional laboratories must be developed or – instead of this – soil properties that can act as suitable **substitutes** must be used. The first procedure – sensing pH buffer directly in the field – has been studied, but not yet realized in an on-the-go mode (Viscarra Rossel and McBratney 2003). As for substitute soil properties, the **cation-exchange-capacity (CEC)** is a prime candidate. It predominantly depends on the clay and organic matter in soils. Both clay and organic matter particles have a net negative electric charge. Hence the higher the clay and organic matter content is, the more cations can be absorbed and thus also be exchanged with the soil water.

The cation-exchange-capacity and the clay content have rather high correlations to the electrical conductivities of soils (see Sect. 5.2.2.1). And the organic matter content can quite precisely be sensed via infrared reflectance (Sect. 5.3). Hence, substituting the traditional buffer pH measurements by techniques that can be recorded on-the-go and thus provide site-specific information is feasible. It probably is reasonable to concentrate on the cation-exchange-capacity alone as a substitute property for the buffer pH. Because this soil property includes already effects of the clay and of the organic matter.

Concerning the spatial resolution, the advantages of on-the-go sensing of soil water pH are obvious. Yet measurement errors might occur. Assuming that the soil water pH indications from conventional laboratories are precise, the question is, how well the on-the-go sensed records of ion-selective electrodes compare to these.

The comparison in Fig. 9.5 is based on the commercialized sensing technique of Fig. 9.4 and furthermore to georeferenced points in fields where exactly samples for analysing in laboratories were taken. This means that the advantages of on-the-go sensing in the spatial resolution are not considered. The on-the-go recorded data that were used without any further processing (Fig. 9.5, top) were only fairly correlated to the soil water pH from the laboratories. The correlation was considerably improved by field-specific data shifts in such a way that a regression slope of 1 was obtained (Fig. 9.5, bottom). This **post-calibration** removed field specific biases. However, for the time being, these data shifts still require a few georeferenced laboratory measurements from each field. So calibration is an important point.

**Simultaneous sensing** and mapping of several ions that are essential for crop nutrition in the same field operation would be an interesting and challenging objective. Experiments in this direction by using different ion-selective electrodes in **naturally moist soil** samples at the same time were simulated in a laboratory by Adamchuk et al. (2005). The ion-selective electrodes corresponded to those listed in Sect. 9.2.1. After converting the voltage output of the electrodes to the respective ion activities, the results were compared to those from methods of conventional soil laboratories. The average correlations of 15 soils from sandy, loamy and clayey fields were:

- for soil water pH $\quad r^2 = 0.93 - 0.96$
- for available potassium $\quad r^2 = 0.61 - 0.62$
- for nitrate nitrogen $\quad r^2 = 0.41 - 0.51$

Since all experiments were conducted in naturally moist soils, it was assumed that primarily the low water content resulting from this caused the less successful

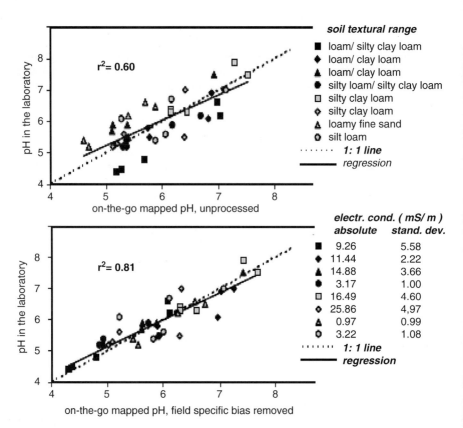

**Fig. 9.5** Comparison between water pH measured conventionally in laboratories or sensed by ion-selective electrodes on-the-go (Fig. 9.4) using naturally moist soil samples. The graphs are based on records from eight fields within six states of the USA. The data in both graphs are from the same fields, but differ in the processing as explained in the text (Extracted from Adamchuk et al. 2007, altered)

results for potassium and nitrogen. From Fig. 9.3 it can be seen that potassium and especially nitrate nitrogen react on the water content in the samples. But the coefficients of determination ($r^2$) for the soil water pH were even better than those in Fig. 9.5.

In all these experiments, soil phosphorus was not included. Up to now, sensing this nutrient with ion-selective electrodes has not been successful. Reasons for this are dealt with later.

### 9.2.3 Sensing Nitrate in Slurries of Soil

Sensing in slurries or in liquid extractions of soils by **ion-selective electrodes** is being practiced in laboratories. In an on-the-go operational mode, up to now neither

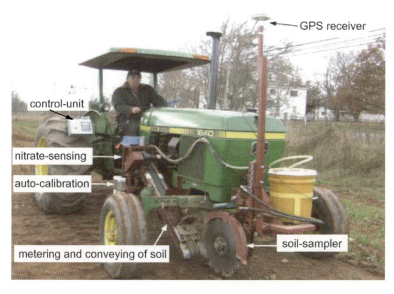

**Fig. 9.6** Technology for sensing of soil nitrate in slurry samples on-the-go by ion-selective electrodes. The soil sampler is in a lifted position (From Sibley et al. 2010, altered)

the use of slurries nor of liquid extractions from soil samples are state of the art in site-specific farming. The reason for this is probably the more complicated sensing procedure compared to direct sensing of naturally moist soil. However, prototypes that can be used to sense nitrate by ion-selective electrodes in soil slurries in an on-the-go mode have been developed (Fig. 9.6).

The soil samples are collected by means of a hydraulic-powered wood saw blade. This device cuts an about 15 cm deep slot into the soil and throws a spray of fine soil onto a belt conveyor. The belt of this conveyor is provided with pockets of an oblong shape. These pockets in the belt thus receive samples from the soil. A scraper above the belt levels all soil samples to the same height. When the belt passes around the tail end roller of the conveyor, the pockets are stretched lengthwise. This stretching of the pockets facilitates their complete emptying and the precise delivery of the samples into a small container, that houses an **electrochemical cell**, the nitrate sensing unit.

Just prior to the discharge of the soil sample, water for the extraction of the nitrate ions is pumped into this container, where the ion-selective electrodes are located. A stirrer within this container is activated and creates a **soil slurry**. And consequently, the nitrate in the soil is rapidly extracted into the watery fraction of the slurry. The ion-selective electrodes sense the activity of the nitrate ions via the voltage, and the georeferenced result is recorded. This design provides an unchanging ratio of naturally moist soil to water (Sibley et al. 2008), which is important for accurate measurements of nitrate ions (see Fig. 9.3). In gravimetric terms, about four times as much water as naturally moist soil is in the slurries (Sibley et al. 2009). Because of the intermittent operation, a sensing signal is obtained every 10 s in a similar way as with soil water pH sensing.

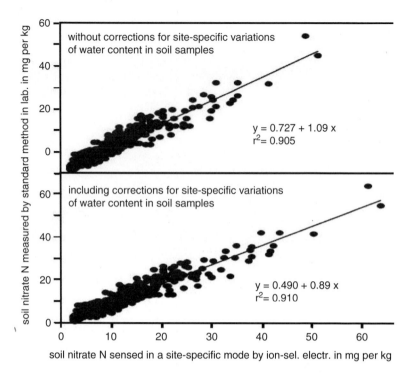

**Fig. 9.7** Soil nitrate N sensed in a site-specific mode either electrochemically or by a standard method in a laboratory. The soil samples came from the same site-specific locations of several fields in Nova Scotia, Canada (From Sibley et al. 2010, altered)

The operation implies that the natural moisture content of the soil samples can vary depending on the location in the field. With a fixed ratio between the naturally moist soil volume and the amount of water added, this also means that small variations in the final water content of the slurry in the electrochemical cell can result. However, these rather small variations in the water content of the slurry due to differences in the moisture of the soil do not deteriorate the **accuracy of sensing**.

This can be seen from comparisons of soil nitrate N that was sensed in a site-specific mode either electrochemically by ion-selective electrodes or by standard methods in a laboratory. These comparisons were made with or without corrections for the varying natural moisture of the soil samples. Both cases resulted in very high correlations to the findings for the same site-specific locations from the standard laboratory methods. Differences that might be due to the corrections for the respective soil moisture of the samples do not show up (Fig. 9.7, top and bottom). This is probably because the water that is added for preparing the slurry counts much more than the natural soil moisture. Yet this does not alter the fact that substantial changes in the water content of the sensed soil have an effect on the output of ion-selective electrodes (Fig. 9.3).

# 9 Site-Specific Fertilizing

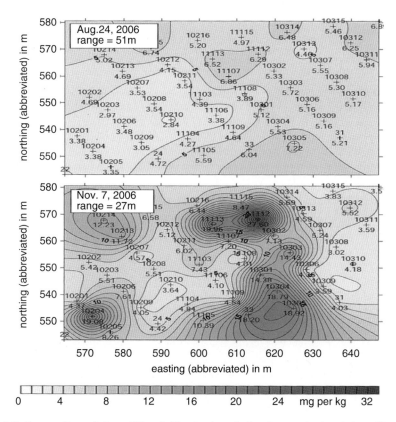

**Fig. 9.8** Contour lines of nitrate-N in a field at the date of wheat harvest (*top*) and about 10 weeks later (*bottom*) as created by the sensing technique shown in Fig. 9.6. *Darker colors* indicate higher amounts of soil nitrate N. For details see the scale. The *plus signs* designate the respective sample locations, the *numbers above them* stand for the identification of samples and the *numbers* below them indicate the respective nitrate N level in mg per kg. The first four numbers of the northings and eastings have been deleted (Extracted from Sibley 2008, altered)

The soil minerals can store a variety of plant nutrients. But there are no minerals that contain nitrate N. Reserves of nitrate N in soils primarily depend on the decomposition of organic matter and the conversion of ammonium that results from this. For the actual plant supply, the main **reserves** of nitrate N are in the soil water. And since the soil water is a very **transient property**, so is the nitrate N in the soil. When soil water moves away from the rooted soil zone – e.g. into the groundwater region – the nitrate N goes with it.

On-the-go soil nitrate sensing allows to track the supply in a site-specific mode, at least for the depth range from which samples are taken. Based on the technique of Fig. 9.6, contour lines about the site-specific nitrate N supply of a field in Nova Scotia, Canada, were mapped several times during the year (Sibley 2008; Sibley et al. 2010). Figure 9.8 shows an extract of the results from the time of wheat harvest to late autumn.

There is on the one hand a **spatially** highly variable nitrate supply within the field, yet in addition on the other hand also a **temporally** rather fast changing situation. At harvest time, the nitrate supply is rather low – probably because of the previous uptake by the crop – and it is fairly uniform. Some weeks later in late autumn, the nitrate content in some parts of the field is much higher. Any similarity in the spatial pattern of the two maps is difficult to detect. The temporal increase in the nitrate content is probably due to mineralization of crop residues. But this increase varies very much spatially. This can result from locally changing amounts of crop residues or from varying conditions for the decomposition. Nitrogen uptake by bacteria that decompose straw may be important. In case much rainfall had taken place in autumn, the nitrate content at the second date probably would have been lower because of leaching. In short, the situation for soil nitrate N is very difficult to predict, spatially as well as temporally.

An adequate response to the spatial problem is site-specific sensing. The best solution for the temporal problem, however, is sensing and controlling in real-time. Up to now, this is not possible with remote sensing. Hence doing this in a proximal way from vehicles and in an on-the-go operational mode is the logical consequence. Yet an important point is that using ion-selective electrodes allows for on-the-go operations, but not in real-time. This is because presently all ion-selective electrodes (Figs. 9.4 and 9.6) rely on intermittent operations with an average time of about 10 s between successive signals. And if a spreader or sprayer of farm chemicals is to be controlled on-the-go with sensing during the same operation, the **time span** before a signal has its effect on the distribution of the fertilizer is even longer than these 10 s. Because the actuator – the device that steadily readjusts the spreader or sprayer based on the signals – in addition needs a fraction of this time interval of 10 s to get effective. So the time interval of 10 s between the signals to get effective can be taken as an absolute lower limit.

Yet spreaders or sprayers operate with travel speeds between 2 m/s (= 7.2 km/h) and about 4 m/s (= 14.4 km/h). The minimum time interval of 10 s hence corresponds to a distance in the direction of travel of 20–40 m. This range is much longer than the conventional distance between the front of a tractor – where the samples might be taken – and the distributing devices of spreaders or sprayers. Of course, these limitations do not hold if mapping takes place and maps are used for controlling subsequent operations.

### 9.2.4   *General Prerequisites and Prospects*

The **first prerequisite** for sensing by ion-selective nutrients is that the respective ion must exhibit an electric potential in a solution or in a slurry. This potential of either cations or of anions depends – among others – on the well known **Hofmeister series**, which ranks the ions according to their hydrophilic or conversely their hydrophobic properties. Within the respective Hofmeister series for anions or for cations, the nitrate anion as well as the hydrogen- and potassium cations are located rather centrally. Thus these ions do not present unusual situations.

But the plant available orthophosphate ions do. These anions stand at the far end of the Hofmeister series, they are strongly hydrophilic. Hence these anions are heavily hydrated. And in turn, the consequence of this is that their electric potential largely has been taken away by hydrogen ions, and thus their free energy is rather small. This makes the sensing of phosphate ions more difficult compared to those from hydrogen-, potassium- and nitrogen, however, not impossible (Kim 2006).

A **second prerequisite** are suitable ion-selective electrodes. The general design is as outlined in Sect. 9.2.1, but differences exist in the treatment of the membranes with special chemicals in order to improve the ability of separating specific ions (Kim et al. 2006, 2007a, b, 2009). The need for frequent calibrations and the wear of some membranes must be considered.

A **third prerequisite** are appropriate extractants that provide for the targeted ions in slurries or solutions from the soil sample. The simplest situation is when water can be the extractant and the content within a naturally moist soil suffices. This situation holds for sensing of the soil water pH. For nitrate, the situation is that water too can be used as an extractant, however, some is added to naturally moist soil to create a slurry. Kim et al. (2007b) searched for an extractant that can be applied to **simultaneously** sense the ions of the three macronutrients nitrate-N, plant available phosphate (orthophosphate) and plant available potassium. The Kelowna extractant – a mixture of acetic acid and ammonium fluoride – can be used simultaneously for all three macronutrients. This solution is employed as a multi-ion extractant in laboratories in British Columbia. It should be noted that the simultaneous sensing of different ions implies that the electrochemical cell is provided with a separate ion-selective electrode for every nutrient.

A **fourth prerequisite** is that there should be no interference between different ions. Theoretically, the composition of the ion-selective electrodes or its membranes aims at excluding any interference. However, there still are limits in this respect. Figure 9.9 shows the situation for simultaneous sensing of the three macronutrients. The respective ion-selective electrodes that were used complied with those mentioned in Sect. 9.2.1 and their membranes – when necessary – were prepared with suitable ligands. The Kelowna solution (see above) was used for extracting the ions. The results are based on 37 different soils located in Illinois and Missouri, USA.

For each of the three graphs in Fig. 9.9, the respective **primary ion** that is to be sensed is on the axis along the bottom. In case of cation sensing (Fig. 9.9, bottom), the voltage rises when the concentration increases – as must be expected. Contrary to this, for the anions, the voltage mainly goes down (Fig. 9.9, top). For the electrodes that pick up nitrogen and potassium as primary ions (top left- and bottom graph), there was no interference by phosphate ions. Furthermore, the nitrate and potash electrodes were not sensitive to potassium and nitrate ions respectively (these graphs are not shown). But the responses of the phosphate electrode were influenced by concentrations of nitrate ions (graph top right) as demonstrated by the changing colour. Even more important probably is the fact that the voltage induced by the phosphate ions did not change unidirectionally when the concentration increased. Simultaneous sensing of several macronutrients in electrochemical cells in an on-the-go mode without doubt is worth to strive for, but it is not yet state of the art in precision farming.

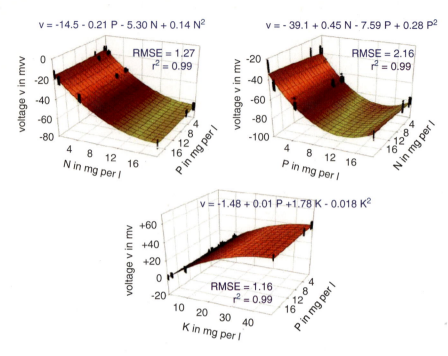

**Fig. 9.9** Simultaneous sensing of ions from nitrate, from plant available phosphate and from plant available potassium by an electrochemical cell with ion-selective electrodes in a laboratory. For each graph, the respective primary ion that is to be sensed is on the front axis along the bottom. And the interfering ion is on the respective right axis of the bottom (From Kim 2006, altered)

Summing up: within the near future, the best prospects in practical applications probably exist for on-the-go soil water pH sensing. The reason for this is not only the fact that this can be done using naturally moist soil and that – contrary to sensing of nitrate – no technique for the creation of a slurry is needed. An additional fact is that recording of the soil water pH is **temporally** simpler than the sensing of nitrate by ion-selective electrodes. Because the soil pH hardly changes over months, but the nitrate content can vary within some days. And the more short-termed the situation is, the more a control system is needed that operates in real-time. Consequently, for the application of nitrogen, a real-time control system is much more important than for the spreading of lime. Yet none of the present sensing systems that use ion-selective electrodes allow real-time control because of the time intervals between the signals.

As for potash and phosphates, the former nutrient is easier to sense by ion-selective electrodes than the latter. However, the site-specific situation for phosphates can – after some preparations – very well be mapped via sensing of the nutrient removal by the previous crop (Sect. 9.1). On soils with clay the same might hold for potash. So it is be reasonable not to use the same sensing method for all nutrients.

## 9.3 Fertilizing Based on Reflectance of Soils

### 9.3.1 General Remarks

Using signals that are derived from optical reflectance and not from the electrochemical potential in soils has distinct advantages for real-time control of fertilizing. Because these signals are almost **immediately available**. The reflectance signals are transferred from the soil to a sensor with the speed of the light. Instead of a signal interval of about 10 s, as with ion-selective electrodes, several signals within 1 s can easily be recorded by a sensor. In practice, some time for processing of the signal and for actuating the control of a spreader or a sprayer must be considered. Yet the operating speed of modern computers and actuators is such that process control within **real-time** for site-specific fertilizing or spraying is feasible and for some applications already is state of the art. This is an important criterion when the situation for the supply of crops varies not only spatially but temporally as well. A prime example for this is the supply of crops with nitrogen, which requires rather fast control responses.

However, the situation for sensing via reflectance in case of **soil nutrients** is not the same as with natural soil properties such as texture (sand, clay, loam), organic matter and water. This is – among others – because the respective amounts differ substantially. In rough gravimetric terms, the constituents that stand for the natural soil properties mentioned above account for many tons per ha, whereas the macronutrients at stake can easily be expressed in kg per ha. This is at least the situation if the macronutrients that are available to crops are considered. With micronutrients, the amounts required might be even below 1 kg per ha.

The **coefficients of determination** between optical reflectances and various soil properties in Table 9.2 are based on **full spectra regression analyses** of soil samples collected either in several areas of the United States (Chang et al. 2001; Lee et al. 2009) or in the Zheijang province of China (He et al. 2007). The natural soil properties listed are rather well correlated to the spectral data. This is in line with some results in Sect. 5.3. About the same applies to the properties that are related to soil pH. But most of the soil nutrients listed in Table 9.2 show a rather poor correlation to the reflectance.

An important point in this respect is, which soil constituents can be regarded as nutrients that are available for plants. A prerequisite is that the respective nutrients are either solvable in water or at least in a chemical **extractant** (Table 9.2) that approximately removes from soil samples the plant available nutrients and hence is used in the laboratories as a standard.

The total N in soils cannot be regarded as being plant available since generally a large part of it is fixed in organic matter. The mineralization of this part can take years or even decades. Most of the mineral N is taken up by crops as nitrate ions, however, plants may also absorb ammonium ions. Up to now, soil sensing methods concentrate on recording the nitrate ions.

The rather discouraging results for spectroscopic sensing of soil nutrients as presented in Table 9.2 should not be regarded as a proof of perpetual failure.

**Table 9.2** Correlations between optical reflectance and various soil properties from spectroscopic experiments in laboratories

| Soil property | Coefficients of determination ($r^2$) | | |
| --- | --- | --- | --- |
| | Chang et al. (2001)[a] | He et al. (2007)[b] | Lee et al. (2009)[c] |
| **Natural properties** | | | |
| Water | 0.84 | – | – |
| Organic carbon | 0.60 | 0.92 | 0.80 |
| Total N | 0.85 | 0.88 | 0.53 |
| Clay | 0.67 | – | 0.76 |
| Silt | 0.84 | – | 0.79 |
| Sand | 0.82 | – | 0.79 |
| **Properties related to pH (extractants in brackets)** | | | |
| pH (water) | 0.55 | 0.87 | 0.68 |
| Ca (ammon. acetate) | 0.75 | – | 0.80 |
| Mg (ammon. acetate) | 0.68 | – | 0.73 |
| **Nutrients, plant available (extractants in brackets)** | | | |
| K (ammon. acetate) | 0.55 | 0.58 | 0.13 |
| P (Mehlich, Olsen) | 0.40 | 0.29 | – |
| Cu (Mehlich) | 0.25 | – | – |
| Mn (Mehlich) | 0.70 | – | – |
| Zn (Mehlich) | 0.44 | – | – |

The wavelength ranges used were
[a]Chang et al. 2001: 1,300–2,500 nm
[b]He et al. 2007: 350–2,500 nm
[c]Lee et al. 2009: 350–2,500 nm

In theory, soil nutrients have an optical fingerprint that is expressed in its reflectance as well as other soil constituents have. However, it may be more difficult to detect their fingerprints because of **interferences**. In case these interferences result from other soil constituents, targeted processing of spectral curves might help to differentiate. However, in addition to interferences from other soil properties, there may be substantial data scattering or **signal noise** due to a too fast response, to inertial reactions in vibrating instruments or many other uncontrolled factors.

A vast number of different techniques for **removing noise** is available and can be used for online data processing such as

- computing weighted moving averages that create a new sequence of data
- smoothing and differentiating of data by simplified least squares procedures (Savitzky-Golay filtering)
- employing first- or second derivatives of full spectra for attenuating the influence of solar angles as well as of the viewing geometry
- using a Fourier analysis or -transformation of full spectra, which is a method of decomposing the signals into sine waves of different frequencies
- transforming a full or partial spectrum by wavelets (= small waves). This method is based on a frequency-time analysis, which supplements the general frequency-amplitude approach.

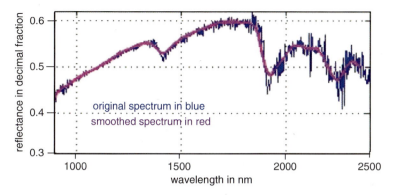

**Fig. 9.10** Soil spectrum as originally recorded and after smoothing or removing noise by discrete wavelets. The three depressions in the spectral curve are caused by water absorbance (Extracted from Ge and Thomasson 2006, altered)

Removing noise can help substantially to get rid of unwanted influences (Fig. 9.10) but must be applied with care in order to avoid loss of important information. In addition to denoising methods, multiple regression analyses can substantially assist in extracting more information from the spectra such as

- partial least squares regressions
- principal component regressions
- stepwise regressions.

Denoising techniques and regression analyses cannot be dealt with here at length. In the future, these methods for processing the sensed data will probably often be used as "**black boxes**" in a similar way as present day computer programs. For details see Brandt (2011), Haykins and Van Veen (2003), Martens and Naes (1992) as well as Savitzky and Golay (1964). The important point is that a careful application of denoising methods and regression analyses makes it possible to extract from spectroscopic spectra exactly the information that is needed for site-specific precision farming.

The first step is using the relation between specific soil properties and the reflectance for the **calibration** of the sensor and its software. The reflectance that is detected in fields of a defined region thus helps to adjust the sensing equipment. The result can be that for soils that are located outside this defined area, the sensing device gets inaccurate. However, this is a matter of how much the soil properties vary geographically. An important question for the future will be, how much generalizing and standardizing of calibration is possible while attempting to use the same calibration for wide areas.

The calibrated sensing equipment can then be subjected to a **validation** process. This validation too is a comparison of soil properties as recorded by traditional procedures against the results that were detected by the spectroscopic sensing method. The difference is that now directly the calibrated instrument is used for the comparison.

Both the calibration as well as the validation can be based on **full reflectance spectra** or on **discrete wavebands**. In case of full spectra, the objective can be simultaneous sensing of several soil properties in one operation. Up to now, the results with simultaneous sensing of several soil properties via full spectra are good or satisfying with some properties, with others they are disappointing (Table 9.2). Possible reasons for this have been dealt with above.

An alternative is to base the first step – the calibration procedure – on full spectra but to extract from this information in a second step the knowledge for sensing single soil properties by discrete wavebands of small width. With this method, the searching of the correlations relies on a rather broad range, the validation is limited to one or a few narrow bands. And these narrow wavebands finally are used for sensing in site-specific applications. Because of the small bandwidths that are involved, this method improves the chance that interferences with spectral ranges that are affected by other soil properties can be avoided. It is probably for this reason that this method of **single property sensing** by discrete wavebands is dominating up to now in practical applications. However, the narrower the wavebands are on which the sensing relies, the more important get effective processing methods for removing noise and for smoothing the data (Fig. 9.10).

An interesting approach for the future might be simultaneous **hyperspectral sensing of several properties**. The term "hyperspectral" in this respect indicates too that only very distinct narrow wavebands are selected for every soil property and that spectral regions in between are disregarded in order to remove the probabilities for interferences.

## 9.3.2 Sensing the Lime Requirement

On-the-go sensing of the soil water pH by ion-selective electrodes in combination with electrical conductivity recording for indicating the buffering properties of soils and hence the lime required is state of the art (Sect. 9.2.2). Yet this method cannot be used to control the lime application in **real-time** because of the time delay that is associated with sensing by ion-selective electrodes. Sensing of the lime requirement by reflectance would remove this time problem.

Sensing the soil water pH alone by reflectance does not suffice. In order to account for the varying **buffering properties** of soils, Viscarra Rossel et al. (2006) used full spectra reflectance sensing techniques in laboratories for Australian soils and processed the data by multivariate analyses with partial least squares regressions. For the estimation of the lime needed, not only soil water pH was the subject of spectral sensing, but several soil properties as *e.g.* cation exchange capacity, organic carbon and texture were considered in the program as well.

The results that were thus obtained depended highly on the spectral range. When solely visible reflectance (400–700 nm) was used, the spectroscopic results were not reliable (Fig. 9.11). The near-infrared reflectance (700–2,500 nm)

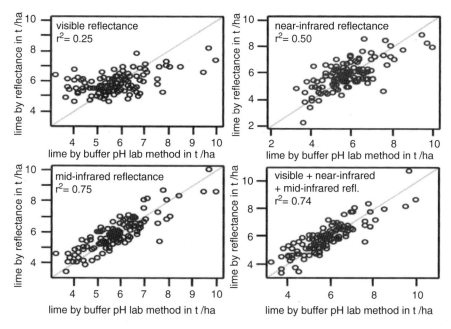

**Fig. 9.11** Comparing the lime requirement as determined in laboratories either by standard methods or by reflectance of full spectra (From Viscarra Rossel et al. 2006, altered)

resulted in an improved yet still not satisfying correlation. But the mid-infrared spectroscopy (2,500–25,000 nm) finally supplied a coefficient of determination ($r^2 = 0.75$) that offered prospects for application. Using all these three spectral ranges in a combined manner together provided a similar result. Consequently, the mid-infrared region *per se* alone should be the best choice. This superiority of sensing in the mid-infrared range is in agreement with results that have been obtained with several natural soil properties (Sect. 5.3.2, Table 5.4). However, it is more expensive to sense in this spectral region than to use visible radiation or the near-infrared range. Even more important is that up to now mid-infrared reflectance needs rather dry soil, which prevents the application for proximal on-the-go sensing.

The soil properties that are important for the control of liming do not present the same conditions on a time scale. With professional farming, checking the soil water pH might be reasonable every 3–5 years. Compared to this, the soil properties that must be known in addition to define the **buffer situation** – like *e.g.* cation-exchange-capacity, texture and organic carbon – remain constant over much longer time periods. Hence a possible approach would also be to sense the soil water pH on a site-specific basis via reflectance in real-time while liming takes place, but to rely for the required buffer corrections on simultaneously used maps. These georeferenced digital maps with data about long-term soil properties would need to be pre-processed for real-time site-specific buffer corrections. Such a combination of real-time sensing on the one hand and simultaneous use of pre-processed georeferenced maps with long-term soil properties on the other hand might be useful for an intelligent control of several field operations.

### 9.3.3 Sensing the Phosphorus Requirement

Phosphorus exists in soils within different compounds, especially in phosphates of calcium, aluminium and iron. These phosphate compounds are mainly a feature of natural soil properties, yet to some extent also originate from fertilizers, hence from human activities. In moderately acid soils as well as in alkaline soils, calcium-phosphates dominate. When the soil pH goes down, the proportion of aluminium- and iron-phosphates increases. All these phosphate compounds have very different optical spectra. They can be sensed spectrally rather easily with a very low classification error (Bogrekci and Lee 2005).

This information, however, only indicates that there are calcium-, aluminium- or iron phosphates in a soil. It discloses nothing about the availability of phosphorus to plants. In fact, the important criterion for the supply of crops and hence for site-specific control of fertilizing is the **plant available phosphorus**. This exists in soil solutions either as hydrogen-phosphate anions ($HPO_4^{2-}$) or as dihydrogen-phosphate anions ($H_2PO_4^-$). The former anion dominates in weakly alkaline soil solutions, and the latter instead in slightly acid situations. Compared to the amounts of phosphorus in the soil phosphates, those of plant available phosphorus in the soil solution usually are very small. This is because phosphorus does not remain in solution for long in soils. The anions that crops extract from the soil solution are usually replenished from calcium-phosphates, provided the soil has a reservoir of these.

In soil laboratories, the plant available phosphorus is defined by chemical extraction. A standard extractant is the Olsen solution of sodium-bicarbonate, though also other extractants are used. For highly fertilized soils, sometimes simply water is recommended as an extractant (Finck 1991).

Sensing the plant available phosphorus by **reflectance** in laboratories traditionally has been done on the basis of **dried soil samples**. However, Maleki et al. (2006) and Mouazen et al. (2006) concluded that the spectral prediction of plant available phosphorus in **fresh, wet soils** is better than in dried samples whilst portending that the sensing takes place for anions in a soil solution. Their hypothesis is that hence this phosphorus fraction in the water phase of the soil correlates well with spectral signals and consequently more water implies also more plant available phosphorus.

The spectra in Fig. 9.12 are based on sensing of fresh, moist soil samples from several fields in Belgium. But since the water content of the sensed soil samples varied, the contents of available phosphorus were presented on the basis of dry soil in order to allow for a precise comparison. Yet this does not alter the original state of the soil samples at the time of sensing. The higher the soil content of available

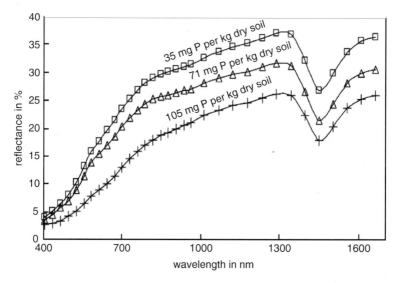

**Fig. 9.12** Average reflectance of 30 spectra of soil samples for three levels of plant available phosphorus, defined by the Olsen extractant. Each sample contains diverse types of soil textures and different soil moisture contents (From Maleki et al. 2006, altered)

phosphorus, the lower the reflectance is. The courses of the spectra resemble those of water sensing by reflectance (see Sect. 5.3.2). More soil moisture also decreases reflectance in a similar way. Hence the question of autocorrelation between available phosphorus and water in soils deserves attention (Maleki et al. 2006). Nevertheless, the **coefficients of determination** ($r^2$) for the relation between the plant available phosphorus based on Olsen extractants and the **full spectra reflectance** from 400 to 1,660 nm were 0.73 and 0.75 and hence rather good. The small difference depended on the data smoothing technique that was used.

This method of sensing available phosphorus by visible and near-infrared spectra has been transferred from laboratory to field application for a site-specific on-the-go operation (Fig. 9.13). The spectral illumination is transferred by an optical fiber cable to the flattened soil underneath a cultivator sweep. A second optical fiber cable leads the reflectance back from this soil to the spectrometer. The cultivator sweep is adjusted for this flat surface sensing along its open bottom plane with a depth of about 15 cm from the soil surface.

For precise results, this **flat surface sensing** needs a very accurate guidance of the illuminating probe in order to get a high and rather constant part of the diffuse reflected light to the spectrometer. For this, the bottom edges of the subsoiler sweep must be held exactly parallel to the soil surface and the illuminating optical probe must be continuously sliding in a slanted position over the measured soil surface. In case the bottom edges of the sweep are not precisely parallel to the soil surface, the illuminating probe is guided somewhere above the sensed soil surface. But any increase in the distance between the illuminating probe and the soil surface reduces

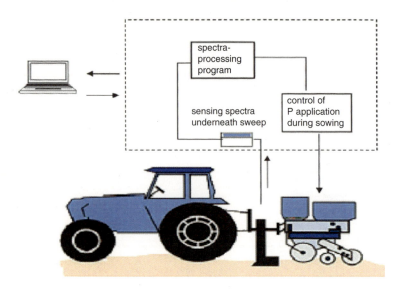

**Fig. 9.13** Scheme for site-specific on-the-go flat surface sensing and control of phosphorus application during sowing (From Maleki et al. 2008a, altered)

the reflectance that is received by the spectrometer and thus impairs the sensing precision (Mouazen et al. 2009).

Furthermore, the **time lag** that is inevitable between the spectral soil scanning and the deposition of the fertilizer deserves attention. For accurate georeferenced site-specific operation, this time lag must be compensated for by a defined distance of the soil scanning device ahead of the fertilizer outlets. This time lag is generated by processes that successively take place, first the soil scanning, then the spectra processing plus signal transportation and finally the reaction of the fertilizer applicator due to the site-specific adjustments by the electric actuator. Each of these processes goes rather fast, but the total succession still takes between 1 and 2 s.

The offset distance in the direction of travel that is needed for the sensor position ahead of the fertilizer outlets in order to compensate for this time lag depends on the tractor speed. Maleki et al. (2008b) found that for each 1 km/h in tractor speed, an offset for the sensor ahead of the fertilizer outlets of 0.5 m was required. This means that the design as outlined in Fig. 9.13 would be adequate only for rather low speeds. For a speed between 7 and 8 km/h it is recommended to install the sensor in front of the tractor. In case this results in excess offset because then the overall time lag is not sufficient to compensate for the distance to the front of the tractor, an **artificial, additional time delay** in the signal transfer can easily correct the situation.

Time lags occur with all sensors that are used in on-the-go field operations. However, as long as the problems can be solved by spatially arranging the machine elements accordingly or by artificial time delays in the signal transfer – as shown above – the precision in the application of agrochemicals still can be good.

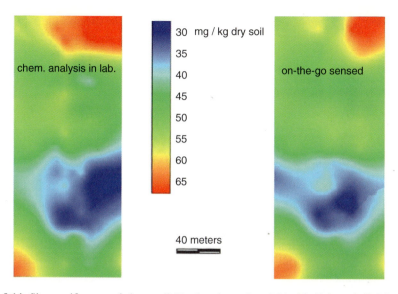

**Fig. 9.14** Site-specific maps of plant available phosphorus for a field with silt loam in Belgium as obtained from chemical analyses in a laboratory (*left*) and from spectroscopic on-the-go sensing (*right*). The data for the chemical analyses as well as for the spectroscopic sensing were recorded at precisely the same georeferenced 126 locations. After that, the maps were created by kriging on a 1 × 1 m grid. So in effect, the two methods of analysing or detecting the phosphorus are only compared for specific spots in the field, they are not at all compared in their ability to resolve an area adequately at reasonable costs for site-specific farming (From Mouazen et al. 2009, altered)

If the soil phosphorus supply is mapped, the time lag situation would be somewhat different. The time lags that occur during the mapping process as well as those that result from the subsequent spreading operation together would have to be considered. In case the time lags, the travel directions and the operating speeds – from mapping and from spreading – are the same, the offset distances cancel each other. Yet there can be many different situations, and georeferenced path control is essential. For mapping, it seems reasonable to combine the flat surface sensing with a cultivating process.

The spatial comparison of detecting plant available phosphorus in Fig. 9.14 by spectroscopic sensing on the one hand and by conventional, chemical analysing in a laboratory on the other hand is based on specific spots in a field. It demonstrates that the spectral on-the-go method (Fig. 9.13) can provide for a similar result in the field as the chemical analysing can in the laboratory – for single spots. This principal precision in detecting the phosphorus is the prerequisite for using the high spatial resolution that is needed for site-specific application. It is obvious that in practice any analysing in laboratories never can provide this spatial resolution because of its high costs.

In field trials with maize in Belgium for which the phosphate fertilisers were either uniformly applied or distributed in a site-specific way according to the method in Fig. 9.13, the site-specific application resulted in a small saving of fertiliser of

4 %. Despite this small saving in fertiliser, the yield with the site-specific application was about 5.8 % higher. This difference in yields was significant (Maleki et al. 2008a).

It should be realized that the results of such field trials depend on many factors, especially in this case on the *a priori* existing supply of plant available phosphorus.

### 9.3.4 Sensing the Potassium- and the Nitrate Requirement

Potassium reserves exist in soils mainly in silicates or as absorbed ions of clay particles, and the amounts can vary greatly with the texture. For crop nutrition, the small amounts of **potassium ions** in the soil water solution or in a soil extractant of the laboratory are important. Detecting these ions by visible and infrared reflectance has been investigated in laboratories (Viscarra Rossel et al. 2006; Lee et al. 2009) Up to now, the results have not been very encouraging. And probably because of this, site-specific on-the-go sensing of plant available potassium has not yet been attempted. So for site-specific application of potassium – according to the present state of the art – either recording the situation via the removal by previous crops (Sect. 9.1) or sensing based on ion-selective electrodes (Sect. 9.2) should be considered as alternatives.

The soil nitrogen exists largely in compounds within the soil organic matter. Plants take up inorganic nitrogen that either results from decomposing of organic matter or from fertilizing as nitrate or as ammonium. Of these, nitrate is generally present in much higher concentrations and more mobile. Hence, nitrate ions are the dominant form of nitrogen used by plants. They exist mainly in the soil water and are not absorbed by soil minerals.

For a long time, sensing **nitrate ions** by soil reflectances in the visible and infrared range has not been successful (Viscarra Rossel et al. 2006). Solely the total N of soils was rather well defined by reflectance of this wavelength range (Table 9.2). But this is not surprising since the total N is highly correlated to the soil organic matter, which can be estimated accurately.

However, the perspectives for reliably sensing of nitrate ions by reflectance have been improved (Jahn et al. 2006; Jahn and Upadhyaya 2006). A combination of three steps has enhanced sensing of soil nitrate:

- using mid-infrared reflectance instead of visible and near-infrared reflectance
- smoothing and decomposing this reflectance by wavelets (see Sect. 9.3.1)
- in this way, detecting a narrow waveband in the mid-infrared range that is especially suited for sensing nitrate ions.

The described steps can be regarded as a **calibration procedure** that leads to a suitable selection, setting and adjustment of the spectroscopic technology. The first step – using mid-infrared reflectance instead of near-infrared reflectance – is a proven way for enhancing the estimation of various soil properties (see Sect. 5.3.2,

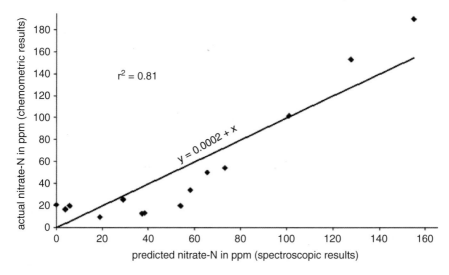

**Fig. 9.15** Sensing nitrate-N in samples of soil paste by mid-infrared reflectance with 7,148 nm wavelength versus results from traditional analyses in a laboratory (From Jahn and Upadhyaya 2006, altered)

Table 5.4). The soil spectra in the mid-infrared range just have more distinct peaks and dips than those in the near-infrared region, and this improves the estimation results. However, a disadvantage is the higher investment for the sensing equipment compared to near-infrared spectroscopy. The second step pertains to the processing of the spectral data. This processing provides the means for the third step, finding the most effective wavelength for sensing nitrate ions.

So the procedure is that the calibration starts with the full mid-infrared spectrum from 2,500 to 25,000 nm wavelength, but the final validation is effected just by a discrete narrow waveband. For soil samples from several Californian fields, the result was a **wavelength of 7,148 nm**. The relation between the predicted nitrate from the spectral sensing with this wavelength and the traditional results from the chemometric analyses in a laboratory is shown in Fig. 9.15.

The relevance of this method of soil nitrate sensing rests not only on the reliability of the correlation between the spectral sensing and the traditional analyses in laboratories (Jahn et al. 2006). Important is also that it allows to simplify the sensing via mid-infrared radiation by employing a **discrete narrow waveband**, which in effect is a drastic reduction of the measuring range. This in turn can provide prospects to reduce the sensing cost for a final site-specific application of mid-infrared radiation.

However, site-specific spectral sensing of soil nitrate-N in an on-the-go manner has not yet been realized in practice. A question is at which **depth** from the soil surface the sensing should take place. Nitrate-N can be relocated rather fast within the soil profile since it moves with the waterfront. And it might be taken up by crops from depths well below the topsoil (Shanandeh et al. 2011). It is for this reason that with the traditional soil analyses in laboratories for nitrogen – the Nmin method – in humid regions the samples are collected and averaged from a 90 cm deep profile (Marschner 2008). Since the relocation of nitrate-N will depend largely on the precipitation, site-specific spectral sensing should at least be adjustable for the depth from the soil surface.

For on-the-go sensing in fields, the mid-infrared radiation would also have to cope with varying soil moisture (Sect. 9.3.2). A discrete narrow waveband might be better suited for this than a full spectrum. However, the verification of this assumption has not yet been provided.

## 9.4 Fertilizing Nitrogen Based on In-Season Crop Properties

Basing the control of fertilizing on **past crop properties** instead of those of the soil is state of the art when the nutrients that were removed by the yields of previous crops are relied on (Sect. 9.1) However, because of the time scale involved, this method lends itself for detecting rather long-term supplies of those nutrients, whose availability temporally does not change fast such as *e.g.* phosphorus.

Contrary to this, basing the fertilizing on **current crop properties** is a method that can provide signals for rather immediate, short-term control and hence with a high temporal precision. The ideal application for this method is associated with online and on-the-go proximal sensing and control in real-time. This in turn means that this method fits well to nutrients, for which the supply of crops changes rather fast. The classical macronutrient for this method hence is **nitrogen**, and so this section is devoted to the application of this nutrient. This limitation, however, does not imply that sensing on the basis of current crop properties might not be useful for other applications as well.

It is obvious that any use of **crop properties** for the control of farming operations is also a control that is oriented on soil properties – at least partly. This is simply because the properties that the plants have are not independent of the soil on which they grow. Therefore a very important advantage of basing the control on properties of the crops is that the signals that are thus obtained depend on an **interaction** between the plants and its soil. This means that *e.g.* the influence of the respective root development on the nutrient uptake is automatically taken care of.

All nutrient extraction methods that are used in soil laboratories so far aim at a simulation of this influence of the soil to crop relation on the availability of nutrients. However, this simulation of the extraction of nutrients by crops in laboratories has to cope with changing situations, e.g. varying root lengths during the development of plants. The best solution to this problem of fertilizing control certainly simply is to use the respective **site-specific soil to crop relation** in the field as it exists by an intelligent sensing system.

Fertilizing based on current crop properties can provide this. It can be oriented at the actual nutrient uptake of crops, assuming that suitable surrogate properties for this are available and selected.

| system | physical criteria | crop criteria |
|---|---|---|
|  Kiel system | special wavelengths from the visible and near-infrared range of the crop's reflectance | chlorophyll-concentr. in the leaves plus total area of leaves (leaf-area-index) |
|  DLR system | chlorophyll-fluorescence of plants, induced via laser-radiation | chorophyll-concentr. in the leaves plus sometimes the canopy-surface |
|  Bornim system | diversion-angle of a pendulum that is dragged along the upper part of the crop | plant-mass, crop-resistance against bending |

**Fig. 9.16** Sensor based systems for site-specific nitrogen top dressing

Three presently used commercialized systems for on-the-go nitrogen top dressing that rely on sensing of crop properties are listed in Fig. 9.16. The names refer to locations of the research groups that originally developed the systems.

The **Kiel system** was initiated from research at the University of Kiel about two decades ago (Heege and Reusch 1996). It relies on sensing and processing the visible and near-infrared reflectance of crop canopies for signals to control the nitrogen application rate. Among the various alternatives (Fig. 9.16), this method is the most frequently used in farming.

The **DLR system** originated from research by Günther et al. (1999) in the Department of Optoelectromics of the DLR Research Center in Wessling near Munich. The signals for the control of nitrogen application are based on on-the-go sensing of the crop chlorophyll fluorescence. More details are outlined in Sect. 9.4.5.

Finally, a mechanical pendulum sensor has been developed in the **Bornim** Institute of Agricultural Engineering by Ehlert et al. (2004a). The deflection angle of a pendulum, which is dragged along the upper part of the crop canopy, is used to control the application rate. This sensing system is dealt with in more detail in Sect. 9.4.6.

Whatever sensing system is used, it will have to rely on quantitative effects of nitrogen supply on specific crop properties. Differences based on species, varieties, soil-conditions and prevailing weather must of course be expected.

In order to understand the functioning of optical nitrogen sensing systems, knowing about the effects of the nitrogen supply on the **chlorophyll content** in the leaves and on the **leaf-area-indices** of crops is fundamental. Because this knowledge helps to find the sources of reliable signals for nitrogen sensing. The leaf-area-index is the relation between the photo-chemically active, one-sided leaf area and the ground surface.

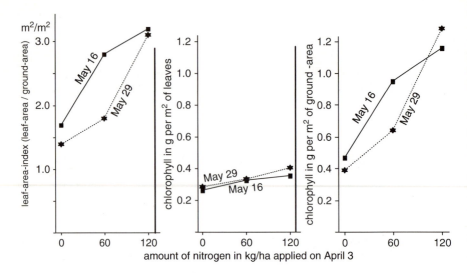

**Fig. 9.17** Effect of nitrogen on the leaf-area-index (LAI) and on the chlorophyll content of winter-wheat in Indiana, USA. The chlorophyll mass per unit of field area in the *right graph* is the product of the leaf-area-index in the *left graph* and the chlorophyll mass per unit of leaf area in the *central graph* (Compiled from data by Hinzman et al. 1986)

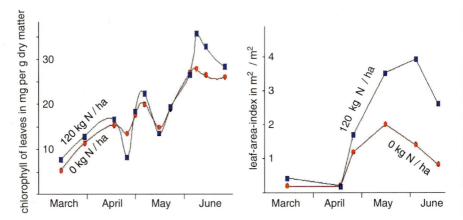

**Fig. 9.18** Effect of nitrogen on leaf-area-index and on chlorophyll content of winter-rape (winter-colza) in Schleswig-Holstein, Germany. For the fertilized plots, the crop was dressed with 40 kg of N per ha respectively at the end of March, at the end of April and at the beginning of May. Hence for these plots, the total amount of N applied in the spring was 120 kg per ha (From Kappen et al. 1998, altered)

Research by Hinzman et al. (1986) with wheat and by Kappen et al. (1998) with rape (colza) revealed the effect of nitrogen on the chlorophyll concentration within the leaves as well as on the leaf-area-index. It can be seen from Figs. 9.17 and 9.18 that nitrogen changes the leaf-area-indices much more than the chlorophyll concentration within the leaves. The dominating effect of nitrogen on biomass production or on leaf-area-indices probably holds for most crops. Belanger et al. (2005) as well as Jongschaap (2001) published similar results for potatoes.

A reasonable conclusion from these results is to use the product of both indicators of nitrogen application as the criterion for sensing. This product of the chlorophyll concentration within the leaves and the leaf-area-index can be defined as the **chlorophyll per unit of ground area**. The curves in Fig. 9.17, right, indicate that this product is sensitive to the nitrogen supply.

Using the chlorophyll concentration per unit of ground area as criterion is analogous to sensing methods that are oriented at defining the yield potential of crops as outlined in Sect. 6.3. However, it should be realized that sensing for **yield potentials** or for **nitrogen fertilizing** may not be quite the same. Because sensing for yield potentials is oriented at detecting the photosynthetic capacity of crops per unit of field area. And sensing for nitrogen fertilizing aims at recording the site-specific nitrogen uptake of the crop. Strong correlations between these sensing objectives can be expected, but differences should be looked forward too as well. Differences can result from the fact that not all the nitrogen in crops is within its chlorophyll. And the yield potential of crops depends on many growth factors, not only on nitrogen.

## 9.4.1 Fundamentals of Nitrogen Sensing by Reflectance

Nitrogen has two effects on the reflectance of a plant canopy (Fig. 9.19). Firstly, it increases the chlorophyll concentration per unit area in the leaves. Thus, more light is absorbed and consequently the reflectance decreases. However, this effect occurs only with the visible light, the photosynthetically active radiation (PAR).

Secondly, as mentioned above, nitrogen has a very pronounced effect on the growth of plant mass and, therefore, on the leaf-area-index of a crop. Theoretically, the higher the leaf-area-index, the more incident solar radiation should be scattered back by the canopy instead of by the bare soil. Yet in reality this is important only in the near-infrared region, since in these wavebands light is barely absorbed by plant pigments. As a result, the effect of nitrogen supply in the near-infrared range is opposite to that in the visible range.

The rather steep slope between the red and the near infrared reflectance is generally denoted as the red edge. It has a concave and a convex part, which meet at the **red edge inflection point**. This point moves to longer wavelengths when the supply with nitrogen is improved.

How can the information of the reflectance curves be processed for the control of the nitrogen supply of crops? Using full-spectrum methods is complicated as a result of the opposite effect of the nitrogen in the visible and the near-infrared wavelength range. Furthermore, these methods are associated with rather high costs for the sensing instruments. Using discrete narrow wavebands instead of these methods can simplify the sensing. The problem however is, finding the most suitable narrow wavebands including the best index that combines these mathematically.

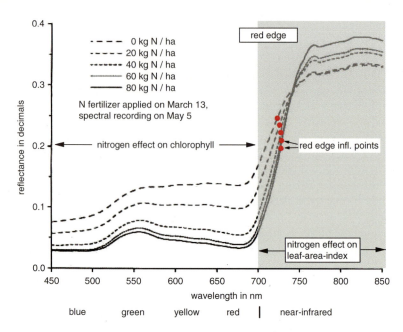

**Fig. 9.19** Reflectance of winter-rye at time of second top-dressing depending on the amount of nitrogen spread 7 weeks earlier. The *curves* are based on natural illumination and *vertical view directions*. The red edge inflection points are not to scale (From Reusch 1997 and Heege et al. 2008, altered)

Two approaches have been used for detecting suitable narrow wavebands or optical indices. The first approach has been to rely on established, well known optical indices and to find out, which of these are the best indicators of the nitrogen supply. So the selection occurs among **standard indices** that originally just were used to detect and assess vegetation (see Sect. 6.2).

The second approach is based on the assumption that – regarding especially nitrogen – these standard indices may disregard or miss the information that can be obtained from very discrete parts of the spectral curves. Consequently, initially full spectra without any interruptions are processed with the objective of finding discrete narrow bands that are the most sensitive to the nitrogen supply. These wavelengths from suitable narrow sections may in a second step be combined to **new indices**, which then can be regarded as **special nitrogen indices** for agricultural crops.

Remote sensing from satellites mainly is done via standard indices, whereas for proximal sensing from farm machines increasingly new indices supply the signals. There are reasons for these differences.

The remote sensed signals from satellites hardly are used solely for sensing the nitrogen status of crops, especially until now not on a site-specific field basis. Predominantly, the remote sensed signals are employed on a whole field basis or for even wider areas for *e.g.* biomass sensing or crop classification. However, if the objective in remote sensing is biomass sensing in combination with detecting the chlorophyll concentration in the leaves (see Figs. 9.17, 9.18, and 9.19), the optical indices that are useful for this may be similar to those that might be suitable for nitrogen sensing.

Any deficiencies that still exist for remote sensing regarding the spatial resolution that is needed for site-specific fertilizing are slowly being removed by technological progress. Hence for the future, site-specific nitrogen fertilizing based on field maps that were obtained by remote sensing of crop canopies from satellites may be feasible. However, this technology will probably be limited to areas where clouds seldom obstruct the transmission of radiation. In areas with maritime climate and hence frequent overcast situations, the concept of site-specific mapping of the nitrogen supply by crop-reflectance sensing from satellites will be difficult to apply. Because here a clear sky may not be available for days or even weeks. This means that the **temporal precision** that is needed for accurate nitrogen sensing often cannot be met when taking into consideration the transient supply by soils.

The situation is quite different with **proximal site-specific nitrogen sensing** from farm machines for on-the-go spreader control. Since in this case the control is done in real-time, the best temporal precision that is possible can be obtained and this irrespective of clouds. Standard reflectance indices as well as new indices – hence special nitrogen indices – can be used.

## 9.4.2 Sensing by Standard Reflectance Indices and Natural Light

Reflectance indices are defined by reflectance bands with special wavelengths and by mathematical combinations of these. Some standard reflectance indices and their formulas are listed in Table 9.3. Basic indices may consist just of **narrow bands**, e.g. from the green, red or near-infrared wavelength range. Instead of these, often also **simple ratios** of reflectances are used such as the ratio of the near-infrared to red or of near-infrared to green.

Other standard approaches are the "**Normalized Difference Vegetation Index**" (**NDVI**) or the "**Soil Adjusted Vegetation Index**" (**SAVI**). Both indices can easily be calculated from the reflectance in the red and near-infrared range. The NDVI is a very frequently used index for remote sensing of vegetation from satellites (see Sect. 6.2). The SAVI is also employed for this. It has the same wavelengths as the NDVI, however, there are additional correction constants in the formula (Table 9.3). The latter reduce the influence of varying soil colours (Huete 1988).

Furthermore, signals for the nitrogen situation could be derived from the point of inflection of the S-shaped curve of the reflectance in the red and adjacent near-infrared range. This **red edge inflection point** (Fig. 6.6 or Fig. 9.19) can be obtained in different ways:

**Table 9.3** Relationships between standard spectral reflectance indices and the nitrogen supply, expressed by coefficients of determination

| Definition | Formulas for wavelengths R (number added is wavel. in nm) | Coefficient of determination ($r^2$) |
|---|---|---|
| Green reflectance | R 550 | 0.910 |
| Red reflectance | R 670 | 0.888 |
| Near-infrared reflectance | R 800 | 0.569 |
| Ratio of near-infrared to red | $\dfrac{R\ 800}{R\ 670}$ | 0.911 |
| Ratio of near-infrared to green | $\dfrac{R\ 800}{R\ 550}$ | 0.943 |
| Normalized difference vegetation index (NDVI) | $\dfrac{R\ 800 - R\ 670}{R\ 800 + R\ 670}$ | 0.914 |
| Soil adjusted vegetation index (SAVI) | $\dfrac{1.5(R\ 800 - R\ 670)}{R\ 800 + R\ 670 + 0.5}$ | 0.907 |
| Red edge inflect. point, numerical differentiation | $\dfrac{d^2 R}{d\lambda^2} = 0$ | 0.932 |
| Red edge inflect. point, approx. formula (REIP) | $700 + 40 \dfrac{(R\ 670 + R\ 780)/2 - R\ 700}{R\ 740 - R700}$ | 0.970 |

The results are based on natural illumination and vertical view directions (From Heege and Reusch 1996)

- by numerically calculating the second derivative of the reflectance curve and determining the wavelength where it is zero,
- by using empirical approximating formulae.

The disadvantage of the first method is that the reflectance data must be available with a high spectral resolution. Therefore, many bands are needed and the method may become expensive. For the second method, several empirical approximating formulas have been proposed (Dash and Curran 2007; Guyot et al. 1988). The red edge inflection point in Table 9.3, bottom and those in the subsequent text (abbreviated as **REIP**) are based on Guyot et al. 1988.

The coefficients of determination in the last column of Table 9.3 refer to experiments with nitrogen fertilization of winter-rye and its spectral results as shown in Fig. 9.19. All spectral indices depend on the nitrogen fertilization. However, there are differences. The 800 nm near-infrared reflectance on its own – surprisingly – shows the weakest relationship. The red edge inflection point indices on the other hand are very closely related to the nitrogen application rate. The best coefficient of determination results from the **red edge inflection point** that is obtained by the approximating formula.

Another index that is closely related to the nitrogen supply is the simple ratio of near-infrared to green (Table 9.3). This agrees with results from Solari et al. (2008) with maize.

The outstanding ability of the red edge inflection point among the various standard spectral indices for indicating the nitrogen uptake of crops was also demonstrated in

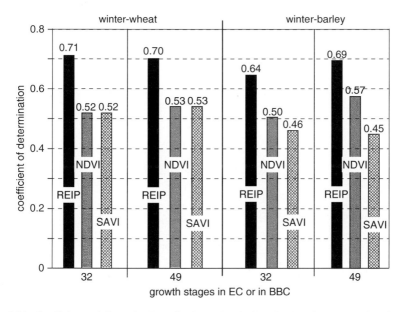

**Fig. 9.20** Coefficients of determination ($r^2$) of standard indices for the nitrogen uptake of small grains. The results are based on reflectance measurements by handheld instruments, natural illumination and vertical view directions. *REIP* red edge inflection point, *SAVI* soil adjusted vegetation index, *NDVI* normalized difference vegetation index (Drawn from data by Schmid and Maidl 2005)

results that were obtained at Munich University in Southern Germany (Fig. 9.20). The red edge inflection point was based on the approximating formula as defined above. However, the coefficients of determinations of the same reflectance indices in Fig. 9.20 differ more than those listed in Table 9.3.

The ratio of near-infrared to green is not listed in Fig. 9.20, though this index too provided good results that were not much inferior to those of the red edge inflection point (Schmid and Maidl 2005).

### 9.4.2.1 Interfering Factors with Standard Indices and Natural Light

Main factors that can interfere with the reflectance are the soil-colour, the zenith-angle of the solar radiation (Fig. 9.21) and eventually cloud-covers. Which of these factors are important, depends on the illumination that is employed for creating the reflectance. If natural illumination is used – which is the rule for remote sensing and also sometimes the choice for proximal sensing – all three factors that are mentioned above are important. When artificial illumination induces the reflectance with proximal sensing and hence the zenith-angles of the sun as well as cloud covers become unimportant, it is only the soil-colour that might interfere.

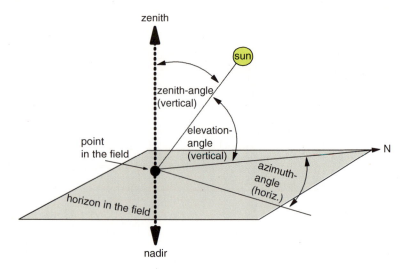

**Fig. 9.21** Geometry of solar radiation

In order to evaluate the influence of the solar zenith-angle, records were taken in Northern Germany on a sunny day in mid-May with a field spectrometer vertically fixed on one spot in winter-barley. During the day, the zenith angle varied from 40° to 80°. Similarly, reflectances were recorded on a day with varying cumulus clouds. Finally, the soil-colour was altered by watering a dry, bright soil and thus creating a dark, wet soil. For further details see Heege et al. 2008 or Reusch 1997.

The effects of the interfering factors zenith angle, cloud cover and soil brightness were defined by coefficients as listed below:

**Factor zenith-angle of the sun**

$$Coefficient\ Cz = \frac{index\ difference\ caused\ by\ altering\ the\ nitrogen\ dose}{index\ stand.\ deviation\ caused\ by\ change\ of\ zenith\ angle}$$

**Factor cloud-cover**

$$Coefficient\ Cc = \frac{index\ difference\ caused\ by\ altering\ the\ nitrogen\ dose}{index\ stand.\ deviation\ caused\ by\ change\ in\ cloud\ cover}$$

**Factor soil-brightness**

$$Coefficient\ Cs = \frac{index\ difference\ caused\ by\ altering\ the\ nitrogen\ dose}{index\ difference\ caused\ by\ wetting\ dry\ soil}$$

The numerators of the coefficients or ratios are given by the index difference due to the change in nitrogen applied. The denominator is the index difference pertaining

9 Site-Specific Fertilizing

**Table 9.4** Evaluating the effect of interfering factors by coefficients and by their geometric means

| Standard spectral indices | Factors and coefficients | | | Geom. mean $\sqrt[3]{CzCcCs}$ |
| --- | --- | --- | --- | --- |
| | Zenith Cz | Clouds Cc | Soil Cs | |
| Reflectance at 550 nm (green) | 4.5 | 0.2 | 6.6 | **1.9** |
| Reflectance at 670 nm (red) | 7.2 | 0.8 | 8.4 | **3.6** |
| Reflectance at 800 nm (near-infrared) | 5.1 | 5.7 | 1.2 | **3.2** |
| Ratio of near-infrared to red reflectance | 7.4 | 4.9 | 7.9 | **6.6** |
| Ratio of near-infrared to green reflect. | 12.9 | 4.2 | 19.0 | **10.8** |
| Normal. differ. veget. index (NDVI) | 11.9 | 2.7 | 12.6 | **7.4** |
| Soil adjusted vegetation index (SAVI) | 11.5 | 7.4 | 32.4 | **14.4** |
| Red edge infl. point, num. differentiat. | 45.9 | 10.7 | 21.8 | **22.0** |
| Red edge inflect. point, appr. formula | 61.9 | 12.7 | 10.1 | **19.8** |

Large coefficients mean low interferences (From Heege and Reusch 1996)

to the change of the interfering factor such as the solar zenith-angle, the cloud-cover or the colour of the soil. Hence the coefficients represent the respective signal to noise ratio. Its suffixes z, c, and s stand for the zenith angle, the cloud cover or the soil colour respectively.

In Table 9.4, the coefficients of the interfering factors are listed. The larger the coefficients are, the lower the interference by the respective noise factor is and hence the more reliable the reflectance index is.

Since **natural illumination** provided the reflectance and consequently all three interference factors are important, it seems reasonable to use the mean of the coefficients as criterion. Instead of arithmetic means, **geometric means** are listed in the right column of Table 9.4. This is because generally geometric means are superior to arithmetic means when normalized results or ratios are averaged. And this is the case with the coefficients.

Regarding the geometric means, the best results – i.e. the lowest average noise – can be obtained with the **red edge inflection point** indices. These indices on the one hand depend very clearly on the nitrogen supply of the plants (Table 9.3) and on the other hand are not much influenced by the interfering factors. The widely used normalized difference vegetation index (NDVI) is on the average much more affected by these interfering factors. The soil adjusted vegetation index (SAVI) as well as the ratio of the near-infrared to green reflectance can be regarded as candidates in intermediate positions. However, when looking at the individual interference factors or its coefficients, it can be seen that the average "noise" ranking of the soil adjusted vegetation index results from the very low influence that in this case is exerted by the factor soil (red soil column in Table 9.4). On the other hand, the outstanding rankings of the red edge inflection point indices originate from the low influence (large coefficients) that the radiation factors zenith angle and clouds have.

If solely **artificial illumination** provides the reflectance, noise from changing solar zenith angles or from varying cloud covers practically is excluded. Hence from the coefficients listed, only the factor soil remains (Table 9.4). However, this factor exerts its influence on the noise situation only if soil is in the field of view of the sensor. In case the sensor views only closed canopies, the factor soil too is excluded. So with artificial illumination and closed canopies, all these noise factors get irrelevant.

### 9.4.2.2 Viewing Directions with Natural Light

The reflectance sensor may have a vertical or an oblique **viewing direction** towards the crop canopy. The above results from Sects. 9.4.2 and 9.4.2.1 refer to vertical directions. Sensors that are operating on sprayer booms usually use this view (Fig. 6.5). And for sensing from aerial platforms or from satellites this vertical view too may be appropriate. However, for an on-the-go control of the widely used centrifugal fertilizer spreaders by proximal sensing, additional considerations deserve attention.

It is essential to have the field of view out of any shades of the field machinery and out of any tramlines, since this affects the reflectance. With vertical viewing, this can be achieved by means of a transverse boom. Sprayers and pneumatic spreaders have this boom anyway, but the dominating centrifugal spreaders do not. And because of this, sensing with oblique views from the top of the tractor's roof sidewards into the canopy is widely practiced. This has an additional effect: the sensor is seeing less soil in case the canopy is not closed. Hence the noise that is caused by soil is reduced (Table 9.4).

However, an oblique viewing direction into the canopy means also that an additional solar radiation factor gets important. Whereas in case of vertical viewing only the **zenith angle** of the solar radiation needs to be considered, with an oblique viewing direction it is necessary to take into account the effect of its **azimuth angle** as well (Fig. 9.21). Because the oblique direction – as opposed to the vertical direction – inherently also has a horizontal component. And this horizontal component is affected in the course of a day by the varying azimuth angle as well as by changing directions of travel of the machinery in the field. An exception from this holds only when the sun is precisely in the zenith (zenith angle=0°). But this happens only in the tropics and even there just at noon. So in most cases, the effects of solar azimuth angles cannot be neglected.

Reusch (2003) has shown that this adverse effect of the azimuth angle on the reflectance from natural light can be practically removed by increasing the number of viewing spots and distributing these evenly around the tractor (Fig. 9.22). He investigated the **correction of the azimuth-effect** that can be obtained by multiple viewing-directions and by averaging the results of these. As a criterion, the ratio of the near-infrared to red reflectance was used.

In case only one direction supplies the signals, the reflectance ratio goes from 18 to 30 % and back to 18 % while the azimuth angle goes from zero to 360° (Fig. 9.23). This wide range for the azimuth angle is encountered while the machinery is operating in different directions within a field. So this range can affect the signals within rather short time spans.

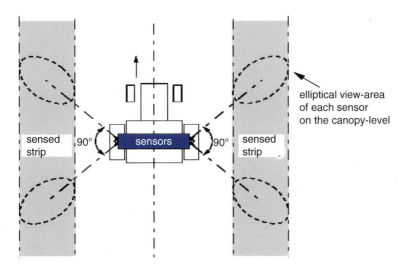

**Fig. 9.22** Geometry for oblique viewing of the crop canopy by two sensors on each side of the tractor. The sensors are positioned on top of the roof of the tractor (From Reusch et al. 2004, altered)

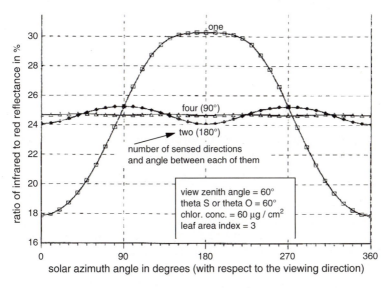

**Fig. 9.23** Effect of solar azimuth angle on the ratio of near-infrared to red reflectance from natural light. The curves demonstrate the results that can be obtained by averaging the signals from multiple directions. The reflectance signals are based on the PROSPECT/ SAIL model. For details to this model see Sect. 6.1 and Fig. 6.3 (From Reusch 2003, altered)

The correcting effect of averaging the results from several directions is drastic. With two opposite directions, the reflectance varies only between 24 and 25 %, and when four equally spaced directions are used, it remains practically constant. It should be noted that the results are based on an optical model (see legend to Fig. 9.23) that does not take into account any other factors than the solar azimuth angle.

This averaging of the signals from four directions is state of the art in practical farming when the sensing is based on natural light.

### 9.4.3 Sensing Nitrogen by Reflectance Based on Artificial Light

A general experience with the Kiel system in Europe when operating on the basis of natural illumination is that the solar elevation angle – the complementary angle to the zenith angle (see Fig. 9.21) – should not be below 25°. This means that early and late hours in the day cannot be used. Yet in regions with maritime- or mountainous influence on the climate, this is just the time of the day when there is less wind. And high wind speeds can seriously reduce the accuracy of fertilizer-spreading, especially with the currently dominating centrifugal spreaders.

Artificial illumination can overcome this problem. It also eliminates the noise associated with any varying irradiance of natural light and with shade effects caused by the machinery or by trees. It allows operating not only during daytime, but at dawn and at night as well.

Technically, the spectral effects of the artificial light source need to be separated from those of any natural light. This **separation** can be achieved by subtracting the reflectance spectrum of solely natural light from the spectrum that results, when in addition the artificial illumination is switched on. The artificial light is applied as **regular flashes** and supplies "On" spectra. In the short intervals between the flashes, the "Off" spectra come from the natural light. The "Off" spectra are subtracted from the respective adjacent "On" spectra. The thus obtained spectral signals result only from the artificial illumination and any effect of varying natural light is eliminated.

In case continuous artificial illumination is applied, separating the effects of the different light sources would also be possible by **modulating the artificial light**. However, this separating technique is more complicated.

In the total cost calculation, the higher investment for artificial illumination might be offset by the capability of treating a larger area because of the longer operating time per day.

It should be mentioned that the **artificial light source** must adequately supply the respective wavelengths. Many commonly used light sources do not, as can be seen in Baille (1993) and Lawrence et al. (2005). If the wavelengths that are needed are provided, there is no reason why the nitrogen effect on the reflectance indices with artificial illumination should be different than with natural light. Commonly

9 Site-Specific Fertilizing

**Fig. 9.24** Control of nitrogen spreading via reflectance sensing based on artificial light with vertical viewing of the canopy from transverse booms in front of the tractor (Crop Circle concept, photo from Wilson J., SoilEssentials Ltd., England)

used light sources include xenon lamps, laser diodes and light emitting diodes (LEDs). As with natural light, the view on the canopy can be in a vertical or in an oblique direction.

For an oblique direction sidewards down from the tractor-roof, there is no need anymore to take care of the effect of any solar azimuth angle. So, contrary to the concept dealt with above in Figs. 9.22 and 9.23, it suffices to use a single sensor on each side of the tractor.

Since shade effects on the signals that result from the machinery too are excluded with artificial illumination, the transverse booms needed for vertical viewing can be shorter (Fig. 9.24) than with natural light. Yet the most compact design still in possible with oblique viewing from the tractor roof (Fig. 9.25).

However, the mounting itself of the sensors probably is less important with artificial illumination. Crucial is that the sensors see what is needed for the control of the spreader. And this depends on the viewing direction too. With an oblique viewing direction, the sensor sees less soil than with a vertically oriented view. This is because the slanted view hides small bare patches by plants. In this respect, the oblique view is advantageous with early development stages and with widely spaced crops.

But because an oblique orientation sees more **biomass** than a vertical view, its reflectance indices also tend to saturate faster. The term **"saturation"** here is used for the relation between the respective reflectance index and the leaf- area-index of the crop. A saturated reflectance index cannot differentiate between leaf-area-indices at high levels. But the ability to indicate high levels of leaf-area-indices with precision is very important for nitrogen sensing. Because an increasing nitrogen supply affects the leaf-area-index of a crop even more than the chlorophyll concentration within its leaves (Fig. 9.17).

**Fig. 9.25** Control of nitrogen spreading via reflectance sensing based on artificial light with oblique viewing of the canopy from the roof of the tractor (From Agri Con GmbH, Jahna, Germany)

In short, sensing by oblique viewing of the canopy particularly needs reliable reflectance indices that do not saturate. This is dealt with in the next section.

### 9.4.3.1 New Reflectance Indices Instead of Standard Indices

The standard spectral indices dealt with so far mostly have been developed for remote sensing of vegetation and landscapes from satellites and aerial platforms. The fact that some of these indices depend on the nitrogen supply of crops does not imply that they are best suited for the detection of this nutrient. The "Normalized Difference Vegetation Index" (NDVI) is widely used as the standard index for assessing the "greening" of the global surface. However, despite this, the potential of the NDVI to indicate the density or the leaf-area-index of the vegetation is rather limited. Therefore Schepers et al. (1998) proposed to use a **green NDVI** instead of the **standard NDVI**. In this green NDVI, the red reflectance in the standard NDVI equation (Table 9.3) is substituted by green reflectance.

It is well known that the standard NDVI saturates around a leaf area index of about 2.0–2.5 (Fig. 6.7). Yet a well developed cereal crop can have a leaf area index that is three to four times higher. A similar saturation effect applies to standard infrared to red ratios. The reason for this is the very high absorption of red light by photosynthesis. Because of this, the light does not penetrate deeply into dense crop canopies and hence is not able to sense higher leaf-area-indices. In the green and in

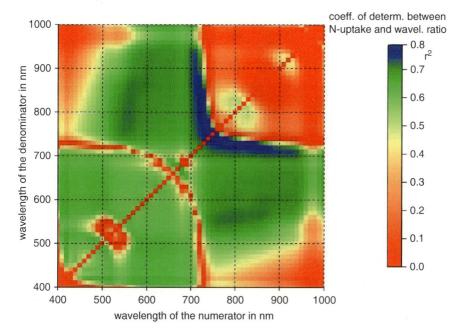

**Fig. 9.26** Matrix showing coefficients of determinations ($r^2$) of simple reflectance ratios for sensing of N in wheat with a growth-stage in EC or BBCH of 31 (From Reusch 2005, altered)

the red edge range the absorption of light by photosynthesis is lower (Fig. 6.1), thus the light penetrates better into the canopy and thus can indicate its biomass. For sensing nitrogen, sensing the biomass of a crop or it surrogate – the leaf-area-index – is at least as important as sensing the chlorophyll concentration within the leaves well. This context explains why among the standard indices, the ratio of near-infrared to green and especially the red edge inflection points provided the best results for nitrogen sensing (Table 9.3).

However, none of the standard indices was developed especially for sensing of nitrogen. Even the best standard index for nitrogen sensing – the red edge inflection point – is essentially a chlorophyll index. Its potential for nitrogen sensing is derived from the fact that chlorophyll is an important carrier of plant nitrogen (Lamb et al. 2002), but it is not the only one.

These considerations led to **systematic searching** in steps of 10 nm within the wavelength range from 400 to 1,000 nm (Reusch 2003, 2005). The searches were confined to simple ratio indices with one wavelength in the numerator and the other one in the denominator. Figures 9.26 and 9.27 show results that were obtained with oblique viewing of winter-wheat canopies and with artificial illumination. The colors stand for coefficients of determination between the nitrogen uptake and the respective wavelength ratios. The locations within the squares of the matrices represent the coordinates of the wavelength ratios.

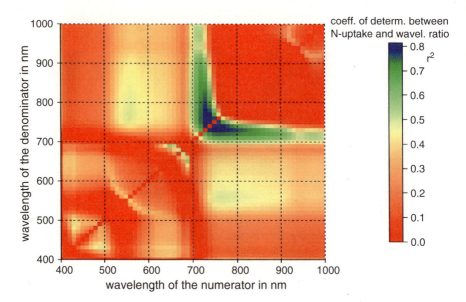

**Fig. 9.27** Matrix showing coefficients of determinations ($r^2$) of simple reflectance ratios for sensing of N in wheat with a growth-stage in EC or BBCH of 38 (From Reusch 2005, altered)

The results are unambiguous: the highest coefficients of determination ($r^2$) and hence the most reliable predictions of the nitrogen uptake are obtained with wavelengths for both the numerator as well as for the denominator of the index ratio between about 720 and 800 nm. In the early development stage of EC or BBCH 31 (Meier 2001), wavelengths somewhat above 800 nm also provided still good results (Fig. 9.26). However, in the later stage of EC or BBCH 38 this is not so. At this stage, the colors in the matrix around 800 nm wavelength already indicate a slight decrease for the coefficients of determination. Müller et al. (2008) as well as Inoue et al. (2012) presented rather similar results of systematic spectral searching in steps of 10 nm in the form of matrices from experiments with vertical viewing and natural light respectively for oilseed rape (Brassica napus) in Germany or for rice (Oriza sativa) in Japan and China.

So the most reliable predictions of the nitrogen uptake come from wavelength ranges, in which also the red edge inflection point is located (Fig. 9.19). However, for reflectance sensing there is a fundamental difference between the red edge inflection point on the one hand and a simple wavelength ratio from the same spectral range on the other hand. The **red edge inflection point** indicates just a small dot on the spectral curve that is defined in nm of wavelength and its respective reflectance. The wavelength ratio instead depends on the slope of the spectral reflectance curve between the respective wavelengths and is dimensionless. Its precise definition in terms of wavelengths for the numerator and the denominator might vary somewhat within the range indicated above. Wavelength pairs that have been proposed for this **red edge ratio index** are *e.g.*:

$\dfrac{R\,730}{R\,780}$ (Reusch 2005) or also $\dfrac{R\,760}{R\,730}$ (Jasper et al. 2009) or identical to this

$\dfrac{R\,760}{R\,730}$ (Erdle et al. 2011) and $\dfrac{R\,740}{R\,780}$ (Müller et al. 2008).

These proposals are based on experiments with winter-wheat (Reusch 2005; Jasper et al. 2009 as well as Erdle et al. 2011) and winter-oilseed rape (Müller et al. 2008). The small spectral differences between these indices probably hardly matter and inverse ratios of the wavelengths indicated can be used as well. Jasper et al. (2009) and Erdle et al. (2011) found that the red edge ratio index was largely unaffected by different varieties, varying seed densities and growth stages.

How do red edge ratio indices compare with the standard indices? And can nitrogen sensing still be improved by using more sophisticated indices, *e.g.* normalized difference indices with wavelengths that come exclusively from the red edge range? The criterion of a normalized difference spectral index is that it relates the difference of two wavelengths to the sum of the same wavelengths (Table 9.3). Would a normalized difference red edge index that relies solely on wavelengths from the red edge still improve the results?

The coefficients of determination ($r^2$) for the prediction of nitrogen uptake by reflectance indices in Table 9.5 refer to winter-oilseed rape. Among the standard indices, the best results again are obtained with the near-infrared to green ratio and especially with the red edge inflection point. Yet still better predictions were supplied by the new indices that rely exclusively on wavelengths from the **red edge range**. Whether **red edge ratio indices** or alternatively **normalized difference indices** from the **red edge range** are used seems to be unimportant (Table 9.5). These good results with various indices from the red edge range are in line with findings that support the significance of this spectral region for sensing of chlorophyll (Fig. 6.8), which is closely related to the nitrogen supply.

The relation between red edge ratio indices and the supply or uptake of nitrogen is nearly linear in most cases (Fig. 9.28). Similar regressions have been obtained with barley, oats, oilseed rape and maize and at different growth stages (Reusch et al. 2010). The sensor readings depend on the growth stages of crops, as does the nitrogen uptake of crops.

There is some influence of various **fungal infections** on the reflectance, as might be expected. Reusch 1997 made trials with barley that was infected by leaf blotch (*Rhynchosporium secalis*), by leaf rust (*Puccina hordei*) and by powdery mildew (*Erysiphe graminis*) and that was either sprayed to remove the diseases or not. The result of the infections on the red edge inflection point index were rather small when compared to nitrogen effects. Yet it must be realized that the severities of fungal infections will affect such results.

**Table 9.5** Standard indices or new indices for nitrogen sensing and results by coefficients of determination ($r^2$)

| Definition | Formulas for wavelengths R (number added is wavel. $\lambda$ in nm) | Coeffic. of determination ($r^2$) |
|---|---|---|
| **Standard indices (commonly used)** | | |
| Near-infrared reflect. | R 850 | 0.16 |
| Ratio of near-infrared to red | $\dfrac{R\,780}{R\,670}$ | 0.63 |
| Ratio of near-infrared to green | $\dfrac{R\,780}{R\,550}$ | 0.72 |
| Normalized difference vegetat. index (NDVI) | $\dfrac{R\,780 - R\,670}{R\,780 + R\,670}$ | 0.55 |
| Soil adjusted vegetation index (SAVI) | $\dfrac{1.5(R\,780 - R\,670)}{R\,780 + R\,670 + 0.5}$ | 0.51 |
| Red edge inflect. point, approx. formula (REIP) | $700 + 40\dfrac{(R\,670 + R\,780)/2 - R\,700}{R\,740 - R\,700}$ | 0.73 |
| **New indices with wavelengths exclusively from the red edge range** | | |
| Red edge ratio index | $\dfrac{R\,780}{R\,740}$ | 0.82 |
| Inverse red edge ratio index | $\dfrac{R\,740}{R\,780}$ | 0.81 |
| Normalized difference red edge indices | $\dfrac{R\,780 - R\,740}{R\,780 + R\,740}$ | 0.82 |
| | $\dfrac{R\,750 - R\,740}{R\,750 + R\,740}$ | 0.82 |

The results are based on winter-oilseed rape, vertical viewing and natural illumination (Arranged from data by Müller et al. 2008)

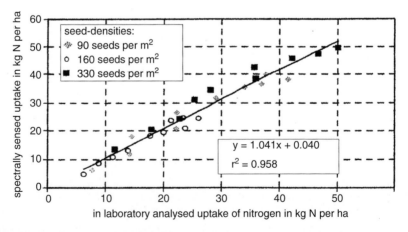

**Fig. 9.28** Accuracy of nitrogen sensing by a red edge ratio index (R760/R730) versus analysing of samples chemically in a laboratory with winter-wheat in Northwest Germany. The sensing was done from the roof of a tractor in an oblique direction with artificial light and at a growth stage of EC or BBCH 31. The field was fertilized with different rates of nitrogen at a growth stage of EC or BBCH 20 (From Jasper et al. 2009)

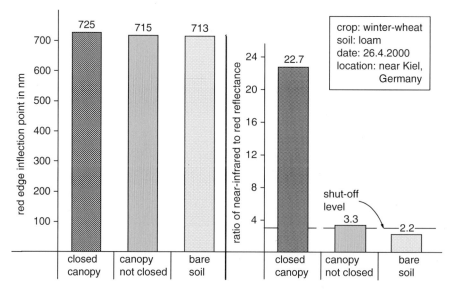

**Fig. 9.29** Soil- and crop canopy effects on spectral indices (From Heege and Thiessen 2002)

## 9.4.4 Soil or Plants in the Field of View

Fertilizer control using reflection from the leaves requires **green plants** in the field of view of the sensor. In the case of patches with little or no plants, the resulting application rate may be inappropriate. The reason for this is that a lack of nitrogen within a canopy as well as bare soil instead of plants turns the signal in the same direction.

A lack of nitrogen within a normally developed canopy should be compensated for by increased fertilization, whereas it certainly is not reasonable to apply high amounts of nitrogen to larger patches of **bare soil**. In rolling landscapes with undulating fields, these patches of bare soil can be the result of uneven winter killing of plants. Therefore, an additional control for completely shutting off the flow of fertilizer is needed.

A reliable indicator for this control is the ratio of near infrared- to visible reflectance (Felton and McCloy 1992). The relation of near-infrared to red already is used for weed control by **spot spraying** in fallow fields or by strip spraying in crops with tramlines (Sect. 6.2). In the case of bare soil instead of a closed canopy, this relation decreases drastically from 20 or more to 3 or less (Heege and Thiessen 2002). For winter cereals in Germany, it seems to be reasonable to stop fertiliser spreading when the near infrared to red relationship drops below 3 (Fig. 9.29). This automatic shut-off does not necessitate additional sensing equipment if infrared and red reflectance signals are recorded anyway. This depends on the spectral indices that are used for the prediction of the nitrogen supply.

With small grains, the Kiel system is rarely used for the first dressing of small cereals. Its application is mostly limited to the second and later dressings. This is mainly because at the time of the first dressing, the crop canopy is not yet closed. Even if an indication of biomass and chlorophyll content could be obtained at the time of the first dressing, the very small plants often do not provide a reliable estimation of the nutrient supply from the soil. Because initially the germinating and emerging plants obtain their nitrogen from the seeds and not from the soil. Therefore, differences in the supply from the soil might not well show up with small grains at a developmental stage below EC or BBCH 25 (less than 4 tillers).

Apart from this, for **not closed crop canopies** it is important to exclude the influence of soil on the reflectance as much as possible in order to avoid mixed signals. Two methods can be used for this, either **narrowing the view** of the vertically oriented sensor to a closed canopy strip exactly above and along a row, or using an **oblique view** that is oriented perpendicular to the row-directions. The first method – narrowing the view – can only be successful if the canopy strip above the row is really closed. The second method – oblique viewing – relies on getting the signals mainly from the upper part of the crop. The more oblique this field of view is, the more the reflectance received comes from the crop instead of the soil.

For most crops, the **minimum growth stage** for avoiding mixed signals is lower with oblique viewing than with vertical viewing. So for small grains, the minimum growth stage with oblique viewing is about EC or BBCH 30, which is the beginning of stem elongation (Meier 2001). With vertical viewing it is about EC or BBCH 32, which is stage node 2 (Schmid and Maidl 2005). A similar effect of the viewing direction on the minimum growth stage that is needed has been observed for maize (Bausch and Diker 2001).

However, oblique- instead of vertical viewing also means that the spectrum might saturate faster with increasing leaf-area-indices of the crops (see Sects. 6.2 and 9.4.3.1). Since this **saturation effect** reduces the sensitivity in measuring, reliable oblique viewing depends particularly on reflectance indices that do not tend to saturate fast. Consequently, accurate nitrogen sensing in lush crops with oblique viewing requires avoiding wavelengths from the main absorption regions, hence from the blue and especially from the red range. With proximal sensing, this practice presently is state of the art. The poor results of the red Normalized Difference Vegetation Index (NDVI) as defined in Tables 9.3 and 9.5 for nitrogen sensing might be even worse when the viewing occurs from an oblique- and not from a vertical direction.

With natural illumination, oblique viewing also increases the noise due to the varying solar azimuth angle. Yet an effective method of eliminating this noise – apart from using artificial light – is averaging signals coming from several and opposite directions (Fig. 9.23). The original Kiel system (Figs. 9.1 and 9:25) now is operating in practice with oblique viewing by sensors positioned on the roof of the tractor relying either on solar or artificial illumination.

## 9.4.5 Sensing Nitrogen by Fluorescence

In addition to the reflected radiance, crop leaves usually emit fluorescent light (see Sect. 6.4). The difference is that reflectance simply results from irradiance that is thrown back, whereas the fluorescence comes from the photosynthetic apparatus of plants as surplus light energy or as a by-product. Artificial induction of fluorescence often is done by pulsed red laser beams.

In the red wavelength range, fluorescent light is emitted from the leaves with two peaks at 680 and 735 nm. It has been shown (Lichtenthaler 1996) that the relationship of the fluorescent intensities at these two peaks is an indicator of the **chlorophyll concentration** in leaves. This indication of chlorophyll concentration is explained by differences in the re-absorption of fluorescent light at these two emission peaks in the red and near-infrared range. The red fluorescent light at 680 nm wavelength is partly re-absorbed by chlorophyll for photosynthesis. This re-absorption depends on the chlorophyll concentration within the leaves. On the other hand, the fluorescent light at 735 nm wavelength is above the range of absorbed light. It therefore is barely reduced by absorption. For that reason, the **ratio** of the **two red fluorescence peaks** can be used to sense the chlorophyll concentration and thus also the nitrogen concentration in the leaves. So the background is chlorophyll estimation from **fluorescence absorption**. But what is the correlation between this fluorescence ratio and the nitrogen supply?

The fluorescence ratio was recorded by means of a handheld instrument (Thiessen 2002). Nitrogen top dressings for winter-cereals and winter-rape were applied three times in the growing season, beginning in early March. Readings were taken at the time of the third dressing and 3–4 weeks after this date. The actual dates for the dressings depended on the crop species, but typical dates for Northern Germany were used (Fig. 9.30).

The basic assumption is that with increasing nitrogen supply by the preceding dressing, the chlorophyll concentration in the leaves should go up and thus the fluorescence ratio should drop. At the time of the third dressing – which is in mid-May for winter-barley or in early June for winter-wheat – this assumption is supported up to a preceding nitrogen supply of 120 kg/ha. Beyond this level of the nitrogen supply, the fluorescence ratio does not drop any more. On the contrary, it rises (Fig. 9.30, top).

A similar trend showed up 3–4 weeks after the third dressing with winter-barley, with winter-wheat and with winter-rape. However, at this time, the fluorescence ratio decreased up to a preceding nitrogen supply of 160 kg/ha (Fig. 9.30, bottom), but above this rate, it also rose. In conclusion, at both dates and with all crops, there was no unidirectional relationship, which is a prerequisite for a simple control system. One might argue, that a unidirectional relationship up to a range of 120–160 kg/ha suffices, since higher pre-applications at earlier dressings seldom occur. So sensing of nitrogen by fluorescence absorption might be a feasible option except perhaps for very high dressing rates. For further results about nitrogen sensing by fluorescence see Thiessen (2002), Schächtl et al. (2005), Thoren (2007), Thoren and Schmidhalter (2009), Thoren et al. (2010).

Contrary to signals from reflectance, fluorescence sensing is never associated with erroneous information from **non-vegetated soil**. Because bare soil does not emit fluorescence. Neither any soil nor its plant residues or its stones influence the signals. The fluorescence ratio is solely based on the chlorophyll concentration in the leaves. This can be a distinct advantage when the crop canopy in not closed.

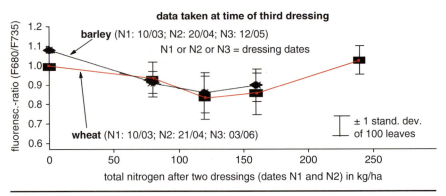

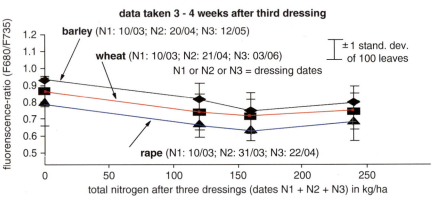

**Fig. 9.30** Chlorophyll-fluorescence and nitrogen supply (From Thiessen 2002, altered)

Therefore, fluorescence signals might be better at early growth stages. Restrictions to this are outlined below. Another consideration is whether for the same reason, fluorescence signals should be preferred for widely spaced crops such as maize, potatoes and sugar-beets, since with these crops the canopy closes fairly late.

However, the fluorescence ratio is not *per se* influenced by the leaf-area-index, in contrast to reflectance indices derived from the red edge and near-infrared range (Fig. 9.19). Since the chlorophyll concentration in the leaves often is inferior to the leaf-area-index as an indicator of nitrogen supply (Fig. 9.17), fluorescence sensing has been combined with means of recording the plant density. A concept for this approach is the combination of fluorescence sensing with a **scanning technique**. Bredemeier and Schmidthalter (2005) as well as Thoren (2007) used such a laser-induced fluorescence sensing combined with the recording of a "biomass index" by scanning the field surface. The surface scanning was obtained via the **number of fluorescence signals** per unit area that was sensed when the plants were irradiated by pulsed red laser beams and thus were induced to emit fluorescence. Hence a relative indication of the site-specific **plant density** per unit area was obtained. The sensors were located on the roof of the tractor and had an oblique field of view

**Fig. 9.31** Simultaneous sensing of the fluorescence ratio (F680/F735) and of estimates of the plant density per unit area. The plant density is sensed by the number of signals per unit of area above a noise level. The inclination of both sensors perpendicular to the direction of travel oscillates with a frequency of 1–2 Hz while driving through the field. Hence the field surface is scanned for signals (From Thoren 2007, supplemented)

(Fig. 9.31). By varying the inclination of this oblique view continuously while driving though the field, the canopy surface was scanned. In this way, signals related to the chlorophyll concentration in the leaf area as well about the vegetated area in the field were recorded.

Yet despite this there are still limits with the present technology of nitrogen sensing by fluorescence. These limits have to do with the irradiation that induces the fluorescence in the photosynthetic apparatus. The irradiance that causes the emission of fluorescence has wavelengths below the near-infrared. In most cases, red radiation is used for the induction. Contrary to this, the indices for nitrogen sensing by reflectance sensing include near-infrared radiation. And there are fundamental differences in the ability of near-infrared or red irradiance to record the development of a canopy.

The near-infrared irradiation is not absorbed, but highly transmitted. The red irradiation is mainly absorbed and barely transmitted. Consequently, the near-infrared irradiance induces mainly **volume-reflectance**. Red radiation causes, for the most part, **surface-reflectance** (Fig. 9.32) or **surface-fluorescence**.

This explains, why near-infrared reflectance responds much better to an increasing **leaf-area-index** of a canopy than red reflectance does (Fig. 9.32, right). Above a leaf area index of 2, the red radiation does not deliver reliable signals. In principle, the same applies to fluorescence that is induced by the red radiation.

In short, nitrogen sensing by fluorescence instead of reflectance is better at early growth stages. But at later development stages, the situation is *vice versa*. Intermediate growth phases can be sensed by either method. So the problem boils down to the question, at which developmental stage of crops, site-specific nitrogen sensing is reasonable or needed.

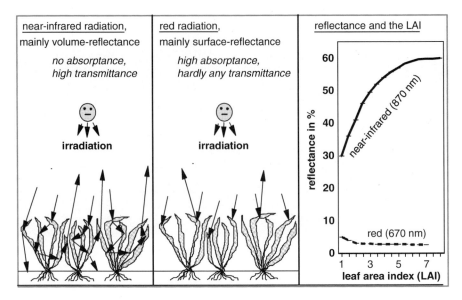

**Fig. 9.32** The optical background of volume and surface reflection (The curves in the right part are from Guyot 1998)

At very early growth stages – *i.e.* below EC or BBCH 29 with small grains – the nitrogen content of the leaves often is not yet a reliable indicator of the supply by the soil. This is, because the seedlings get their initial supply from the seed and not from the soil.

There may be exceptional cases that call for an early nitrogen dressing on a site-specific basis. These cases might arise when local differences in field emergence develop because of uneven fields or irregular sowing and an early site-specific nitrogen dressing is targeted at compensating for this. The compensation might not be based on the nitrogen content in the leaves, but instead rely on using a fluorescence technique for sensing the **plant density** per unit area (Thoren 2007). An alternative to this method could be a site-specific first dressing that is oriented at the height of the plants (see next section).

Up to now, fluorescence sensing for controlling the nitrogen application is rarely used in practical farming. Interesting applications for fluorescence sensing might develop in combination with the detection of fungal diseases.

### 9.4.6 Sensing Nitrogen Based on Bending Resistance or Height

These methods inherently omit the chlorophyll as a control factor for the nitrogen fertilization. Instead, these methods rely on physical properties of crops for the control, either the resistance of the canopy against bending or alternatively the height of the plants.

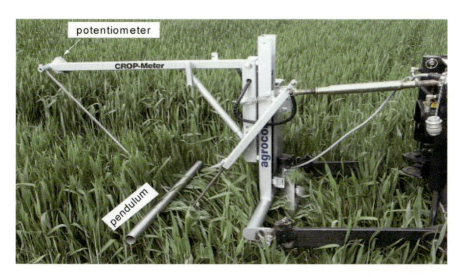

**Fig. 9.33** Sensing the canopy resistance against bending in a small cereal crop by a pendulum that operates in front of a tractor (From Thoele and Ehlert 2010, altered)

The **resistance of the canopy against bending** is sensed via the deflection of a pendulum that operates in front of a tractor (Fig. 9.33). The deflection of this pendulum is indicated by its angle to the vertical via a potentiometer. And the use for site-specific farming is based on the correlation between this angle and the biomass or the **leaf-area-index** of crops. Actually, the coefficients of determination ($r^2$) between the pendulum deflection angles and the biomasses were 0.89 or higher (Ehlert et al. 2003) and those for leaf-area-indices were between 0.64 and 0.91 (Dammer et al. 2008). However, there are limits of precise sensing in rough fields because of vibrations or in areas with varying slope.

This rather simple method of sensing crop properties is state of the art in farming and aims at applications in the site-specific control of both **nitrogen fertilizing** and **fungicide spraying**. The results that have been obtained are encouraging (Dammer et al. 2008; Thoele and Ehlert 2010) despite the fact that this method inherently leaves out any influence of chlorophyll on applications of agrochemicals.

A serious limitation is that it needs advanced growth stages. With small cereals, a growth stage of at least EC or BBCH 35 is essential (Ehlert et al. 2004b), whereas with reflectance sensing the minimum growth stage is about EC or BBCH 30. This means, the earliest time for sensing by the pendulum is when the first flag leaf – still rolled – has appeared, while reflectance sensing can already start when the tillering has ended.

Another concept that also leaves out any direct influence of plant ingredients is based on sensing the apparent **crop height** by an ultrasonic transducer. This method relies on distance recording as shown in Fig. 8.7 and operates similar to well known applications in automobiles as driver aids for parking. Reusch (2009) has shown that an ultrasonic transducer, which is pointing vertically downwards from a front mounting of a vehicle towards the crop canopy, simultaneously can indicate the distance to the soil surface as well as to the top of the canopy. Hence from the difference of these distances, it can sense the apparent crop height, which in turn is well correlated to the biomass dry matter. In experiments with winter-wheat and growth stages between EC or BBCH 21 and stepwise up to EC or BBCH 69, the biomass was indicated with coefficients of determination ($r^2$) between 0.79 and 0.94. Contrary to sensing via canopy resistance against bending, the biomass was also precisely recorded in early developmental stages between BBCH 21 and 32. Hence this rather simple physical method could be used even for the first top dressing of small cereals. However, practical experiences in this direction do not yet exist.

## 9.4.7 Cell Sizes or Resolution

An important question is the spatial resolution in site-specific nitrogen application. In case the resolution that is obtained is too low, and consequently the cell sizes are too large, any site-specific application might be useless (see Chap. 2). However, it is essential to differentiate between

- the cell sizes for site-specific **distributing** (spreading or spraying)
- the cell sizes for site-specific **sensing** of the crop status during fertilizing
- the cell sizes when **recording the heterogeneity** of crops is done.

An adequate cell size in site-specific distributing is the final aim. But having a distributing cell size that is below that of sensing is useless. To be effective, the cell sizes that hold for site-specific distributing must be as large or larger than those that are sensed. Usually the cell sizes in proximal sensing are much smaller and averaged during processing because of the speed and frequency with most optical methods. Yet a prerequisite of having adequate cell sizes both for distributing and for sensing during fertilizing is knowledge about the heterogeneity of crops.

Recording the **heterogeneity** of crops in respect to the nitrogen supply can easily be done by proximal reflectance sensing of crop canopies with a high frequency. This method can provide more than 100 basic signals per second for small sampling areas and the field-transects thus recorded or mapped are very accurate if a precise georeferencing method is used (Thiessen 2002). The semivariograms that are obtained from the transects can supply the upper limits of cell sizes or of cell side lengths as outlined in Fig. 2.5. Exceeding this **upper limit of cell side length** deteriorates the precision in site-specific management. The upper limit of cell side length sometimes is called the mean correlated distance.

Results from this method of recording the heterogeneity by sensing of red edge inflection indices and hence of the nitrogen supply of small cereal crops in Schleswig-Holstein, Germany are listed in Table 9.6.

The upper limits of cell side lengths vary between 5.9 and 139.0 m. The reasons for this wide span are difficult to find out. The geographic areas in Schleswig-Holstein that are involved are

## 9 Site-Specific Fertilizing

**Table 9.6** Heterogeneity of small cereal crops in three regions of Schleswig-Hostein, Germany

| Field and crop | Date | Upper limit of cell side length |
|---|---|---|
| **Eastern hilly, loamy region of Schleswig-Holstein** | | |
| Achterkoppel, winter-wheat | 23rd of March | 33.5 m |
| | 27th of April | 28.4 m |
| | 31st of May | 13.2 m |
| Achterkoppel, winter-barley | 4th of April | 5.9 m |
| | 23rd of April | 32.8 m |
| | 30th of May | 34.1 m |
| Achterkoppel, winter-rye | 16th of April | 75.0 m |
| Kronskoppel, winter-barley | 26th of April | 11.8 m |
| | 11th of May | 14.7 m |
| Viehkoppel, winter-wheat | 27th of April | 8.6 m |
| | 26th of May | 12.0 m |
| Niedeel, winter-wheat | 4th of April | 13.4 m |
| | 17th of April | 7.2 m |
| | 7th of May | 7.5 m |
| | 30th of May | 7.4 m |
| | 13th of June | 32.9 m |
| Niedeel, winter-barley | 16th of April | 17.0 m |
| | 26th of May | 99.1 m |
| *Average for Eastern hilly, loamy region* | | **25.3 m** |
| **Central flat, sandy region of Schleswig-Holstein** | | |
| Olenhoebek, winter-wheat | 8th of May | 46.9 m |
| | 7th of June | 27.3 m |
| Olenhoebek, winter-barley | 23rd of April | 29.4 m |
| | 19th of May | 33.7 m |
| Olenhoebek, winter-rye | 11th of April | 18.0 m |
| *Average for central, flat, sandy region* | | **31.1 m** |
| **Flat, loamy region of Fehmarn Island, Schleswig-Holstein** | | |
| Ostenfeld, winter-wheat | 6th of May | 116.1 m |
| | 4th of June | 139.0 m |
| Hohlblöcken, winter-barley | 6th of May | 28.6 m |
| | 4th of June | 32.0 m |
| *Average for flat, loamy region of Fehmarn* | | **78.9 m** |

The criterion of the heterogeneity is the upper limit of cell side length as defined in Fig. 2.5. The results are from vehicle based, site-specific sensing for N with red edge inflection indices and a vertically viewed sampling area of 1.5 m$^2$ (From Thiessen 2002, altered)

- the eastern, hilly, loamy region with mounds and depressions in the fields
- the flat, sandy, central region
- the very flat, loamy region of Fehmarn island.

From the topography of the fields, the highest heterogeneity should be expected from the rolling, hilly fields in the eastern region and the opposite from the very flat agricultural areas in Fehmarn. The average upper limits of cell side lengths confirm these expectations, however, there is much overlapping of individual data.

The results from different times within the growing season differ widely for the same crops, and there is no uniform trend discernible. The same holds for differences in the results between the crop species. In short, there seem to be some factors involved that are not listed.

From the **averages** of the upper limits of cell side length (Table 9.6), it can be concluded that the working widths of most present day fertilizer spreaders probably comply with the site-specific application needs for nitrogen. The lowest average limit of cell side length is about 25 m. Although the spreading widths of modern **centrifugal spreaders** exceed this range, the overlapping that occurs with this method of distributing must be considered. The actual working widths of centrifugal spreaders that are obtained after taking into account the overlapping are usually not above this range.

For modern **pneumatic spreaders**, the working width can be above 25 m. However, with many of these spreaders, it is technically feasible to practice sectional control and thus to supply the right and left section with separate rates of fertilizer. Hence with this method too it would be possible to stick to the needed limits of cell side length.

Very high resolutions and thus small limits of cell side lengths can be realized with **sectional rate control** for the outlets or nozzles of either pneumatic spreaders or of sprayers. This method is state of the art for spot spraying of herbicides in fallow fields with dryland farming (Fig. 6.5). With proper reflectance indices and variable control devices, it could be used for site-specific nitrogen application as well and would allow for an upper limit of cell side length that is as low as 0.5 m.

Hence it can be concluded that principal technical limits for nitrogen application with resolutions that correspond to the needs of crops probably do not exist. The overriding question is: what resolution is needed in order to correspond to the heterogeneity of the crop? Most present day machinery for site-specific nitrogen application is designed for lateral cell side lengths that comply with the working widths of conventional centrifugal spreaders (Figs. 9.24 and 9.25).

The really sensed widths or areas usually account only for a small fraction of the fertilized width or area (*e.g.* Fig. 9.22). The inaccuracy that might result from this is partly offset by averaging of optical signals that are received in the direction of travel. The finally obtained **cell side lengths** in the direction of travel are much smaller than the working width in fertilizing.

So the presently realized resolution in distributing results in **cell-shapes** that correspond to rectangles whose long side is oriented perpendicular to the direction of travel. Theoretically it might be better to have squared cells. However, the present rectangular cells with long sides perpendicular to the direction of travel result from technical possibilities with conventional spreading machinery.

It must be mentioned that any results of heterogeneity sensing (Table 9.6) depend on the **sample areas** that are recorded. This is because of the averaging effect that automatically takes place when the sample area is increased. Thus the resolution that is used for recording the heterogeneity of crops greatly influences the obtained upper limit of cell side length (Solie et al. 1996). An interesting question to the

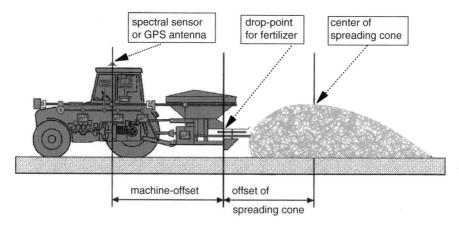

**Fig. 9.34** Offset distance between locations of signal reception and distribution of fertilizer, consisting of a machine-offset plus a spreading cone-offset (From Griepentrog and Persson 2001, altered)

method of determining the heterogeneity of small cereal crops is this: is the sample area of 1.5 m$^2$ that was used (Table 9.6, legend) adequate? This thinking reduces the question about the resolution in site-specific distributing to a problem of selecting the suitable sample area. And the latter will certainly depend on the crop species and on the nitrogen flux within a soil.

### 9.4.8 Distance- and Time Lag in Site-Specific Application

The signals for site-specific fertilizing normally are received by a process-controller that is located within the tractor-cabin. These signals are supplied either by the spectral sensor or – in case of mapping – by the GPS system that correlates the mapped data to the geographic location within the field. The fertilizer, however, is distributed some distance away from the tractor. This **offset distance** is composed of a machine offset plus an offset of the fertilizer position in the field that is due to the spreading (Fig. 9.34). Accurate site-specific fertilizer placement is not possible without compensating for this offset distance.

In addition to this offset distance, a **time-lag** in the control of site-specific spreading or spraying must be considered. This time-lag is the result of response-delays in the control adjustments for the spreading or spraying. With mounted centrifugal spreaders that feed the spinning discs by gravity, this time-lag is approximately 2 s. When the feeding of the spinning discs is done by conveyor belts, the time lag increases to about 3.8 s.

And finally with mounted pneumatic spreaders, the time-lag is between 4 and 5 s (Griepentrog and Persson 2001). In most cases, sprayers for liquid applications have rather low time-lags that are between 1 and 2 s (Bennur and Taylor 2010).

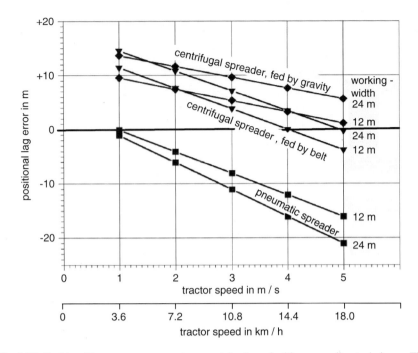

**Fig. 9.35** Positional lag error versus tractor speed for three fertilizer spreading techniques. The positional lag error is the difference between the offset distance and the time-lag distance (From Griepentrog and Persson 2001, altered)

For precise site-specific application, the distance that the machinery moves in the field within the time-lag is important. If this **time-lag distance** does not match the total offset-distance, an error in the site-specific placement of the fertilizer results. Any difference between the offset-distance and the time-lag distance here is called **positional lag error**, which can be positive or negative.

The aim is to have a positional lag error close to zero. In case the positional lag error is positive – *i.e.* the total offset-distance is larger than the time-lag distance – a correction can rather easily be obtained by purposely increasing the response time of the controller.

If the positional lag error is negative, the travel speed can be reduced in order to decrease the time-lag distance. However, this deteriorates the labor efficiency of field operations. So the question is, which positional lag errors occur with the presently dominating distributing techniques and their typical time lags.

As can be seen from Fig. 9.35, the mounted **centrifugal spreaders** have positional lag errors that are positive except for very high tractor speeds. Yet these positive lag errors can rather easily be removed by adjusting the control devices for longer response- or default times. Contrary to this, the mounted **pneumatic spreaders** have negative positional lag errors with the exception of very low tractor speeds. These errors can only be avoided by driving more slowly.

The positional lag error for **spraying** of liquid fertilizers (not shown in Fig. 9.35) corresponds approximately to the situation for centrifugal spreaders. This is because for sprayers the time-lags or response-times are on a low level as well and hence any positive positional lag errors that exist can be removed by adjusting the control devices for longer response times.

## *9.4.9 Sensed Signals and the Control of Nitrogen Application*

### 9.4.9.1 Agronomic Background

The site-specific signals that indicate the crop properties cannot be used directly for the control of the nitrogen application. These signals just provide information about the chlorophyll content of the leaves, the biomass or the leaf-area-index. The property-substitutes must be converted into fertilizer application rates.

In early **growing stages** of crops, the general consensus is that when crop properties indicate a low nitrogen supply, the application rate should be increased and *vice-versa*. With sensing by reflectance indices this means that when *e.g.* the red edge inflection point or the red edge ratio (Table 9.5) go up, the application rate should go down. This basic relation is rather easily implemented into the control-algorithm, however, there is much more needed. The site-specific indications that modern crop sensors supply do not at all spare the need for a detailed agronomic knowledge about the reasonable use of nitrogen fertilizer by taking into account

- the crop species and varieties
- the soil properties
- the growth stages
- the water supply
- the effects of rotations, crop-residues, manure and nitrogen-mineralization
- the use of the product (food, feed, fiber *etc.*)
- the costs of fertilizers and product prices.

The list is not complete and just should demonstrate that about every situation in a field is a unique one. In order to simplify the search for the best application rate, it has been proposed to use integrating factors. One such **integrating factor** is the final crop yield. Several of the items listed above affect the crop yield. And if this yield were known precisely at the time of fertilizing, taking this into account would simplify the control needs definitely.

The problem is that the final crop yield cannot be predicted precisely enough at the time of fertilizing. This is because the final crop yield is affected by the weather in the growth stages that still lie ahead. The uncertainties in the weather-forecasting beyond a few days do not allow a precise prediction of the expected final yield.

So in view of the abundance of factors that have to be considered, it might be asked how generally the rates for in-season nitrogen application are defined. The usual procedure in practical farming is that the farmer or his consultant inspects and

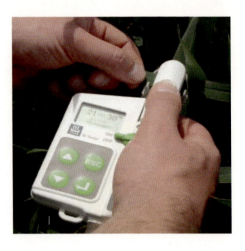

**Fig. 9.36** Measuring the transmittance of the youngest wheat leaf for precise calibrating of site-specific nitrogen application (From Agri Con GmbH, Jahna, Germany)

considers the local situation of the crop and makes a decision. Small handheld meters that can indicate the chlorophyll content of plant leaves at a few spots in the field (Fig. 9.36) and specific schemes of advisory services are often used to assist in defining the proper rate. Yet the decision still relies largely on the knowledge and experience of the farmer or consultant. And this decision is indispensable for a uniform application in the whole field as well as for a site-specific application. The difference is just that for a uniform application the decision is about the constant rate in the whole field, whereas for a site-specific application a minimum rate, a maximum rate – hence the limiting rates – and the control line between these must be defined.

### 9.4.9.2 Limiting Rates and Slope of the Control-Line

Entering the minimum- and maximum rate into the algorithm is easily done with modern process-controllers. For the control-line that extends between these **limiting rates**, its relation to the reflectance index must be defined. Presently, the dominating practice is to assume a simple linear relation between the points along the control-line and the reflectance index, *e.g.* the red edge inflection point or the red edge ratio. This simplifying assumption may not be far off, though there are indications that a still more precise definition of the relation between the nitrogen needed and the reflectance indices is possible (Holland and Schepers 2010).

The **control-line** stands for the relation between the reflectance index and the nitrogen rate (Fig. 9.37). As a linear line, its course is defined by the slope and an intercept. Its course can also be determined by the position of the minimum- and of the maximum rate in the graph that has the reflectance index and the nitrogen rate as ordinates.

Hence a logical method of defining the control line is to direct the field of view of the reflectance sensor on one spot in the field that needs only the minimum rate

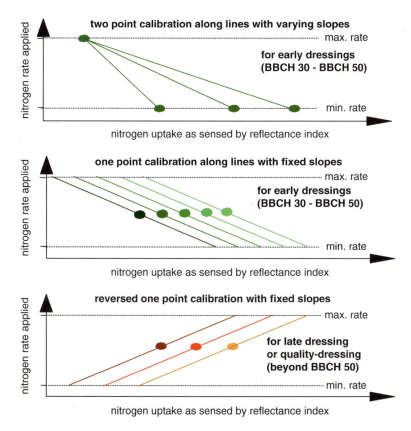

**Fig. 9.37** Principal control-algorithms for site-specific nitrogen application based on reflectance sensing and limiting rates. The course of the linear control lines is defined by the minimum rate, the maximum rate and by the slope of the lines. The developmental stages for small cereals are indicated in BBCH units. These are identical to EC units or Zadoks units (From Thiessen 2001, 2002, altered)

and then on another place for that the maximum rate seems adequate. At these two points, the respective reflectance indices of both limiting rates are obtained for the algorithm. The site-specific control should then logically take place along the line that connects these two extreme points (Fig. 9.37, top). A prerequisite for accuracy with this **two point calibration concept** is that the farmer carefully selects the best spots for the minimum- and maximum rate in the field.

In order to ease the calibration work for the farmer, a **one point calibration concept** has been introduced and now has become a standard method for main crops (Fig. 9.37, center). With this concept, a **default slope** of the linear control line is in the program of the algorithm. The default slope has been obtained from numerous field experiments with the respective main crop. This allows to calibrate the sensor by just recording the reflectance on one spot or on one short strip within the field and assigning an estimated nitrogen rate to it. Within this spot or short strip, the farmer can enhance his estimate about the nitrogen supply and hence the needed rate by transmittance measurements from the youngest leaf of some plants (Fig. 9.36). This too supplies information about the chlorophyll content.

With all these site-specific calibration methods, the farmer does not know well at the outset of the fertilizing operation how much nitrogen finally the whole field will get. Because this depends on a sum of many small applications that are not known in advance. Yet increasing the sample area that correlates reflectance and estimated nitrogen rate allows to correct this at least partially. So instead of a spot or short strip within the field, some farmers take a **full transect** in a typical region and relate its mean reflectance index to an estimated nitrogen rate. But also in this case, the control line is based on a one point calibration concept.

Whenever default slopes from past experiences are not available, the two point calibration method is indispensable. This method also is essential if more flexibility in the site-specific application is desired. In case knowledge about a locally different reaction of a crop on nitrogen fertilizer is available, this flexibility can be necessary. But how to get information about this local reaction to nitrogen?

The use of **nitrogen-rich strips** within fields aims at getting information about this. These strips are applied at the start of the growing season in one or several small areas of the field with the objective to test the effect of the nitrogen. The term "N-rich" indicates that the crop within this strip really has sufficient nitrogen. Since it is known how much nitrogen was applied, the comparison of the strips with the adjacent non-rich plants can inform about the slope that the control line should have. If the reaction to more nitrogen is small, the slope to the horizontal should also be small and *vice versa*.

But on what scale or resolution should this information from N-rich strips be applied? The use can be oriented at **field scales** and hence assist in getting the right slope of the control line for an individual field. Yet nitrogen-rich strips can also be applied in such a way that the slope of the control line is continuously adjusted while the tractor with the spreader or sprayer is moving through the field. The implementation of such a continuously adjusting system for the slope of the control line needs a thin enriched strip or **transect within each pass** of the fertilizing machine. By sensing the reflectance precisely along this narrow strip and referencing it to a standard, the signals for adjusting the slope of the control line in an on-the-go mode are obtained. Thus the N-rich concept is based on **site-specific cell scales**. The result is that the site-specific control of nitrogen application occurs in a combined **dual mode**. The control is based firstly on the nitrogen supply that the plants have, but secondly also on the site-specific reaction to more nitrogen. Both the nitrogen supply as well as the reaction to more nitrogen are sensed by reflectance. For details to this concept of dual mode site-specific sensing of the nitrogen supply as well as the response to nitrogen see Thiessen (2001, 2002).

However, any calibration method – whether used on a field scale or on a site-specific scale – represents a reaction that is based on the weather conditions of the past. But the application control aims at reactions in the future. This temporal difference implies errors that can result from varying weather and its effect on changing soil or crop conditions (Roberts et al. 2011).

Another concept of a site-specific adjustment for the slope of the control line would be to take into account the effect of different soils. It is well known that – *ceteris paribus* – the efficiency of nitrogen use also depends on the **texture of soils** (Tremblay et al. 2012). On sandy soils, the nitrogen use efficiency often is not as good as on soils with more clay or silt. And since a surrogate of texture – the soil conductivity – can rather easily be mapped, a feasible approach would be to orient the slope of the control line on site-specific field maps of this soil property. The result would be a site-specific **soil conductivity based slope** of the control line. This approach is not state of the art, but it would be rather easy to implement such an additional control mode.

### 9.4.9.3 Controls for Improving the Quality of Products

A completely different situation arises for small cereals when it comes to a **late dressings** that might be applied after heading, *i.e.* after BBCH stage 50. The objective of such late applications is not to increase the yield mass, which hardly is possible at this stage any more. Instead, the application of small amounts of nitrogen in this developmental period aims at raising the protein content of the kernels. Accordingly, this rather late application is denoted as "quality-dressing". However, this "quality-dressing" needs biologically active plants. Therefore, the more dark and green the canopy is, the more effective this late application can be, and the higher the nitrogen rate beyond BBCH 50 should be. This explains why, for quality-dressing, the basic controlling mode goes in the opposite direction: with rising red edge inflection point or red edge ratio index, the application rate is increased (Thiessen 2001, 2002; Reckleben and Isensee 2004). Hence, compared to earlier dressings, the slope of the control line is inverted (Fig. 9.37, bottom).

Proper quality-dressings of nitrogen for small cereals result in a higher protein contents, which can improve the value of the grain if it is used for bread or for feed. The advantage of applying this quality-dressing in a site-specific mode instead of a uniform way for the whole field is that this can provide for less **variation in the protein content** of the harvested grain. This holds especially for small cereals from fields with varying soil texture (Reckleben 2003).

Still another situation exists if the objective is to produce **malting barley** for beer breweries. For this product, the ideal crude protein concentration is not high, but rather low, namely about 10.7 % of the dry matter. Site-specific nitrogen application too can assist in getting close to this protein content. Any late dressings after heading can be left out completely, since high levels of protein content deteriorate the quality in this special case. The approach for keeping the protein content at this rather low level is reflectance sensing for the control of nitrogen application at the earliest possible growth stage, *i.e.* at about BBCH 30 – BBCH32 (Hopkins et al. 2007; Pettersson and Eckersten 2007; Söderström et al. 2010). The site-specific control provides the means for limiting the application and for evening out variations in the protein content within a field that – without using this technique – might develop as a result of differences in soil texture.

Similar concepts of improving the quality of plant products by correcting an uneven supply that is due to varying soil conditions via site-specific control of the nitrogen supply are conceivable for many crops, *e.g.* potatoes, sugar-beets and vegetables.

### 9.4.9.4 Control by a Sufficiency Index

This method for the control of site-specific nitrogen application via reflectance sensing was – starting with maize – developed by Holland and Schepers (2010). It is based on defining normalized reflectance indices by using a generalized plant growth function. The premise is that the plant growth function depends on the nitrogen supply. The **normalized reflectance indices** are given by a **sufficiency index** that represents the ratio of the locally sensed reflectance index to the same index that stands for plants of the same crop that have no nitrogen limitation. Hence:

$$\text{sufficiency index} = \frac{\text{locally sensed reflectance index of crop}}{\text{reflectance index of non}-N\text{ limited crop}}$$

For suitable reflectance indices that can define the sufficiency index see Table 9.5.

The corresponding **generalized plant growth function** is assumed to be a second order polynomial with downward cavity, thus a quadratic function of the nitrogen that was applied (Fig. 9.38). The basic concept is that the sufficiency index is a surrogate for the ability of the crop to grow, thus for its vigor, and that the nitrogen rate given should be related to it.

It is essential to differentiate between on the one hand the nitrogen supply of a crop in the past that provides the sufficiency index (Fig. 9.38) and on the other hand the nitrogen rate that should be applied. Both nitrogen parameters need to be detected on a site-specific basis. The first parameter, the nitrogen supply of the past, is directly represented by the sufficiency index. The second parameter, the nitrogen to be applied, must be derived from the sufficiency index.

Holland and Schepers (2010) provide a mathematical deduction of the nitrogen rate that should be applied. The result of this mathematical deduction is that the site-specific nitrogen rate to be applied can be described by a rather simple equation:

$$\text{Nitrogen rate in kg per ha to be applied, } N_{appl} = C\sqrt{\frac{1-\text{sufficiency index}}{\text{delta sufficiency index}}}$$

The term in the numerator, (1 – sufficiency index), is represented by the respective vertical distance in the green area of Fig. 9.38 for a given *ex ante* nitrogen supply. And the delta of the sufficiency index is the vertical difference between the intercept of the plant growth function and maximum of the latter. The term C stands for factors such as *e.g.* nitrogen from mineralization of soil organic matter and organic fertilizers or from previous legumes. This term can also take into account the fact that the agronomic optimal nitrogen rate in practice should be replaced by

# 9 Site-Specific Fertilizing

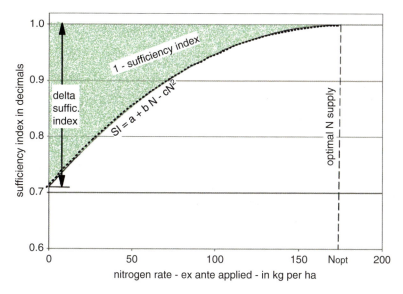

**Fig. 9.38** The sufficiency index SI as defined by the *ex ante* applied nitrogen rate via a generalized plant growth function (From Holland and Schepers 2010, altered)

an **economically optimal nitrogen rate**, which is a little lower because of the costs of the fertilizer. So an interaction by the farmer with his local situations and experiences is possible via the term C.

Compared to the use of a limiting rate method from the previous sections, the control by the sufficiency index does not simply follow a straight line but depends on the respective plant growth function. This can be regarded as an improvement, since the plant growth function defines the response of the crop to nitrogen probably better. The control via the sufficiency index does not exempt the farmer from deciding about minimum- and maximum rates. The difference is that with the control by the sufficiency index these rates just limit the application. The maximum- and minimum rate do not any more define the course of the control line in between as they do with the limiting rate method (Fig. 9.37).

There remains one problem for all algorithms and all present fertilizing strategies for site-specific control as well as for whole fields: they are based on informations that are related to the past of the crop. But what is needed is an adequate supply for some weeks in the future. This fact presents uncertainties since the weather and especially the water supply of the crop are not known *a priori*.

## 9.4.10 Interactions Between Nitrogen and Water

Water stress is one of the most common limitations in plant production. When the water supply is inadequate, nitrogen transport from the soil to the leaves is hindered.

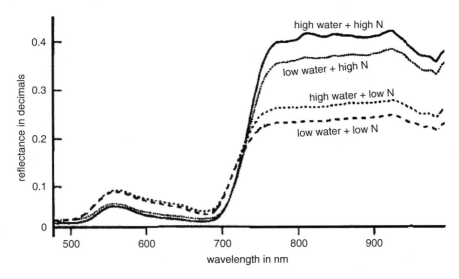

**Fig. 9.39** Spectral reflectance of pepper (Capsicum annuum) with four treatments concerning the nitrogen- and water supply (From Filella and Penuelas 1994, altered)

If the whole field is too dry, any application of nitrogen might be useless, except in cases, when a remedy by artificial irrigation or by natural rain soon can be anticipated. But this is reasoning about the water supply in the **future**.

Reasoning about the water supply of the crop in the **past** leads to other considerations. Since the signals of the past nitrogen supply of the crop have been derived from the plants appearance, it could be possible, that these signals were influenced by the previous water supply. This is indeed an important point.

Penuelas et al. (1996) have shown that a lack of water as well as an insufficient nitrogen supply have similar effects on the visual appearances of wheat. In both cases, the plants tend to have **xeromorphic characteristics**: the cell elasticity is reduced, the cell walls are thicker as well as more rigid and the cellulose content of the leaves is higher. But how are the effects on the reflectance?

Starting with the not **decomposed reflectance curves** of the visible and near-infrared region, the prospects of separating the effects of nitrogen from those resulting from the water supply do not look good. The reason for this is that the influences of nitrogen deficiency and lack of water have similar effects on the general course of the reflectance curves in the visible and near-infrared range. A rising water stress results in an increase of the visible- and in a decrease of the near-infrared reflectance. The same effect is caused by a lack of nitrogen (Fig. 9.39).

Fortunately, modern reflectance sensing generally has moved away from signals derived from the course of not decomposed curves, which extend over long ranges of wavelengths. Reflectance sensing for site-specific plant production has gone from red-green-blue (RGB) digital imaging to broad band recording and presently to hyperspectral narrow band signals. And further decomposing of the reflectance

**Table 9.7** Correction of nitrogen rate depending on the water situation

| Nitrogen supply | High | High | Low | Low |
|---|---|---|---|---|
| Water supply | High | Low | High | Low |
| Correction needed | No | No | No | **Yes** |

curves by using first or second derivatives, Fourier- or wavelet analyses should not be excluded (see Sect. 9.3.1).

The use of the red edge inflection point or of red edge ratios near this point for nitrogen sensing (Table 9.5) is an example for this trend of **targeted decomposing** of the reflectance curve. Several authors as *e.g.* Filella and Penuelas (1994), Yang and Su (2000) as well as Shiratsuchi et al. (2011) found that the red edge inflection points or red edge ratios were not much influenced by the water supply. Among the various spectral indices that were used for nitrogen sensing, the red edge inflection points and especially the red edge ratios were the least affected by crop water stress.

However, even if the red edge inflection points and particularly the red edge ratios are less influenced by water stress, this does not completely remove the problem. These indices allow only for a look at the supply of the crop in the past days or weeks. The actual sensing is directed at providing the nitrogen for coming weeks. And it can be taken for granted that when the crop cannot transpire because of a starting or ongoing drought, it will not be supplied with nitrogen. Hence the current supply of nitrogen as well as of water should be sensed separately and simultaneously by a **dual sensing strategy**. The prerequisites for doing this are good. Fortunately, the best wavelengths for sensing nitrogen on the one hand and water on the other hand are well separated within the spectrum (see Sects. 6.5 and 9.4.3.1).

How should the site-specific nitrogen rate be adapted to the water situation? Only the extreme cases can be outlined here. These extreme cases are described by respectively high or low supply of either nitrogen or water (Table 9.7).

In places of high nitrogen supply, the crop needs no fertilizer; therefore the water situation is of no avail for the fertilizing strategy. This also applies, when the sensing signals indicate a low nitrogen supply, while the water situation is good. The alert situation is, when both the nitrogen as well as the water supply are signaled as low. Without any information about the water situation, a high site-specific nitrogen dose would be applied. The dual sensing of nitrogen and water would prevent wasteful and perhaps environmentally harmful nitrogen application in this case. However, this concept of dual sensing is not yet state of the art.

Since it probably can be assumed that the water supply of crops varies temporarily as much as the nitrogen supply does and hence calls for short-termed control reactions, the best concept would be tractor based **on-the-go dual sensing**. An alternative would be a dual control via a combination of mapping the water supply of the crop by signals from satellites and tractor based on-the-go nitrogen sensing. This could be a reasonable solution provided the transmission of the radiation is neither impeded by the atmosphere nor by clouds and the frequency of remote recording suffices. Unfortunately, the prime candidate for water sensing, namely infrared radiation, does not get through clouds. Hence the concept of remote sensing based on infrared reflectance would be risky for regions with maritime climate.

The restrictions that result from clouds do not exist for remote sensing via **radar**. However, remote water sensing of soils or crops via radar still is in an experimental state. This applies especially to water sensing in crops, which would be preferable, because it automatically takes into account any interaction between soils and roots on the supply of plants.

## 9.4.11 Benefits, Costs and Economics

### 9.4.11.1 Benefits

The potential benefits from site-specific nitrogen application can vary between

- higher yields in masses of product per ha
- savings for nitrogen fertilizer
- better conditions for harvesting the crop
- better qualities of the harvested product
- less leaching of nitrogen into ground-waters.

Which of these possible benefits dominates, might depend on the crop, the variations in soil properties and weather, the control algorithm as well as on the time within the growing season when the site-specific technique is used. Hence for comparisons of uniform- versus site-specific nitrogen applications, a variety of results might evolve.

Higher **yields** in masses of product per ha can be expected if the average nitrogen rate for site-specific application is about the same – and hence not lower – than with uniform fertilizing. In field experiments with site-specific and uniform application, this is not easy to obtain, since the average rate with site-specific fertilizing is not known precisely beforehand. Another prerequisite for higher yields is site-specific application within a rather early vegetation stage, *e.g.* EC or BBCH 30–40 with small grains.

The comparisons of site-specific and uniform applications in Fig. 9.40 are based on approximately the same average nitrogen rate and early dressings of winter-wheat. The site-specific applications were controlled by reflectance sensing. It can be seen that for the location Raguhn the yields hardly were influenced. The reasons for this probably are uniform soil properties or low rainfall in the growing season. The other 16 locations resulted in a **mean yield increase of 4.1 %**. However, the spread around this average result is remarkable and presumably due to the many factors that affect the yields of crops.

**Savings for nitrogen fertilizers** can be based on the area that is treated or on the product that is harvested. The latter approach is probably preferable, because the product is the objective. This leads to defining the efficiency of mineral N use, *e.g.* in kg of grain per kg of N applied.

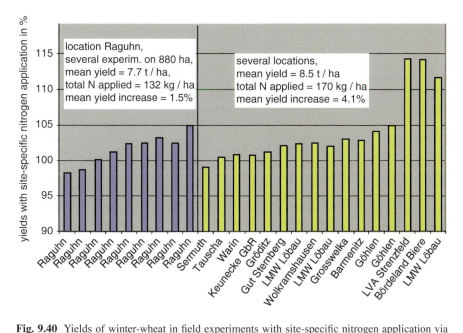

**Fig. 9.40** Yields of winter-wheat in field experiments with site-specific nitrogen application via the Kiel system of reflectance sensing versus uniform nitrogen fertilization. The results of uniform nitrogen application are set to 100 %. The average nitrogen rates within each field with site-specific and with uniform application are about the same. The names below the columns represent the locations of the fields in Germany (From Werner et al. 2004, altered)

Higher yields that are obtained with the same level of total nitrogen application (Fig. 9.40) indicate an improvement of mineral N use. However, it should be realized that the **efficiency of N use** depends on the yield level. This is because of the law of the diminishing returns (Mitscherlich 1922; Holland and Schepers 2010). Hence a precise indication of the efficiency of N use can only be provided from experiments in which the yields from uniform N fertilizing on the one hand and from site-specific application on the other hand do not differ significantly. Such results can be expected from experiments in which the N rate with site-specific application is lower than with uniform fertilizing. The experiments that are listed in Table 9.8 comply with this.

The site-specific control is either based on topography (Griepentrog and Kyhn 2000), on reflectance sensing (Havrankova et al. 2008) or on the resistance of the canopy against bending (Thoele and Ehlert 2010). It can be seen that all site-specific methods improve the efficiency of nitrogen use (Table 9.8). Especially the control based on topography clearly enhances the nitrogen use. This suggests that in hilly fields a combined control both by a surrogate for the nitrogen supply and in addition via mapped topography might make sense.

A key question in this respect is whether site-specific nitrogen control should aim at **higher yields** in masses of product per ha or at **lower expenditures** for

**Table 9.8** Efficiency of mineral nitrogen use expressed in kg grain per kg N

| Crop | Mean yield in t/ha | Total mineral N used in kg per ha | | Efficiency of min. N use in kg grain per kg N | |
|---|---|---|---|---|---|
| | | Unif. rate | Var. rate | Unif. rate | Var. rate |
| Griepentrog and Kyhn (2000), *experiments in Schleswig-Holstein, Germany* | | | | | |
| W. wheat | 11.0 | 201 | 149 | 54 | 74 |
| W. barley | 9.5 | 191 | 147 | 50 | 64 |
| W. wheat | 10.8 | 206 | 131 | 53 | 81 |
| Havrankova et al. (2008), *experiments in Bedfordshire, United Kingdom* | | | | | |
| W. wheat | 7.5 | 221 | 205 | 34 | 37 |
| W. wheat | 7.3 | 221 | 205 | 33 | 36 |
| Thoele and Ehlert (2010), *mean of 13 experiments in Eastern Germany* | | | | | |
| W. cereals | 6.2 | 156 | 138 | 43 | 49 |
| **Average efficiency of mineral N use** | | | | **44.5** | **56.8** |

The results are based on experiments in which the yield level with both site-specific and uniform application of nitrogen is approximately the same. However, the site-specific variable rate application is limited to the second dressing and does not apply to any first dressing

nitrogen fertilizer. Thriwakala et al. (1998) dealt with this question on the basis of a simulation model for whole fields with varying fertility. The latter was defined on the basis of the *a priori* given nitrogen supply, *e.g.* from mineralization of organic matter. The result was that for whole fields with a high mean fertility the site-specific application control should be programmed for higher yield averages. And for fields with a low fertility it should be targeted for lower mean fertilizer expenditures. The problem is that the fertility as defined above does not only depend on soil properties but on weather as well.

The **efficiency of nitrogen use** is a very important criterion for the environmental impact of plant production. This is not only because inefficient nitrogen use causes higher fertilizer costs per unit of the product. In humid areas, inefficient nitrogen use by crops also results in more nitrate that seeps into groundwaters (Schepers et al. 1995; Schepers 2008). This side-effect of nitrogen fertilization has generated the most serious environmental problem that is caused by agrochemicals, namely too high nitrate contents in drinking waters. In the human digestive system, most nitrate is reduced to nitrite. This holds especially for babies. The nitrite in turn lowers the ability of the red blood cells to carry oxygen. So a serious disease due to the lack of oxygen may develop, called "blue baby syndrome" or methemoglobinemia. As a safeguard, government directives have set **limits for drinking water** in large parts of the world. For many countries, the limit is 50 mg of $NO_3^-$ ions per one liter of water, which too meets the proposal by the World Health Organization of the United Nations (WHO 2006). The USA set the limit at 10 mg nitrate N per one liter of water (EPA 2011). This corresponds to 45 mg of $NO_3^-$ ions per liter water. So there is rather good agreement in the objective.

However, the actual situation is that despite small recent improvements, these limits quite often are exceeded (Nolan and Stoner 2000; Sprague et al. 2009,

Umweltbundesamt 2011; Wolter 2004). In agricultural areas, the main cause for this is fertilization. Reducing this by generally accepting lower yields is no worldwide solution of the problem when taking into consideration that still about 800 million people are starving. Instead, technologies that provide a high efficiency of nitrogen use should be aimed at. Because nitrogen that is taken up by a crop is prevented from leaching into groundwaters. Hence more precision in application is needed. Important means for improving the efficiency of nitrogen use are site-specific applications (Table 9.8) as well as temporally split dressings. In other words, the application should be at the right place, in the right time and with the right rate. The thus obtained higher efficiency in nitrogen use results in two-sided advantages: both higher yields or lower nitrogen expenses and in addition better qualities of water resources.

For small grains, also the effect of site-specific nitrogen application on the **conditions for harvesting** the crop should be considered. Nitrogen application for high yielding small grains can result in crop canopies that at harvest time are **lodging** partly. This nuisance for harvesting is mainly caused by disregarding site-specific differences in soil qualities, hence by a local oversupply with nitrogen. Site-specific nitrogen application is oriented at local differences in the condition of the crop, therefore can remove this local oversupply and thus can homogenize the canopy. If the site-specific control is well adjusted, the lodging problem is gone.

However, there is a homogenizing effect via site-specific fertilizing also if no lodging occurs. This is due to the fact that this technique provides more **even shoots**. At harvest time of small grains, this means for shoots of a low order that the ears are located higher from the soil level. Hence harvesting by a combine is facilitated even if no grain lodges (Feiffer et al. 2005).

For combines there exists a relationship between throughput in tons of harvested product per time unit on the one hand and losses in % of the grain on the other hand. Generally it holds that when – *ceteris paribus* – the throughput or the travel speed is increased, the losses get higher and this in a more than proportional manner. Farmers try to keep the losses at a level of approximately 1 % of the small grain harvest and adjust the travel speed accordingly. Under this premise, the site-specifically fertilized and hence more uniform crop allows for a higher throughput. In extensive trials by Feiffer et al. (2005) in Eastern Germany, the field capacity of combines in ha/h for strips of small grains that had been homogenized by site-specific nitrogen application was on the average about 10 % higher than for strips that had been fertilized uniformly. With average costs for combine operations including the driver of 100 euros per ha, this means a saving of 10 euros per ha.

Another potential benefit from site-specific nitrogen application can result from better qualities of the harvested product, especially a more precise control of its **protein content** (see Sect. 9.4.9.3). Perspectives in this direction might exist for many crops, not only for small grains. However, much more knowledge about control algorithms for site-specific quality control is needed.

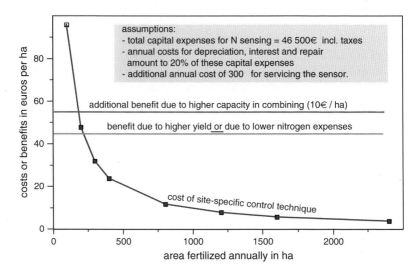

**Fig. 9.41** Costs of the control technique *versus* benefits of site-specific nitrogen application for winter-wheat. It is assumed that the site-specific application – depending on the control setting – provides either for a yield increase or alternatively for a saving of nitrogen expenses. With present market prices for grain or for nitrogen fertilizer, both alternatives amount to approximately the same financial benefit, hence both are represented by the *green line*. The homogenizing effect of site-specific nitrogen application on the crop results in a higher capacity of the combine. This benefit amounts to 10 euros per ha (*brown line*). For details see Sect. 9.4.11.1

### 9.4.11.2 Comparing Benefits and Costs

The costs of site-specific in-season nitrogen application result primarily from the capital expenditures for the equipment that is needed for the control of the spreading machine. These expenditures amount to about 46,500 euros for an implement that is designed for reflectance sensing by means of artificial light (Fig. 9.25) including operational terminal, software, a handheld transmittance meter to assist the calibrating procedure and taxes. If natural- instead of artificial light is used, the total expenditure for the corresponding items is about 31,000 euros. However, the annual area that can be treated with the latter outfit might be smaller too because of limitations in operational time.

The control costs per ha that result from the equipment depend heavily on the annual area that is treated (Fig. 9.41).

As for benefits that may result from higher **yields**, an average increase of 4 % is assumed (Fig. 9.40, right). With a yield level of 8.5 t/ha, the corresponding absolute increase of 0.34 t/ha represents a benefit of approximately 45 euros per ha, assuming that the unprocessed small grains in the field are valued at 130 euros per t.

Alternatively, the site-specific control can be used in such a way that no yield increase, but instead a **saving of nitrogen** fertilizer results. From the average efficiencies in Table 9.8, bottom line, it can be calculated that for the same yield of

8.5 t/ha, the saving would amount to 44 kg of N per ha. Since the expenses per kg N for fertilizers – with the exception of anhydrous ammonium – are around 1 euro, the monetary savings would add up to 44 euros per ha. Hence the financial value of the eventual yield increase on the one hand or of the possible saving in nitrogen fertilizer instead of this on the other hand hardly differs and therefore is represented in Fig. 9.41 by a common green line.

The benefits via higher yields or via lower expenses for nitrogen plus those via the higher **capacity of the combine** cross the cost curve for the control technique for an annually fertilized area of about 175 ha. So this area should be exceeded for an economic investment. Since a fertilizer spreader or sprayer with a width of 20 m and a speed of 10 km/h theoretically covers 20 ha per hour, this area might even be handled per day. With large farms or by means of contract management, annual areas of 1,000 ha or more can easily be treated per machine set. Under these circumstances, the financial advantage of site-specific application as represented by the difference between the sum of the benefits and the cost of the control technique (Fig. 9.41) can go up to 40–50 euros per ha. This explains why in practice site-specific nitrogen application has become a leading technique in precision farming.

In addition, there may be benefits from better **qualities** of the harvested products and from less **leaching of nitrogen** into ground-waters. Future experience will show how these benefits should be evaluated. And more knowledge about the best control algorithms will help to fully exploit the possibilities with various crops.

## 9.5 Summary

The prospects for the realization of **site-specific fertilization** concepts differ substantially among nutrients:

- for phosphorus, the application according to a mapped removal by previous crops is promising. An alternative to this might be the control of fertilizing via flat surface-soil-sensing by visible and near-infrared reflectance.
- as for potassium, too fertilizing proportional to the mapped removal by previous crops has good prospects. However, contrary to phosphorus this holds not for sandy soils.
- concerning nitrogen, excellent prospects exist for sensing the supply and hence the need during distributing via the reflectance of the crop. This limits the application to the respective growing period. Yet this temporal limitation is inevitable as long as efficient nitrogen fertilizing necessitates "immediate feeding" of the crop.
- for the control of soil pH via calcium or magnesium, sensing and mapping the situation by means of ion-selective-electrodes is a promising technique.

With many applications, a site-specific control that takes into account several soil- or crop properties simultaneously might be worth to aim at in the future.

# References

Adamchuk VI (2008) On-the-go mapping of soil pH using antimony electrodes. ASABE annual international meeting, Providence, June 29–July 2, 2008. Paper No. 083995

Adamchuk VI, Lund ED, Sethuramasamyraja B, Morgan MT, Dobermann A, Marx DB (2005) Direct measurement of soil chemical properties on-the-go using ion-selective electrodes. Comput Electron Agric 48:272–294

Adamchuk VI, Lund ED, Reed TM, Ferguson RB (2007) Evaluation of an on-the-go technology for soil pH mapping. Precis Agric 8:139–149

Baille A (1993) Artificial light sources for crop production. In: Varlet-Grancher C et al (eds) Crop structure and light microclimate. Institut National de la Recherche Agronomique, Paris, pp 107–120

Bausch WC, Diker K (2001) Innovative remote sensing techniques to increase nitrogen use efficiency of corn. Commun Soil Sci Plant Anal 32(7 and 8):1371–1390

Belanger MC, Viau AA, Samson G, Chamberland M (2005) Determination of a multivariate indicator of nitrogen imbalance (MINI) in potato using reflectance and fluorescence spectroscopy. Agron J 97:1515–1523

Bennur PJ, Taylor RK (2010) Evaluating the response time of a rate controller used with a sensor-based variable rate application system. Appl Eng Agric 26(6):1069–1075

Bogrekci J, Lee WS (2005) Spectral measurement of common soil phosphates. Trans Am Soc Agric Eng 48(6):2371–2378

Brandt A (2011) Noise and vibration analysis. Signal analysis and experimental procedures. Wiley, Chichester

Bredemeier C, Schmidthalter U (2005) Laser-induced chlorophyll fluorescence sensing to determine biomass and nitrogen uptake of winter wheat under controlled environment and field condition. In: Stafford JV (ed) Proceedings of the 5th European conference on precision farming. Wageningen Academic Publishers, Wageningen, pp 273–280

Chang CW, Laird DA, Mausbach MJ, Hurburgh CR (2001) Near-infrared reflectance spectroscopy – principal components regression analyses of soil properties. Soil Sci Soc Am J 65:480–490

Dammer KH, Wollny J, Giebel A (2008) Estimation of leaf-area-index in cereal crops for variable rate fungicide spraying. Euro J Agron 28:351–360

Dash J, Curran PJ (2007) Evaluation of the MERIS terrestrial chlorophyll index. Adv Space Res 39:100–104

Ehlert D, Hammen V, Adamek R (2003) Online sensor pendulum-meter for determination of plant mass. Precis Agric 4:139–148

Ehlert D, Dammer KH, Völker U (2004a) Application according to plant mass. Landtechnik 59(2):76–77 (in German)

Ehlert D, Schmerler J, Voelker U (2004b) Variable nitrogen fertilization in winter wheat based on a crop density sensor. Precis Agric 5(3):263–273

EPA (2011) Drinking water contaminants. United States Environmental Protection Agency. Update 20 Sept 2011 www.epa.gov/drink/contaminants/#List

Erdle K, Mistele B, Schmidhalter U (2011) Comparison of active and passive spectral sensors in discriminating biomass parameters and nitrogen status in wheat cultivars. Field Crops Res 124:74–84

Feiffer A, Kutschenreiter W, Feiffer P, Rademacher T (2005) Harvest of small grains – clean, safe, fast. A guide about combining. DLG Verlag, Frankfurt, p 126 (in German)

Felton WL, McCloy R (1992) Spot spraying. Agric Eng 73:9–12

Filella I, Penuelas J (1994) The red edge position and shape as indicators of plant chlorophyll content, biomass and hydric status. Int J Remote Sens 15(7):1459–1470

Finck A (1991) Fertilization. Verlag Ulmer, Stuttgart (in German)

Ge Y, Thomasson JA (2006) Wavelet incorporated spectral analysis for soil property determination. Trans ASABE 49(4):1193–1201

Griepentrog HW, Kyhn M (2000) Strategies for site-specific fertilization in a highly productive agricultural region. In: University of Minnesota, Precision Agriculture Center (ed) Proceedings of the 5th international conference on precision agriculture, Minneapolis, July 2000

Griepentrog HW, Persson K (2001) A model to determine the positional lag for fertilizer spreaders. In: Grenier G, Blackmore S (eds) Proceedings of the 3rd European conference on precision agriculture, Montpellier. Agro Montpellier, Ecole Nationale Superieure Agronomique, pp 671–676

Günther KP, Dahn HG, Lüdeker W (1999) Laser-induced-fluorescence, a new method for "precision farming". In: Bill R et al. (eds) Sensorsysteme in precision farming. Workshop, University of Rostock, 27–28 Sept 1999, Rostock. Institut für Geodäsie und Geoinformatik, pp 133–144

Guyot G (1998) Physics of the environment and climate. Wiley, New York, p 49

Guyot G, Baret F, Major DJ (1988) High spectral resolution: determination of spectral shifts between the red and infrared. Int Arch Photogramm Remote Sens 11:750–760

Havrankova J, Godwin RJ, Rataj V, Wood GA (2008) Benefits from applications of ground based sensing systems in winter wheat nitrogen management in Europe. In: 2008 ASABE annual international meeting, Providence, Paper no. 083560

Haykins S, Van Veen B (2003) Signals and systems, 2nd edn. Wiley, New York

He Y, Huang M, Garcia A, Hernandez A, Song H (2007) Prediction of soil macronutrient content using near-infrared spectroscopy. Comput Electron Agric 58:144–153

Heege HJ, Reusch S (1996) Sensor for on-the-go control of site-specific nitrogen top dressing. In: International meeting in Phoenix. American Society of Agric Engineering, St. Joseph, Paper No. 961018

Heege HJ, Thiessen E (2002) On-the-go sensing for site-specific nitrogen top dressing. In: ASAE international meeting/CIGR XVth world congress, Chicago. ASAE, St. Joseph, Paper No. 021113

Heege HJ, Reusch S, Thiessen E (2008) Prospects and results for optical systems for site-specific on-the-go control of nitrogen top dressing in Germany. Precis Agric 9:115–131

Hinzman LD, Bauer ME, Daughtry CST (1986) Effects of nitrogen fertilization on growth and reflectance characteristics of winter wheat. Remote Sens Environ 19:47–61

Holland KH, Schepers JS (2010) Derivation of a variable rate nitrogen application model for in-season fertilization of corn. Agron J 102(5):1415–1419

Hopkins BG, Stephens SC, Shiffler AK (2007) Optical sensing for nitrogen management. In: Western nutrient management conference, Salt Lake City, 2007, vol 7, pp 98–105

Huete AR (1988) A soil adjusted vegetation index. Remote Sens Environ 25:295–309

Inoue Y, Sakaiya E, Zhu Y, Takahashi W (2012) Diagnostic mapping of canopy nitrogen content in rice based on hyperspectral measurements. Remote Sens Environ 126:210–221

International Fertilizer Association (2007) In: Wichmann W (ed) World fertilizer use manual, Paris, France. http://www.fertilizer.org/ifa/HomePage/LIBRARY/Our-selection2/World-Fertilizer-Use-Manual

Jahn BR, Upadhyaya SK (2006) Development of mid-infrared-based calibration equations for predicting soil nitrate, phosphate and organic matter concentrations. In: ASABE annual international meeting, Portland, 9–12 July 2006, Paper No. 061058

Jahn BR, Linker R, Upadhyaya SK, Shaviv A, Slaughter DC, Shmulevich I (2006) Mid-infrared spectroscopic determination of soil nitrate content. Biosyst Eng 94(4):505–515

Jasper J, Reusch S, Link A (2009) Active sensing of the N status of wheat using optimized wavelength combination: impact of seed rate, variety and growth stage. In: van Henten EJ, Goense D, Lockhorst C (eds) Precision agriculture '09. Wageningen Academic Publishers, Wageningen, pp 23–30

Jongschaap REE (2001) Integrating remote sensing information in dynamic simulation models: sensing nitrogen status in a potato crop. In: Grenier G, Blackmore S (eds) Proceedings of the 3rd European conference on precision agriculture. Agro, Ecole Nationale Superior Agronomique, Montpellier, pp 923–927

Kappen L, Hammler A, Schultz G (1998) Seasonal changes in the photosynthetic capacity of winter rape plants under different nitrogen regimes in the field. J Agron Crop Sci 181:179–187

Kim HJ (2006) Ion-selective electrodes for simultaneous real-time analysis of soil macronutrients. Ph.D. thesis, University of Missouri, Columbia

Kim HJ, Hummel JW, Birell SJ (2006) Evaluation of nitrate and potassium ion-selective membranes for soil macronutrient sensing. Trans ASABE 49(3):597–606

Kim HJ, Hummel JW, Sudduth KA, Birell SJ (2007a) Evaluation of phosphate ion-selective membranes and cobalt-based electrodes for soil nutrient sensing. Trans ASABE 50(2):415–425

Kim HJ, Hummel JW, Sudduth KA, Motavalli PP (2007b) Simultaneous analysis of soil macronutrients using ion-selective electrodes. Soil Sci Soc Am J 71(6):1867–1877

Kim HJ, Sudduth KA, Hummel JW (2009) Soil macronutrient sensing for precision agriculture. J Environ Monit 11:1810–1824

Lamb DW, Steyn-Ross M, Schaare P, Hanna MM, Silvester W, Steyn-Ross A (2002) Estimating leaf nitrogen concentration in ryegrass (Loliun spp.) using the chlorophyll red edge: theoretical modelling and experimental observations. Int J Remote Sens 23(18):3619–3648

Lawrence KC, Bosoon P, Heitschmidt G, Windham WR (2005) LED lightning for use in multispectral and hyperspectral imaging. ASAE, St. Joseph, Paper No. 053073

Lee KS, Lee DH, Sudduth KA, Chung SO, Kitchen NR, Drummond ST (2009) Wavelength identification and diffuse reflectance estimation for surface and profile soil properties. Trans ASABE 52(3):683–695

Lichtenthaler HK (1996) Vegetation stress: an introduction to the stress concept in plants. J Plant Physiol 148:4–14

Lund ED, Adamchuk VI, Collings KL, Drummond PE, Christy CD (2005) Development of soil pH and lime requirement maps using on-the-go soil sensors. In: Stafford J (ed) Precision agriculture '05. Wageningen Academic Publishers, Wageningen, pp 457–464

Maleki MR, Van Holm L, Ramon H, Merkx R, De Baerdemaeker J, Mouazen AM (2006) Phosphorus sensing for fresh soils using visible and near infrared spectrometry. Biosyst Eng 95(3):425–436

Maleki MR, Mouazen AM, De Ketelaere B, Ramon H, De Baerdemaeker J (2008a) On-the-go variable-rate phosphorus fertilisation based on a visible and near-infrared soil sensor. Biosyst Eng 99:35–46

Maleki MR, Ramon H, De Baerdemaker J, Mouazen AM (2008b) A study of the time response of a soil-based variable rate granular fertiliser applicator. Biosyst Eng 100:160–166

Marschner H (2008) Mineral nutrition of higher plants, 2nd edn. Academic Press, Amsterdam

Martens H, Naes T (1992) Multivariate calibration. Wiley, Chichester

Meier U (2001) Growth stages of mono – and dicoledoneous plants, 2nd edn. BBCH Monograph, Federal Biological Research Centre for Agriculture and Forestry, Braunschweig, http://www.bba.de/veroeff/bbch/bbcheng.pdf

Mitscherlich EA (1922) The law of the diminishing returns. Z Pflanzenernähr Dung 1(2):49–84 (in German)

Mouazen AM, De Baerdemaker J, Ramon H (2006) Effect of wavelength range on the measurement accuracy of some selected soil properties using VISNIR spectroscopy. J Near Infrared Spec 14(3):189–199

Mouazen AM, Maleki MR, Cockx L, Van Meirvenne M, Van Holm LHJ, Merckx R, De Baerdemaeker J, Ramon H (2009) Optimum three-point linkage set up for improving the quality of soil spectra and the accuracy of soil phosphorus measured using an online visible and near infrared sensor. Soil Till Res 103:144–152

Mulla DJ, McBratney AB (2000) Soil spatial variability. In: Summer ME (ed) Handbook of soil science. CRC Press, Boca Raton, pp A321–A352

Müller K, Böttcher U, Meyer-Schatz F, Kage H (2008) Analysis of vegetation indices derived from hyperspectral reflection measurements for estimating crop canopy parameters of oilseed rape (Brassica napus L.). Biosyst Eng 101:172–182

Nolan B, Stoner J (2000) Nutrients in groundwaters of the conterminous United States. Environ Sci Technol 34(7):1156–1163

Osmond DL, Kang J (2008) Soil facts. Nutrient removal by crops in North Carolina. North Carolina State University. Cooperative Extension Service. AG-439-16W

Penuelas J, Filella I, Serrano L (1996) Cell wall elasticity and water index (R970nm/R900nm) in wheat under different nitrogen availabilities. Int J Remote Sens 17(2):373–382

Pettersson CG, Eckersten H (2007) Prediction of grain protein in spring malting barley grown in northern Europe. Eur J Agron 27:205–214

Potash Development Association (2006) Principles of potash use. Leaflet 8. The Potash Development Association, Laugharne. www.pda.org.uk

Reckleben Y (2003) Differences in yield and protein content with site-specific treatments for wheat. Landtechnik 58:252–253 (in German)

Reckleben Y, Isensee E (2004) Influences on protein content and yield with small cereals. Landtechnik 59:242–243 (in German)

Reusch S (1997) Development of an optical reflectance sensor for recording the nitrogen supply of agricultural crops. Doctoral dissertation, University of Kiel, Kiel, Forschungsbericht Agrartechnik der Max-Eyth-Gesellschaft Agrartechnik im VDI 303 (in German)

Reusch S (2003) Optimization of oblique-view remote measurement of crop N-uptake under changing irradiance conditions. In: Stafford J, Werner A (eds) Precision agriculture. Papers from the 4th European conference on precision agriculture, Berlin, 16–18 June 2003. Wageningen Academic Publishers, Wageningen

Reusch S (2005) Optimum waveband selection for determining the nitrogen uptake in winter wheat by active remote sensing. In: (a) Stafford J (ed) Precision agriculture '05. Wageningen Academic Publishers, Wageningen, pp 261–266 (b) extended paper from Precision agriculture '05, the 5th European conference of precision agriculture, held in Uppsala, 2005

Reusch S (2009) Use of ultrasonic transducers for online biomass estimation in winter-wheat. In: van Henten EJ, Goense D, Lockhorst C (eds) Precision agriculture '09. Wageningen Academic Publishers, Wageningen, pp 169–175

Reusch S, Jasper J, Link A (2004) Advanced concept for research on variable rate nitrogen application based on remote sensing. In: Mulla DJ (ed) Proceedings of the 7th international conference on precision agriculture and other resources management, University of Minnesota, St. Paul, 25–27 July 2004

Reusch S, Jasper J, Link A (2010) Estimating crop biomass and nitrogen uptake using CROPSPEC™, a newly developed active crop canopy sensor. In: Khosla R (ed) 10th international conference on precision agriculture, Denver, 18 July 2010

Roberts DC, Wade Brorsen B, Taylor RD, Solie JB, Raun WR (2011) Replicability of nitrogen recommendations from ramped calibration strips in winter wheat. Precis Agric 12:653–665

Savitzky A, Golay MJE (1964) Smoothing and differentiation of data by simplified least squares procedures. Anal Chem 36(8):1627–1639

Schächtl J, Huber G, Maidl FX, Sticksel E, Schulz J, Haschberger P (2005) Laser-induced chlorophyll fluorescence measurements for detecting the nitrogen status of wheat (*Triticum aestivum* l.) canopies. Precis Agric 6:143–156

Schepers JS (2008) Potential of precision agriculture to protect water bodies from negative impacts of agriculture. Landbauforschung – vTI Agric Forest Res 58(3):199–206

Schepers JS, Varvel GE, Watts DG (1995) Nitrogen and water management strategies to reduce nitrate leaching under irrigated maize. J Contam Hydrol 20:227–239

Schepers JS, Hagopian DS, Varvel GE (1998) Monitoring crop stresses. In: Illinois fertilizer conference proceedings, 26–28 Jan 1998, report 16:3. http://frec.cropsci.uiuc.edu/1998/report16/index.htm

Schmid A, Maidl F-X (2005) Optimizing site-specific crop management by contact-free sensing of the heterogeneity of the canopy. In: IKB Dürnast, IKB-Teilprojekt 9, Abschluss – symposium, TU München-Weihenstephan: 23 (in German). http://ikb.weihenstephan.de/ikb2/deutsch/symposium/pdf/schmid.pdf

Sethuramasamyraja B, Adamchuk VI, Marx DB, Dobermann A, Meyer GE, Jones DD (2007) Analysis of an ion-selective electrode based methodology for integrated on-the-go mapping of soil pH, potassium, and nitrate contents. Trans ASABE 50(6):1927–1935

Shanandeh H, Wright AL, Hons FM (2011) Use of soil nitrogen parameters and texture for spatially-variable nitrogen fertilization. Precis Agric 12:146–163

Shiratsuchi L, Ferguson R, Shanahan J, Adamchuk V, Rundquist D, Marx D, Slater G (2011) Water and nitrogen effects on active canopy sensor vegetation indices. Agron J 103:1815–1826

Sibley KJ (2008) Development and use of an automated on-the-go soil nitrate mapping system. Ph.D. thesis, Wageningen University, Wageningen

Sibley KJ, Adsett JF, Struik PC (2008) An on-the-go soil sampler for an automated soil nitrate mapping system. Trans ASABE 51(6):1894–1904

Sibley KJ, Astatkie T, Brewster G, Struik PC, Adsett JF, Pruski K (2009) Field scale validation of an automated soil nitrate extraction and measurement system. Precis Agric 10:162–174

Sibley KJ, Brewster GR, Astatkie T, Adsett JF, Struik PC (2010) In-field measurement of soil nitrate using an ion-selective electrode. In: Sharma MK (ed) Advances in measurement systems. InTechOpen, pp 1–27, InTech Europe University Campus STeP RiSlavka Krautzeka 83/A51000 Rijeka, Croatia. http://www.intechopen.com/books/advances-in-measurement-systems

Söderström M, Börjesson T, Pettersson CG, Nissen K, Hagner O (2010) Prediction of protein content in malting barley using proximal and remote sensing. Precis Agric 11:587–599

Soil Science Society of America (2008) Measuring nutrient removal, calculating nutrient budgets. In: Logsdon S, Clay D, Moore D, Tsegaye T (eds) Soil science: step-by-step field analysis, Madison, WIS, USA. www.soils.org

Solari F, Shanahan J, Ferguson R, Schepers J, Gitelson A (2008) Active sensor reflectance measurements of corn nitrogen status and yield potential. Agron J 100(3):571–579

Solie JB, Raun WR, Whitney RW, Stone ML, Ringer JD (1996) Optical sensor based field element size and sensing strategy for nitrogen application. Trans Am Soc Agric Eng 39(6):1983–1992

Sprague LA, Mueller DK, Schwarz GE, Lorenz DL (2009) Nutrient trends in streams and rivers of the United States, 1993–2003: U.S. geological survey. Scientific investigations report 2008–5202. U.S. Geological Survey, Reston

Swedish Institute of Agricultural Engineering (1988/89) In: Sundell B (ed) Annual report 1988/89. Meddelande nr 427, Uppsala, 12

Thiessen E (2001) Experiences with sensor-controlled nitrogen-application. Landtechnik 56:278–279 (in German)

Thiessen E (2002) Optical sensing-techniques for site-specific application of agricultural chemicals. Doctoral thesis, Department of Agricultural Systems Engineering, University of Kiel, Kiel. VDI-MEG Forschungsbericht Agrartechnik 399 (in German)

Thoele H, Ehlert D (2010) Biomass related nitrogen fertilization with a crop sensor. Appl Eng Agric 26(3):769–775

Thoren D (2007) Laser induced chlorophyll-fluorescence for detecting the N content, the biomass and the plant density – technology, sensing in the field, effect of light conditions. Doctoral dissertation, Chair of Plant Nutrition, Technical University Munich, Munich (in German)

Thoren D, Schmidhalter U (2009) Nitrogen status and biomass determination of oilseed rape by laser-induced chlorophyll fluorescence. Eur J Agron 30:238–242

Thoren D, Thoren P, Schmidhalter U (2010) Influence of ambient light and temperature on laser-induced chlorophyll fluorescence measurements. Eur J Agron 32:169–175

Thriwakala S, Weersink A, Kachanowski G (1998) Management unit size and efficiency gains from nitrogen fertilizer application. Agric Syst 56(4):513–531

Tremblay N, Bouroubi YM, Belec C, Mullen RW, Kitchen NR, Thomason WE, Ebelhar S, Mengel DB, Raun WR, Francis DD, Vories ED, Ortiz-Monasterio I (2012) Corn response to nitrogen is influenced by texture and weather. Agron J 104(6):1658–1670

Umweltbundesamt (2011) Data on the environment. Environment and agriculture. Bonn. www.umweltdaten.de/publikationen/fpdf-1/4129.pdf

Viscarra Rossel RA, McBratney AB (2003) Modelling the kinetics of buffer reactions for rapid field predictions of lime requirements. Geoderma 114:49–63

Viscarra Rossel RA, Walvoort DJJ, McBratney AB, Janik LJ, Skemstad JO (2006) Visible, near infrared, mid infrared or combined diffuse reflectance spectroscopy for simultaneous assessment of various soil properties. Geoderma 131:59–75

Werner A, Jarfe A, Leithold P (2004) Use of modules in practice. Management system für den ortsspezifischen Pflanzenbau. Verbundprojekt pre agro. Herausgegeben vom KTBL. Abschlussbericht, Kap 2, pp 51–65 (in German)

WHO (2006) Protecting groundwater for health. Managing the quality of drinking water sources. In: Schmoll O, Howard G, Chilton J, Chorus I (eds) World health organization. IWA Publishers, London

Wolter R (2004) Nitrate contamination of surface- and groundwater in Germany – results of monitoring. In: Bogena H, Hake JF, Vereeken H (eds) Water and sustainable development. Schriften des Forschungszentrums Jülich, Reihe: Umwelt/Environment 48, pp 71–79

Yang CM, Su MR (2000) Analysis of spectral characteristics of rice canopy under water deficiency. In: Proceedings of the Asian conference on remote sensing. Session agriculture and soil, Taipei, 4–8 December 2000

# Chapter 10
# Site-Specific Weed Control

**Roland Gerhards**

**Abstract** Spatial and temporal variations in weed seedling distributions in arable fields are analysed. It is described how weed distributions can be assessed by manual grid sampling and by using sensor technologies from the near range. The potential for herbicide savings using site-specific weed management in different crops is calculated. Two different approaches for site-specific weed control are presented. First, an offline-approach based on georeferenced weed distribution maps and secondly a real-time approach combining sensor– and patch spraying technologies. The decision rules for patch spraying should take into account density, coverage and yield loss effects by weed species, its growth stages and costs of weed control. Herbicide savings using precision weed control varied from 20 to 70 %. Real-time patch spraying is the most economic treatment followed by map-based site-specific weed control. Uniform herbicide applications and uncontrolled treatments gave the lowest economic return. Several studies showed that weed species distribution remained stable over time when site-specific herbicide applications were realized based on economic weed thresholds.

**Keywords** Direct injection system • Distribution • Image analysis • Mapping • Multiple field sprayer • Patches • Shape features • Site-specific control

## 10.1 Introduction

Weed seedling distribution changes spatially and temporally within agricultural fields. It often presents itself in **aggregated patches** of varying size or in stripes along the direction of cultivation (Marshall 1988; Gerhards et al. 1997; Christensen

R. Gerhards (✉)
Department of Weed Science, University of Hohenheim, 70593 Stuttgart, Germany
e-mail: gerhards@uni-hohenheim.de

and Heisel 1998). The variation in weed seedling population has often been ignored for weed management decisions since techniques to assess the weed seedling distribution in acceptable time were not available. Many studies were conducted to apply post-emergence herbicides in winter wheat and maize based on georeferenced maps of the weed seedling distribution (Nordmeyer and Niemann 1992; Tian et al. 1999; Gerhards and Christensen 2003). Herbicide use with this map-based approach was reduced some 40–50 %. With a large within-field variation in weed occurrence, patch spraying that is based on the need for weed control reduces costs, herbicidal pollution of the environment and the risk of herbicide residues in the food chain (Dammer et al. 2003; Timmermann et al. 2003; Gerhards and Oebel 2006).

In many studies, weed species were grouped into grass weeds, annual broadleafs and perennial weeds. Perennials such as bindweed (*Convolvulus arvensis*) and thistle (*Cirsium arvense*) were found to be highly aggregated in arable crops with less than 20 % of the field being infested. Grass weeds covered on average 30–40 % of the fields at infestation levels higher than the economic thresholds and annual broadleaves between 20 and 90 % (Timmermann et al. 2003; Gerhards and Oebel 2006).

Site-specific weed management needs patch sprayers as well as automatic and real-time sensors for weed detection. The objective of this study is to describe the state-of-the-art and evaluate current **patch spraying systems**.

## 10.2 Weed Mapping

Weed seedling distribution in the field was usually assessed using discrete weed mapping or continuous-area sampling (Rew and Cousens 2001). In most studies, discrete **weed mapping** was applied in a regular sampling grid that was established in the field. The side length of the squared grid varied from a few meters up to approximately 50 m and depended on the width of the spray boom used for site-specific herbicide application. Density and/or coverage of emerged weed seedlings were counted and measured prior to and after post-emergence herbicide application in a sampling frame placed at all grid intersection points.

**Efficacy of weed control** was determined relating weed density after post-emergence herbicide application to prior herbicide application. Different mapping programs have been applied to characterize spatial distribution of weeds within fields. Maps differed based on the interpolation method that was applied, the area sampled and the distance between sampling points (Isaaks and Srivastava 1989; Johnson et al. 1995; Rew and Cousens 2001; Gerhards et al. 1997). Geostatistics and interpolation methods were applied to overcome the problem of discontinuities between adjacent sampling points that result from grid sampling. Interpolated weed maps were reclassified based on weed infestation levels (Gerhards et al. 1997). Most weed patches will be detected when sampling grids are not wider than 6 × 6 m (Gerhards and Oebel 2006).

A weed treatment map was derived from the weed distribution maps using weed control thresholds to provide a decision rule for the patch sprayer (Fig. 10.1).

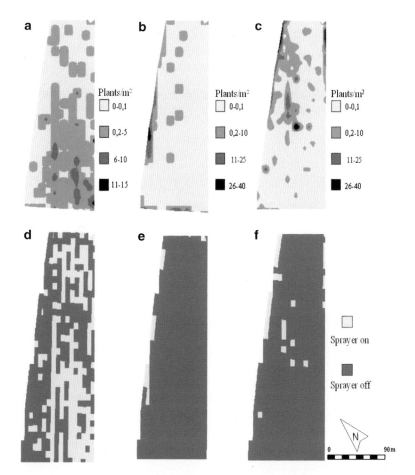

**Fig. 10.1** Distribution of different weed species (**a–c**) in a 3 ha spring barley field in the year 2003 and application maps as a decision rule for the patch sprayer (**d–f**). Maps were created according to economic weed thresholds for all three weed species classes (Gerhards et al. 1997)

Different methods to continuously record in-field variation of weed distributions were to surround and record the borders of aggregated patches of weed species such as wild oats (*Avena fatua*) using a data logger connected to a differential global positioning system (DGPS) (Colliver et al. 1996) or to map weed patches during harvest operations (Barroso et al. 2005).

A major step towards a practical solution for site-specific weed management was the development of sensor technologies and of differential global positioning systems (DGPS) to automatically and continuously determine in-field variation of **weed seedling populations**. Airborne remote sensing was found to be capable for detection of high density weed patches of wild oats (*Avena fatua* L.) and of sterile oats (*Avena sterilis* ssp. *ludoviciana* Durieu) in wheat as well as of infestations of perennial weed species (Lamb and Brown 2001). However, weed seedling

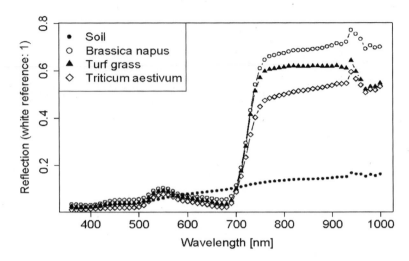

**Fig. 10.2** Reflectance curves for soil (*filled dots*) and for different plant species with the typical steep incline (red edge) between 680 and 750 nm wavelength (Weis and Sökefeld 2010)

populations and weed infestations lower to or equal to the economic weed thresholds could only be detected with sensors from a low distance because of the limited spatial resolution of the sensors and the small area of weed seedlings. Many scientists therefore mounted the sensors on the tractor. Three groups of optical sensors have so far been applied for weed detections:

- spectrometers
- fluorescence sensors
- digital cameras with subsequent image analysis.

## 10.2.1 Spectrometers

Intact green plants transform the incoming light by their chlorophyll pigments, which absorb mostly red as well as violet and blue light. Only some part of the green and most of the near-infrared light is reflected. The spectral reflectance of plants has a minimum in the visible wavelengths of about 650 nm and increases considerably towards the invisible near infrared above 700 nm (Fig. 10.2).

The steep part of the curve is called the "red edge" (Guyot et al. 1992). Important properties of plants, such as chlorophyll content, leaf-area-index (LAI), biomass and water status, age, plant health levels can be derived from the position of the red edge (REP). The spectral curves of different plants have a similar nonlinear shape, but the soil curve in Fig. 10.2 is linear. The local extremes of the plant curves are within the green band (550 nm, maximum), the red band (660 nm, minimum) and near-infrared (750 nm, maximum) (Weis and Sökefeld 2010).

Vrindts and de Baerdemaeker (1997) as well as Biller (1998) used spectrometers to detect weeds between the crop rows or before sowing and after harvesting the crop by measuring the reflectance in the green, red and near-infrared light wave bands. Green leafs were characterised by a high reflectance in the green and near-infrared and a low reflectance in the red spectrum compared with the reflectance curve of bare soil. A few commercial products for weed control with optoelectronic equipment exist that use this spectral information; *e.g.* DetectSpray® (evaluated by Biller 1998) and WeedSeeker® (used by Sui et al. 2008).

## *10.2.2 Fluorescence Sensors*

After exposing green plants with radiation for a specific amount of time, leafs emit radiation of a longer wavelength as the excitation light. The intensity of this radiation named fluorescence highly depends on the leaf properties and on the physiological state of plants (Cerovic et al. 1999).

UV-induced **chlorophyll fluorescence** has also been applied to discriminate plant species based on the characteristic leaf structure. Longchamps et al. (2010) measured a range of fluorescence spectra of maize, grass-weeds and broadleaved weeds under greenhouse conditions with natural illumination. They classified the three plant species groups based on their distinct spectral signatures with a recognition rate above 90 %. Tyystjärvi et al. (1999) developed a method called fluorescence fingerprinting with which leafs are exposed to a series of different spectra and intensities of light to record changes in the fluorescence of chlorophyll a. The emitted light curve could be used to identify plant species with an accuracy of more than 90 % under laboratory conditions. Later, Tyystjärvi et al. (2011) applied a similar approach under field condition and achieved 90 % recognition in maize and weeds when plants were shaded for 1 s before measuring.

In our working group, the MiniVeg® sensor (Fritzmeier Umwelttechnik) has been used in field and greenhouse studies to map the spatial distribution of weed species in arable crops. Red- plus far red fluorescence was induced by a red laser. When the laser hit plants, fluorescence was induced and recorded in the processor of the sensor. Due to the high frequency of measurements (500 s$^{-1}$), **plant density** highly correlated with the number of hits. As crop density was rather homogeneous within the fields at early growth stages, variations of the hit-number correlated with the weed density. Therefore, weed distributions maps could be derived from the sensor measurements when a GPS-receiver was mounted on the sensor vehicle (Fig. 10.3). Blackgrass (*Alopecurus myosuroides* Huds.) was the dominant weed species in the winter wheat field sampled in 2008. Forty-four percent of the area remained untreated when site-specific weed control was applied in this field using a weed control threshold. The MiniVeg®-sensor provided 75 % correct decisions compared to manual weed countings.

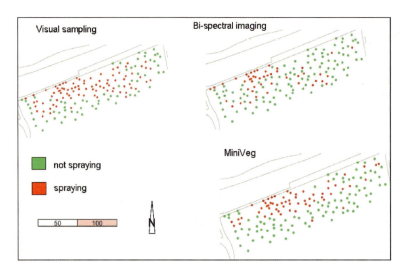

**Fig. 10.3** Weed distribution maps derived from visual countings, bi-spectral imaging and MiniVeg® measurements in a 5.6 ha field at Ihinger Hof in winter wheat in autumn 2008. For details to bi-spectral imaging see the next section (After Gerhards et al. 2012)

## 10.2.3 Digital Image Analysis Based on Shape Features

### 10.2.3.1 The Sensing and Processing Concept

A very promising approach for weed detection and identification is a continuous ground-based detection method based on **image analysis** (Weis et al. 2008). With this method, weeds and crops are segmented from digital images in real-time using a bi-spectral camera system connected to DGPS. Weed species as well as crops are identified and counted based on automatic classification of **shape features**. The entire system for site-specific weed control consists of three parts, which are separated and can communicate via interfaces:

1. A bi-spectral camera system with practical suitability for the field connected to DGPS.
2. An image processing and classification component including a weed/crop-database used for the classification.
3. A GPS-controlled patch sprayer.

First, laboratory studies were conduced to analyze reflection properties of green plants, different soils (organic, sandy, stony, wet, dry), stones and organic mulch using a video spectrometer (Fig. 10.4). The reflectance of all objects mentioned above within the field of view of the video spectrometer in the spectral band between 338 and 925 nm was visualized using grey levels from 0 (black) to 255 (white).

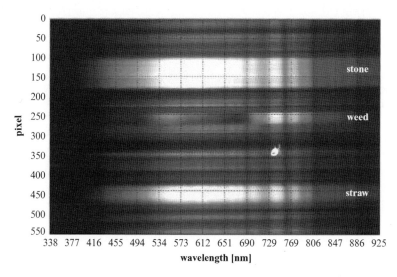

**Fig. 10.4** Video spectrometer image with typical reflections of a stone, weed and straw in the spectrum of 338–925 nm with soil as background (Sökefeld et al. 2007)

Figure 10.4 shows an image of a typical video spectrometer scene. Nearly over the entire analyzed spectrum, stones show a strong reflection. The reflection of dead organic material like straw is very similar to the reflection of stones. Living plants reflect moderately below 570 nm and have a strong reflection above 690 nm. The characteristic decrease of reflection between 610 and 690 nm is typical for green plants because of the absorption of this band by chlorophyll.

Based on these studies, it was concluded that a normalized difference between images above 700 nm and images between 610 and 690 nm would result in high quality images with a strong contrast between green plants and background. A bi-spectral camera was developed computing differential images of the infrared and red wavebands (Figs. 10.5 and 10.6). The resulting images were saved on the hard disc of a computer for further processing. The **bi-spectral camera** allows real-time detection and identification of weed species in arable crops. The speed of the image analysis is high enough to use the cameras for online weed control in combination with a field sprayer.

In the subsequent text, the processing of the data is dealt with in an abbreviated and condensed manner. Details to this are explained in the literature cited.

A **circular closing operator** of size five was used to connect most leafs of a single plant. For the extracted plants, various numerical features were computed that reflect the form of the plant species:

- Region-based: these features are based on the region pixels, which are defined as a connected set of pixels. Examples are the size, compactness, minimum and maximum diameter and several statistical measures (statistical moments (Jähne 2001), Hu moments (Hu 1962))

**Fig. 10.5** Principle of a bi-spectral camera system for the pixel congruent acquisition of two images in different spectral bands (Sökefeld et al. 2007)

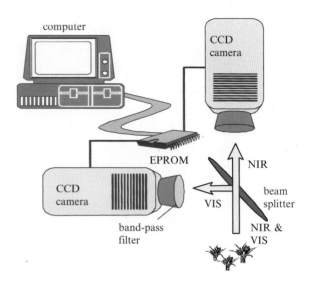

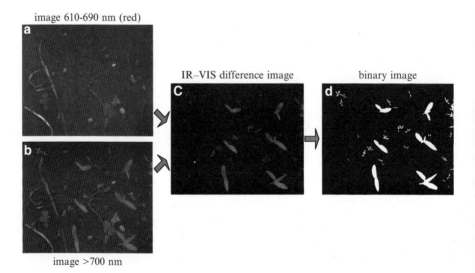

**Fig. 10.6** Difference image (**c**) calculated from the infrared (**b**) and red image (**a**) to remove soil, stones and mulch; followed by binarization (**d**) using automatic thresholding (Sökefeld et al. 2007)

- Contour-based: these features were calculated from a contour representation of the region. Fourier descriptors and curvature scale space representation (CSS) were calculated (Mokhtarian et al. 1996).
- Skeleton-based: the skeleton of a region is a "thinned" representation of the region (Soille 2003). Several numerical features as well as structural ones can be derived from the skeleton.

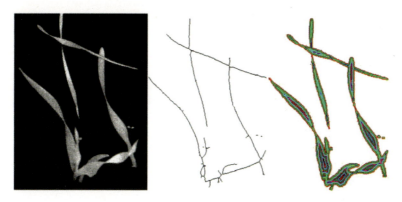

**Fig. 10.7** Gray level image of an object with – from *left* to *right* – overlapping leafs, skeleton and distance transform (Weis and Gerhards 2007)

The **skeleton** is the "middle line" of the objects as shown in Fig. 10.7, center. A distance transform is computed for the leafs, assigning each pixel a value for the minimum distance to the border of the leaf (Fig. 10.7 right). All border pixels have the value one, all others inside the region get a higher value. The combination of the skeleton with the distance transform leads to a distance function that describes the "thickness" of the leaves. Statistical values (maximum, mean, variance, number of skeleton pixels) can be derived which were found to discriminate especially grasses from broadleafs.

Every **feature set**, consisting of more than a 100 features, is associated with a class. The class assignments are determined from training sets of images. To be able to reuse the training sets, all training samples are stored in a database. The database contains the segmented images and the feature sets as well as the class assignments. A few examples of the images stored in the database are shown in Fig. 10.8.

An **image database** was created for six crops (sugar beet, wheat, barley, maize, peas and oil seed rape) and 40 weed species. In the database, prototypes for the different classes are stored. The images are split up into segments each containing only one plant of known class. This allows the images to be re-used for the development of new feature extraction algorithms and classifiers. A comparison of different image segmentation approaches and feature sets can be achieved using the database.

### 10.2.3.2 Identification Results and the Classification of Plant Species

For the identification of weed species, a knowledge-based image analysis system was used (Gerhards and Oebel 2006; Oebel et al. 2004; Sökefeld and Gerhards 2004). First, shape features were extracted and calculated from all plants in the image. Those features were used to discriminate and classify plant species. In order to test the accuracy of the **classification algorithm**, images taken in the field were analyzed visually and by the image analysis system. Between 400 (maize) and

**Fig. 10.8** Classification examples: each classified region denoted by a *color* and a *number* for the class. The latter is defined by a so-called EPPO code (Weis et al. 2008)

**Table 10.1** Automatic classification of plant species in maize (Zea mays) using digital image analysis (data of 400 images) (Sökefeld et al. 2007)

|  | Identification in % (identification figures in horizontal lines add up to 100 %) | | | | |
| --- | --- | --- | --- | --- | --- |
|  | Maize (Zea mays) | Grass weeds | Lambsquart. (Ch. album) | Other broadleafs | Sum |
| Maize | 100 | 0 | 0 | 0 | 100 |
| Grass weeds | 0 | 90 | 4 | 6 | 100 |
| Lambsquarters | 10 | 0 | 90 | 0 | 100 |
| Other broad-leaved species | 0 | 1 | 1 | 98 | 100 |
| Total |  |  |  | 94 |  |

2,100 (winter wheat) images of unknown plants were taken for the testing. The **automatic classification** resulted in an average of about 90 % correct identification when weed species and crops were grouped into 4–5 classes. The results of the automatic classification in maize and winter wheat are presented in Tables 10.1 and 10.2. As must be expected, the identification in % depended on the plant species that were grouped together.

The classification results of the images and the corresponding GPS data were used for weed mapping and site-specific herbicide application. During the past 3 years, the camera system was used for weed identification in more than 100 ha of cereals,

**Table 10.2** Automatic classification of plant species in winter wheat (Triticum aestivum) using digital image analysis (data of 2,100 images) (Sökefeld et al. 2007)

| | Identification in % (identification figures in horizontal lines add up to 100 %) | | | | | |
|---|---|---|---|---|---|---|
| | Wheat (Tritic. aest.) | Grass weeds | Catchweed (Galium aparine) | Mayweed (Matricaria chamonilla) | Other broad-leafs | Sum |
| Wheat | 80 | 13 | 7 | 0 | 0 | 100 |
| Grass weeds | 0 | 100 | 0 | 0 | 0 | 100 |
| Catchweed | 0 | 0 | 92 | 0 | 8 | 100 |
| Mayweed | 0 | 0 | 0 | 100 | 0 | 100 |
| Other broadleafs | 0 | 0 | 20 | 0 | 80 | 100 |
| Total | | | | | 86 | |

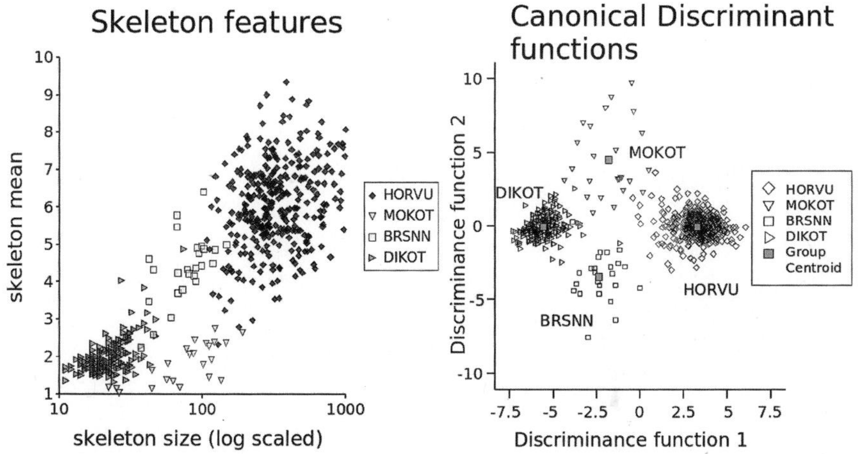

**Fig. 10.9** (*left*) Two dimensions of the feature space: skeleton mean and size; (*right*) the first two discriminant functions. The classes are: *HORVU* Hordeum vulgare, *MOKOT* grass weeds, *BRSNN* Brassica napus, *DIKOT* broad-leaved weeds (Weis and Gerhards 2007)

maize, sugar beet, rape-seed and peas. No problems with the camera technology arose due to vibration of the field vehicle, dust and moisture.

A different dataset was classified including images of four plant species groups: Winter barley (HORVU), grass-weeds (MOKOT), rape (*Brassica napus* BRSNN) and other broadleafs (DIKOT). Plant species groups that were in different growth stages are partly overlapped. This led to a high variation in the features (Fig. 10.9).

Approximately 40 different **classifiers** including Bayes functions, nearest neighbor, classification trees were applied to classify the dataset. All of them performed better than 95 % (correct classification rate) in a 10-fold cross validation. The main result of this test was that the type of classifier was less important than the selection of the right features and grouping of plant species into meaningful classes.

The shape based approach has problems in situations when the plants are in late growth stages and overlap each other. A proposal for the handling of these situations was made that uses a structural description for the separation of the objects into parts (Weis and Gerhards 2007).

## 10.3 Temporal and Spatial Dynamics of Weed Population

The temporal and spatial stability of weed populations are important for map based site-specific herbicide applications that take place in subsequent seasons of a rotation. Any **pre-emergence site-specific applications** rely on this, they would not be possible without some stability of weed populations between subsequent crops. Yet also post-emergence spraying in subsequent crops might be map based.

The actual **dynamics of weed populations** are influenced by the biological characteristics of weed species, by farming practices such as tillage, crop rotation, time of seeding, by harvesting competitiveness of the crop and direct weed control methods as well as by soil parameters (Mortensen et al. 1998; Nordmeyer and Niemann 1992; Timmermann et al. 2002). The major weed species have developed specific adaption- and **survival strategies** to persist in cropping systems (Radosevich et al. 1997). Those strategies include the production of a high number of seeds over a long period of time and seed dormancy (*e.g.* lambsquarter *C. album*). In addition, successful weed species have the capacity to survive under variable environments based on high phenotypic and genetic plasticity to invade new sites (*e.g.* velvetleaf *Abutilon theophrasti*). Many weeds are able to strongly compete for space, light, water and nutrients with the crops by high growth rates and efficiency in using water and nutrients. Several weeds produce mature seeds in a much shorter time than crops so that the seeds are spread long before a dense crop stand has been established (*e.g.* gallant soldier *Galingsoga parviflora*). Other weed species, such as thistle (*Cirsium arvense*) and quackgrass (*Agropyron repens*) have the ability to persist and spread via seeds and vegetative reproduction tissues. Those **perennial weeds** can emerge much faster than annual plants. These are only few reasons for spatial and temporal dynamics of weed populations.

Nordmeyer and Niemann (1992) found that blackgrass (*Alopecurus myosuroides*) populations mostly occurred at locations in the field where the clay content was relatively high. Timmermann et al. (2002) reported that the crop rotation had a long-term effect on weed density and weed species composition. In fields that had been planted with 50 % maize in the rotation more than 20 years ago, the density of lambsquarter (*C. album*) was still much higher than in fields with a high percentage of winter annual grains in the rotation. The crop rotation had also a very strong effect on the organic matter content. Fields that had been planted with potatoes were lower in the organic matter content than fields where mostly grains were planted. The difference in organic matter content again had a strong influence on the weed species composition. Catchweed (*Galium aparine*) predominantly occurred in fields with high organic matter contents (Timmermann et al. 2002).

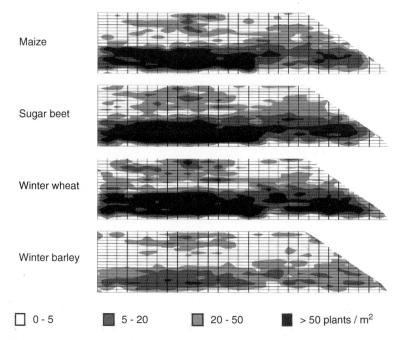

**Fig. 10.10** Distribution of field violet (*Viola arvensis*) in maize, sugar beet, winter wheat and winter barley in a 5 ha arable field at Dikopshof Research Station near Bonn, Germany (Modified after Krohmann et al. 2002)

Krohmann et al. (2002) studied the dynamics of weed seedling distribution over 5 years in a rotation of maize, sugar beet, winter wheat and winter barley and in continuous maize. They found that **weed distribution maps** obtained in maize and sugar beet were suitable for site-specific weed control in winter wheat and winter barley (Fig. 10.10).

Ritter and Gerhards (2008) reported that populations of blackgrass (*Alopecurus myosuroides*) did not significantly change in density, location and size when site-specific weed control methods were applied over a period of 8 years in a rotation of winter annual cereals, maize and sugar beet. In all of the three fields studied, **weed seedling distribution** was heterogeneous. Density was higher in maize and sugar beet than in winter cereals. High density patches with densities higher than 25 plant per $m^2$ consistently recur over the years at the same areas in the fields. **Weed density reduction** due to herbicides and other weed control methods was satisfying in each year indicating that site-specific weed control methods are sustainable for long-term weed suppression. Herbicide savings with blackgrass (*A. myosuroides*) ranged from 50 % in sugar beet to 75 % in winter barley.

Ritter and Gerhards (2008) also studied weed **population dynamics** of catchweed (*Galium aparine*) and blackgrass (*A. myosuroides*) under the influence of site-specific weed management. Most of the tested population parameters were weed

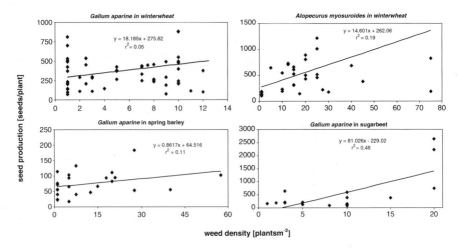

**Fig. 10.11** Weed density and seed production of catchweed (Galium aparine) and blackgrass (Alopecurus myosuroides) in various crops (Ritter and Gerhards 2008)

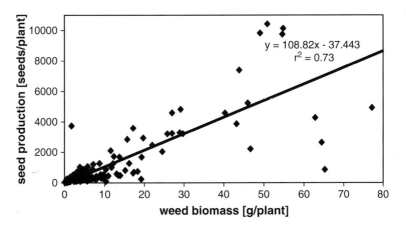

**Fig. 10.12** Correlation of weed biomass and seed production of catchweed (Galium aparine) and blackgrass (Alopecurus myosuroides) over all crops (Ritter and Gerhards 2008)

density dependent. It was presumed that individual weeds without competition evolve better and produce more seeds, but this study showed opposing results. With increasing weed density, weed biomass and fecundity increased in this study (Figs. 10.11 and 10.12). All findings support that weed density has to be considered in weed management strategies.

An understanding of fundamental weed population biology would improve our ability to develop site-specific management decisions. **Weed populations models** have been applied to quantify the effects of site-specific weed management practices

(Paice et al. 1997). However, the mechanism of weed patch stability is rather untapped. A few results show that efficacy of weed control methods (see Sect. 10.2, top) was lower in weed patches than at low density locations (Mortensen et al. 1998). Krohmann et al. (2002) found that the persistence of weed populations was also attributed to weed seedlings that emerged after weed control methods had been applied. Those individuals were able to produce viable seeds in maize and sugar beet but not in winter wheat and winter barley. The authors assume that competition of the crop was higher in winter annual grains and therefore late emerging weed seedlings were suppressed.

Few studies have attempted to quantify spatial stability of **weed patches** in agricultural fields. If weed patches were consistent in density and location over years, maps from 1 year could be used to direct sampling plans and to regulate weed control methods in subsequent years. Wilson and Brain (1991) found that the pattern of blackgrass (*A. myosuroides* Huds.) patches persisted during a 10 year study. Persistence of patches was attributed to the poor ability of blackgrass to colonize new locations when effective herbicides were applied. The pattern of patches was most stable in fields planted to cereals. Pester et al. (1995) observed significant stability for velvetleaf (*Abutilon theophrasti*) populations using Pearson, Spearman rank and chi-square correlation analysis to quantify year by year relationships between weed density at individual X, Y-coordinates of the sampling grid in four fields. Walter (1996) also used the chi-square correlation method and found that field violet (*V. arvensis* Murr.), common lambsquarters (*C. album* L.) and prostrate knotweed (*Polygonum aviculare* L.) distributions were stable in cereal grain fields over 3 years.

Gerhards et al. (1996) studied the spatial stability of velvetleaf (*A. theophrasti* Medik.), hemp dogbane (*Apocynum cannabinum* L.), common sunflower (*Helianthus annuus* L.), yellow foxtail (*Setaria glauca* L.) and green foxtail (*Setaria viridis* L.) over 4 years (1992–1995) in two fields in eastern Nebraska. The first field was planted to soybean in 1992 and corn in 1993, 1994 and 1995. The second field was planted to corn in 1992 and 1994 and soybean in 1993 and 1995. Weed density was sampled prior to post-emergence herbicide application at approximately 800 locations per year in each field on a regular 7 m grid. The same locations were sampled every year. Weed density at locations between the sample sites was determined by linear triangulation interpolation. Weed seedling distribution was significantly aggregated with large areas being weed free in both fields. Common sunflower, velvetleaf and hemp dogbane patches were very persistent in the east–west and north–south directions and in location as well as in area over the 4 years in the first field. Foxtail distribution and density continuously increased in each of the 4 years in the first field and decreased in the second field. A Geographic Information System was used to overlay maps from each year for a species. This showed that 36 % of the sampled area was free of common sunflower, 62.5 % was free of hemp dogbane and 11.5 % was free of velvetleaf in the first field, but only 1 % was free of velvetleaf in the second field. The persistence of broadleaf weed patches observed in this study suggests that **weed seedling distributions** mapped in 1 year are good predictors of future seedling distributions.

Heijting et al. (2007) found strong spatial correlations for cockspur (*Echinochloa crus-galli*), lambsquarter (*C. album*), goosefoot (*C. polyspermum*) and black nightshade (*Solanum nigrum*) in 3 years continuous maize cultivation. They attributed spatial and temporal stability of weed populations to their high recruitment capacity.

Summing up, it can be concluded that in many cases the weed maps from 1 year might provide the site-specific control basis for either pre-emergence or post-emergence herbicide applications in next years. So it might be reasonable to use the results of one sensing operation initially for an simultaneous in-season real-time application followed by map based site-specific sprayings in next crops.

## 10.4 Site-Specific Weed Control

The weed population can vary spatially as well as on a species basis. Precise application of herbicides thus has two objectives: **adapting the mass to the spatial weed density** as well as **adjusting the formulation to the plant species**. Hence site- and species-specific control might be needed.

For this purpose, a **multiple field sprayer** was developed (Fig. 10.13). Each of the three sprayer circuits led to a boom width of 21 m divided into 7 sections of 3 m. Each sprayer circuit and each boom section was turned on and off separately via solenoid valves. This sprayer allowed a separate control of each hydraulic circuit according to information from herbicide application maps. The application rate was regulated from 200 to 290 l ha$^{-1}$ over the whole boom width of each circuit by pressure variation (Gerhards and Oebel 2006).

Another approach for site- and species specific weed control is to employ sprayers with an integrated **direct injection system**. With such injection sprayers, herbicides and carrier (water) are kept separate. According to the indications of the weed treatment map (off-line application) or the sensor data (on-line application), the herbicides are metered into the carrier and mixed immediately before entering the nozzles.

In both scenarios, short **reaction times** of less than 1 s and adequate mixing of the herbicides into the carrier are basic requirements for high weed control efficacy. Figure 10.14 shows an experimental direct injection system with one injection point for each boom section (3 m each). With this configuration, a lag time of 4–7 s was obtained. A shorter lag of approximately 1.0 s can be achieved by the injection of the herbicide at each nozzle.

Site- and species specific weed control was performed in cereals, sugar beet, maize and oil-seed rape resulting in significant areas that remained untreated with herbicides (Table 10.3). Combinations of weed mapping and application technologies for site- and species specific weed control increased the potential for herbicide savings. From the results it is obvious that the herbicide savings can be considerably enhanced when the site-specific application is supplemented with species-specific spraying.

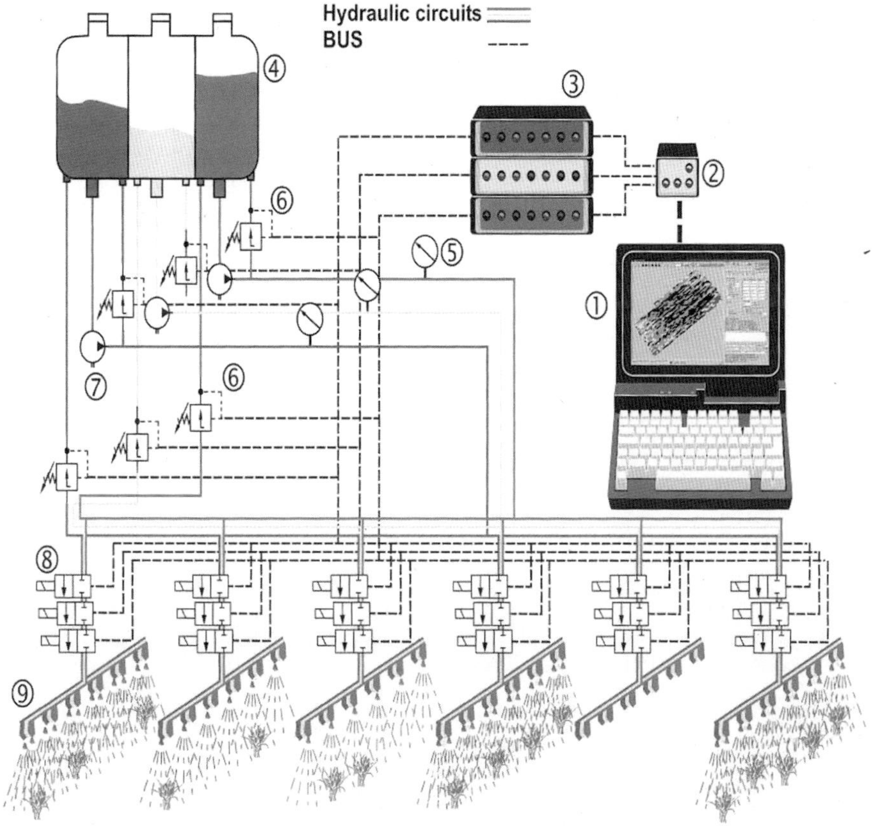

**Fig. 10.13** Schematic configuration of the multiple sprayer (Gerhards and Oebel 2006) with: *1* board computer with application map, *2* control unit for spray computer, *3* spray computer, *4* tanks, *5* manometers, *6* pressure valves, *7* pumps, *8* solenoid valves, *9* boom sections with nozzles

Efficacy of site-specific weed control (see Sect. 10.2, top) attained on the average 85–98 % and was similar to uniform herbicide applications.

Knowledge of spatial and temporal variability of weed populations offers large potentials for precise control methods and thus for using less herbicides resulting in less herbicide residues in the environment and food chain. Site-specific weed control methods can be realized when automatic sensor technologies for weed detection and patch spraying technologies are combined with precise decision algorithms.

In addition to this practical benefit, weed mapping helps to understand weed-crop interactions and population dynamics of weed species. It allows quantifying yield effects of different weed infestations in the fields and modelling the spatial and temporal variability of weed populations under different crop management systems.

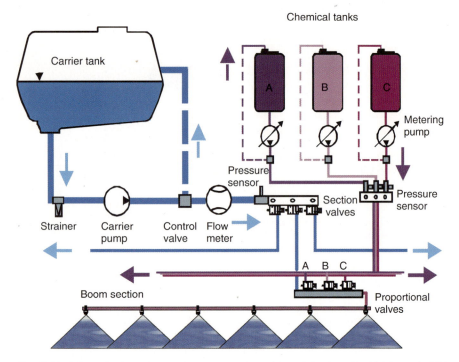

**Fig. 10.14** Schematic view of the hydraulic system for the direct injection of herbicides (Sökefeld et al. 2005)

## 10.5 Outlook and Perspectives

The potential of savings in herbicides by site-specific as well as by species-specific application is impressive (Table 10.3). The technical expenditures for this approach probably will be lower when – instead of expensive techniques for multiple spraying systems – concepts for direct injection will become state of the art (Figs. 10.13 and 10.14). Hence developments in this direction will be important.

Important will also be, whether the attempts of **orienting herbicides at weeds** or instead of this **adapting crops to herbicides** will prevail. The techniques dealt with so far in this chapter mainly aim for an orienting of herbicides at weeds, on a site-specific basis as well as species-directed. Yet the advances that have been made in adapting crops to herbicides are impressive as well. **Genetically modified crops** that are herbicide resistant are widely used in both North- and South America. They allow using singular and uniform chemical formulations of herbicides that in the past destroyed crops but effectively remove all weeds. Recent developments have proved that these **herbicide resistant crops** must not necessarily result from transgenic methods, for which there exist objections in the public of some European countries. These herbicide resistant crops can be developed by traditional breeding methods as

**Table 10.3** Relative untreated area using two different patch spraying systems compared to a uniform herbicide application across the whole field (Gutjahr et al. 2012)

| | | % untreated area | | | |
| --- | --- | --- | --- | --- | --- |
| | | Site-specific control only with uniform tank mixture | Site-specific as well as species-specific weed control with multiple-tank sprayer | | |
| | Field | | Broadleaf- weeds | Grass-weeds | Catchweed (*G. aparine*) |
| Winter wheat | 1 | 64 | 82 | 81 | 96 |
| | 2 | 4 | 99 | 11 | 15 |
| | 3 | 40 | 68 | 52 | 100[a] |
| | 4 | 63 | 67 | 96 | 96 |
| | 5 | 89 | 94 | 96 | 100[a] |
| | 6 | 76 | 88 | 89 | 93 |
| | 7 | 2 | 85 | 14 | 40 |
| | 8 | 40 | 88 | 44 | 89 |
| | 9 | 6 | 10 | 71 | 95 |
| | 10 | 2 | 66 | 2 | 73 |
| | 11 | 23 | 25 | 98 | 84 |
| | Mean | 37 | 70 | 59 | 80 |
| | | | Broadleaf- weeds | Grass-weeds | Perennials |
| Maize | 12 | 0 | 0 | 30 | 100[a] |
| | 13 | 0 | 0 | 2 | 98 |
| | 14 | 0 | 3 | 14 | 100[a] |
| | 15 | 23 | 37 | 90 | 74 |
| | Mean | 6 | 10 | 34 | 93 |

[a]It was assumed that farmers always apply herbicides against all three weed- species groups in uniform application.

well. The new "Clearfield" crop varieties verify this. However, one fundamental problem may arise with herbicide resistant crops. Their outgrowth can appear in subsequent crops of a rotation as weeds that cannot be removed by herbicides any more. This might be important especially with rape (colza).

And still another alternative should be considered, *i.e.* future weed control by small **robots** that might loiter through fields (Chap. 11, Fig. 11.6). These robots might remove weeds either by applying herbicides or by hoeing. It would be feasible to program the robots by precise georeferencing in such a way that they can differentiate between weeds and plants of the crop. This can be achieved by meticulously georeferencing all positions within a field, where the seeds were placed during the planting operation. This might not be feasible with small cereals or grass crops, yet it is possible with more widely spaced crops such as maize, beans and beets. In case of control by herbicides, the robots could by and large apply these only on weeds and leave out any deposition on plants of the respective crop. Hence there would be neither a need for herbicide resistant crops nor for selective herbicides. If the control is done by hoeing, this could include the removal of weeds that result from herbicide resistant outgrowth.

Actually, an **intelligent combination of control techniques** might be sensible. With herbicide resistant crops, any species-specific application can be put aside. But a site-specific application still might substantially save herbicides and reduce the impact on the environment. And with crops that are not resistant to herbicides, it might be wise to supplement the site- and species specific application as dealt with in the sections above with occasional or partial weed control by robots. Because it will hardly be possible to take into account all kinds of weeds in a species- specific application.

Anyway, the variety of feasible control techniques will largely allow to dispense with **soil cultivation** as a means for weed killing.

# References

Barroso J, Ruiz D, Fernandez-Quintanilla C, Leguizamon ES, Hernaz P, Ribeiro A, Dias B, Maxwell BD, Rew LJ (2005) Comparison of different sampling methodologies for site-specific management of Avena sterilis. Weed Res 45:165–174

Biller RH (1998) Differentiating between plants and targeted application of herbicides (in German). Forschungs-Report 1:34

Cerovic ZG, Samson G, Morales F, Tremblay N, Moya I (1999) Ultraviolet-induced fluorescence for plant monitoring: present state and prospects. Agronomie 1:543–578

Christensen S, Heisel T (1998) Patch spraying using historical, manual and real-time monitoring of weeds in cereals. J Plant Dis Prot, Special Issue XVI:257–263

Colliver CT, Maxwell BD, Tyler DA, Roberts DW, Long DS (1996) Georeferencing wild oat infestations in small grains: accuracy and efficiency of three survey techniques. In: Roberts PC et al (eds) Proceedings of the 3rd international conference on precision agriculture, Minneapolis, pp 453–463

Dammer KH, Böttger H, Ehlert D (2003) Sensor-controlled variable rate application of herbicides and fungicides. Precis Agric 4:129–134

Gerhards R, Christensen S (2003) Real-time weed detection, decision making and patch spraying in maize, sugarbeet, winter wheat and winter barley. Weed Res 43:1–8

Gerhards R, Oebel H (2006) Practical experiences with a system for site-specific weed control in arable crops using real-time image analysis and GPS-controlled patch spraying. Weed Res 46:185–193

Gerhards R, Pester DY, Mortensen DA (1996) Characterizing spatial stability of weed populations using interpolated maps. Weed Sci 45:08–119

Gerhards R, Sökefeld M, Schulze-Lohne K, Mortensen DA, Kühbauch W (1997) Site specific weed control in winter wheat. J Agron Crop Sci 178:219–225

Gerhards R, Weis M, Gutjahr C, Schulz J, Jancker H (2012) Research on automatic weed recognition in crops via the sensor-system MiniVeg® (in German). In: ATB-Computer Bildanalyse in der Landwirtschaft, Workshop 2011, Universität Hohenheim-Ihinger Hof

Gutjahr C, Sökefeld M, Gerhards R (2012) Evaluation of two patch spraying systems in winter wheat and maize. Weed Res 52:510–519

Guyot G, Baret F, Jacquemoud S (1992) Imaging spectroscopy for vegetation studies. In: Toselli F, Bodechtel J (eds) Spectroscopy: fundamentals and prospective applications. Kluwer Academic Publishers, Dordrecht, pp 145–165

Heijting S, Van Der Werf W, Stein A, Kropff MJ (2007) Are weed patches stable in locations? Application of an explicitly two-dimensional methodology. Weed Res 47:381–395

Hu MK (1962) Visual pattern recognition by moment invariants. IRE T Inf Theory 8(2):179–187

Isaaks EH, Srivastava RM (1989) An introduction to applied geostatistics. Oxford University Press, New York, 561 p

Jähne B (2001) Digital image processing, 5th edn. Springer, Berlin

Johnson GA, Mortensen DA, Martin AR (1995) A simulation of herbicide use based on weed spatial distribution. Weed Res 35:197–205

Krohmann P, Timmermann C, Gerhards R, Kühbauch W (2002) Causes for persistence of weed populations (in German). J Plant Dis Prot, Special Issue XVIII:261–268

Lamb DW, Brown RB (2001) Remote sensing and mapping of weeds in crops. J Agric Eng Res 78:117–125

Longchamps L, Panneton B, Samson G, Leroux G, Thériault R (2010) Discrimination of corn, grasses and dicot weeds by their UV-induced fluorescence spectral signature. Precis Agric 11(2):181–197

Marshall EJP (1988) Field-scale estimates of grass populations in arable land. Weed Res 28:191–198

Mokhtarian F, Abbasi S, Kittler J (1996) Robust and efficient shape indexing through curvature scale space. In: Pycock D (ed) Proceedings of British Machine Vision Conference 1996, BMVC, British Machine Vision Association, Edinburgh, pp 53–62

Mortensen DA, Dieleman JA, Johnson GA (1998) Weed spatial variation and weed management. In: Hatfield JL, Buhler DD, Stewart BA (eds) Integrated weed and soil management. Ann Arbor Press, Chelsea, p 293

Nordmeyer H, Niemann P (1992) Possibilities of targeted site-specific application of herbicides based on weed distribution and soil variability (in German). J Plant Dis Prot, Special Issue XIII:539–547

Oebel H, Gerhards R, Beckers G, Dicke D, Sökefeld M, Lock R, Nabout A, Therbourg R-D (2004) Site-specific weed control by georeferenced image processing in an offline (and online) TURBO mode. First practical experiences (in German). J Plant Dis Prot, Special Issue XIX:459–465

Paice MER, Day W, Rew LJ, Howard A (1997) A simulation model for evaluating the concept of patch spraying. Weed Res 43:373–388

Pester DY, Mortensen DA, Gotway CA (1995) Statistical methods to quantify spatial stability of weed populations. North Cent Weed Sci Soc 50:52

Radosevich S, Holt J, Ghersa C (1997) Weed ecology – implication for weed management, 2nd edn. Wiley, New York, 589 p

Rew LJ, Cousens RD (2001) Spatial distribution of weeds in arable crops: are current sampling and analytical methods appropriate. Weed Res 41:1–18

Ritter C, Gerhards R (2008) Population dynamics of *Galium aparine* L. and *A. myosuroides* (Huds.) under the influence of site-specific weed management. J Plant Dis Prot, Special Issue XXI:209–214

Soille P (2003) Morphological image analysis, 2nd edn. Springer, Heidelberg

Sökefeld M, Gerhards R (2004) Automatic weed mapping by digital image processing. Landtechnik 59(3):154–155

Sökefeld M, Hloben P, Schulze Lammers P (2005) Development of test bench for measuring of lag time of direct nozzle injection systems for site-specific herbicide application. Agrartechnische Forschung 11(5):145–154

Sökefeld M, Gerhards R, Oebel H, Therburg RD (2007) Image acquisition for weed detection and identification by digital image analysis. In: Stafford J (ed) Precision agriculture '07, vol 6, The Netherlands. 6th European Conference on Precision Agriculture (ECPA). Wageningen Academic Publishers, Wageningen, pp 523–529

Sui R, Thomasson JA, Hanks J, Wooten J (2008) Ground-based sensing system for weed mapping in cotton. Comput Electron Agric 60:31–38

Tian L, Reid JF, Hummel JW (1999) Development of a precision sprayer for site-specific weed management. Trans ASAE 42(4):893–900

Timmermann C, Gerhards R, Kühbauch W (2002) Causes of yield differences with arable crops (in German). J Agron Crop Sci 187:1–9

Timmermann C, Gerhards R, Kühbauch W (2003) The economic impact of the site-specific weed control. Precis Agric 4:249–260

Tyystjärvi E, Koski A, Keränen M, Nevalainen O (1999) The Kautsky curve is a built-in barcode. Biophys J 77:1159–1167

Tyystjärvi E, Norremark M, Mattila H, Keranen M, Hakala-Yatkin M, Ottosen C (2011) Automatic identification of crop and weed species with chlorophyll fluorescence induction curves. Precis Agric 12:546–563

Vrindts E, de Baerdemaeker J (1997) Optical discrimination of crop, weed and soil for on-line weed detection. In: Stafford J (ed) Precision agriculture 1997, 1st European Conference on Precision Agriculture, vol 2, Technology, IT and Management. BIOS Scientific Publishers, Warwick, pp 537–544

Walter W (1996) Temporal and spatial stability of weeds. In: Brown H (ed) Proceedings of 2nd International Weed Congress, Copenhagen, Denmark, pp 125–130

Weis M, Gerhards R (2007) Feature extraction for the identification of weed species in digital images for the purpose of site-specific weed control. In: Stafford J (ed) Precision agriculture '07, vol 6, The Netherlands. 6th European Conference on Precision Agriculture (ECPA). Wageningen Academic Publishers, Wageningen, pp 537–545

Weis M, Sökefeld M (2010) Detection and identification of weeds. In: Oerke E-C, Gerhards R, Menz G, Sikora RA (eds) Precision crop protection – the challenge and use of heterogeneity. Springer, Dordrecht, pp 119–132

Weis M, Ritter C, Gutjahr C, Rueda-Ayala V, Gerhards R, Schölderle R (2008) Precision farming for weed management techniques. Gesunde Pflanzen 60:171–181

Wilson BJ, Brain P (1991) Long-term stability of distribution of Alopecurus myosuroides Huds. Within cereal fields. Weed Res 31:367–373

# Chapter 11
# Site-Specific Sensing for Fungicide Spraying

Eiko Thiessen and Hermann J. Heege

**Abstract** Especially in humid moderate climates, high yields require the application of fungicides. Its site-specific application based on the biomass of crops is state of the art. Yet this technique does not take into account that fungal infections in most cases start and spread out from small, initial spots within a field. So a sensing technique to detect these initially small infected spots would be of great importance for saving fungicides, for reducing damage to crops as well as to the environment and for allowing higher driving speeds in uninfected areas. Reflectance indices of visible and near-infrared light as well as indices of fluorescent light are candidates for detecting these spots.

Detecting the fungi in early stages of infection (= latency stage) can be important for a successful treatment, because stopping the infection after this time gets more difficult. In this latency stage, the diseases might not yet be visible by human eyes. Fungal diseases often change the physiological state of plants either by means of the photosynthesis or by the formation of secondary metabolites like phenols. These changes can be detected in the smartest way by optical sensing. Hereby fluorescence is a sensitive method with the potential of detecting changes before infections are visible by human eyes.

**Keywords** Fluorescence • Fungal infection • Optical sensing • Site-specific

E. Thiessen (✉) • H.J. Heege
Department of Agricultural Systems Engineering, University of Kiel,
24098 Kiel, Germany
e-mail: ethiessen@ilv.uni-kiel.de

## 11.1 The Situation for Site-Specific Fungicide Applications

It is a well known fact that fungal crop infections in most cases start with a few small and very discrete **patches**. At the beginning of a fungal epidemic, these few tiny infected loci within a field are hard to detect. Their visual patterns depend very much on the respective fungi, on the infection stage, on crop development as well as on varieties and on the weather. Therefore, traditionally crop epidemiologists have been **detectives of patterns** with microscopes as manual tools.

The early detection of an infection is crucial, because it allows timely counteractions, hence to eliminate further spreading of the disease and to prevent serious damage to the crop. It is also a prerequisite for site-specific application. Because at the far end of a fungal epidemic and much damage, a more uniform spatial distribution of the disease will exist.

If the challenge of an early and site-specific detection of infected patches is met, there probably are no other farming operations where treating the whole field uniformly can be so far off the actual needs. However, these challenges also mean that any successful site-specific application of fungicides must comply with extremely high spatial and temporal resolutions. So to come up to the prerequisites for site-specific spraying against fungi is not easy, yet the prospects might justify the efforts.

As a consequence of this situation, two different approaches for site-specific application of fungicides have evolved. The first approach is based on a **full-area preventive concept**. Since for practical applications the detection of the initial tiny infected loci is not yet solved, still the whole area is treated in a precautionary manner. The timeliness for these largely prophylactic applications is oriented at local epidemic forecasts of extension services. However, there might still be a site-specific control of the application rate. This control might be based on more general crop properties such as biomasses or leaf-area-indices.

The second approach aims at **discrete spot spraying** of the few just infected loci and if possible not beyond these. This approach – if well conceived and executed – would allow a radical reduction of fungicide use. However, whereas the full area preventive concept with site-specific application is state of the art, the discrete spot spraying of fungicides still is in an experimental stage. And especially the latter concept needs sprayers that allow for separate section- or even better separate nozzle control. Modern sprayers that use **direct injection** of the pesticides into the water close to the nozzles instead of premixed batches provide the technical prerequisites for such a resolution in the application (Vondricka and Schulze Lammers 2009).

## 11.2 The Full-Area Preventive Concept Based on Biomass

It is general experience that dense, lush crops are more susceptible to fungal infections than thin and less developed canopies. On a site-specific basis, the biomass densities or leaf-area-indices of crops vary as the soils do. With the usual

# 11 Site-Specific Sensing for Fungicide Spraying

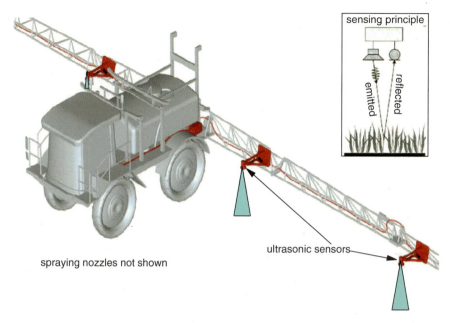

**Fig. 11.1** Section control of a sprayer by ultrasonic sensing of biomass. The *insert* shows the sensing principle (From Reusch 2009 and Agri Con GmbH, Jahna, Germany, altered)

practice of applying a uniform fungicide rate across the whole field, any site-specific differences in crop densities are disregarded. Yet since in many cases the leaf-area is the object of a fungal attack, it seems logical to apply the same concentration of fungicides per unit of leaf surface area. This is the rationale for fungicide application according to site-specific crop densities or its surrogates, the biomasses or leaf-area-indices.

Several methods of detecting the site-specific biomasses or leaf-area-indices are available and have been dealt with:

- proximal or remote **reflectance** sensing (Fig. 6.7)
- mechanical sensing of the **bending resistance** (Fig. 9.33)
- sensing by **ultrasonics** (this section Fig. 11.1 and Sect. 9.4.6).

The growth stages at which fungal infections occur can be very different. For small grains, depending on the respective fungi, infections can develop at almost any growth stage. But the sensing methods listed above differ in their capabilities to detect well at various growth stages. Only the ultrasonic method can reliably sense biomass at any growth stage (Fig. 11.2).

With reflectance sensing, because of soil effects in less developed canopies, the **minimum growth stage** of small grains is about EC or BBCH 30. This is when tillering has ended. Reflectance indices with wavelengths from the red edge range should be preferred in order to avoid limitations at advanced growth stages for lush crops.

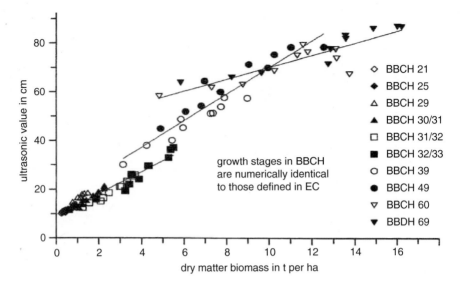

**Fig. 11.2** Sensing the aboveground biomass of winter wheat at various growth stages by ultrasonics (From Reusch 2009, altered)

Finally, sensing leaf-area-indices or biomasses via the bending resistance of crops requires even a minimum growth stage of about EC or BBCH 35 (Ehlert et al. 2004). This is when the first flag leaf – still rolled – has appeared.

The **savings in fungicides** that can be realized with a site-specific control based on biomass via a full-area preventive concept will vary with the heterogeneity of a crop. When the sensing in ten fields with small cereals and growth stages between EC or BBCH 39 and 71 was based on the bending resistance, savings in pesticides between a minimum 7 % and a maximum of 48 % were realized (Ehlert and Dammer 2006). The average saving in fungicides was 23.1 %. The savings did not affect the grain yields.

## 11.3 The Discrete-Spot Sensing Concept Based on Reflectance

### 11.3.1 The Pinpointing Approach and the Field of View

Theoretically, the savings in fungicides could be much higher than indicated above if only at the start of a fungal epidemic the infected loci were treated. This **pinpointing approach** still is in an experimental stage.

Site-specific spraying against fungi is much more difficult than site-specific fertilizing of nitrogen, though the sensing might be based in both cases on the reflectance of the canopy. Because usually nitrogen deficiencies occur in much larger patches or areas than fungal infections in their initial stages do. Hence for sensing fungal infections adequately, much smaller optical **fields of view** are needed. This will be difficult to obtain with sensing from satellites. But for proximal sensing from field based machines, principally it is possible to provide the small field of view by a narrow angle of vision and by a short distance to the canopy. In experiments with infections of yellow rust (*Puccinia striiformis*) and leaf blotch (*Septoria*) in winter-wheat, Moshou et al. (2011) used a field of view with 20 cm diameter on the canopy level for sensing from a sprayer-boom. However, small fields of view must be accompanied by many records in order to scan a field adequately. Already because of this, sensing for a discrete-spot effect will be more expensive than the conventional reflectance sensing for nitrogen fertilization.

Another point to consider is the fact that the **top leaf** of a plant in crops generally is not infected because it has only recently unfolded and incubation periods for the fungi hold (West et al. 2003). Hence in field operations, the viewing should be such that the top leaf is omitted as far as possible. Since with most crops – especially with small grains – the top leaf initially has a vertical position, oblique view directions towards the canopy should be avoided. With vertical viewing directions, primarily the horizontally oriented leaves get into the field of view, and the recently unfolded new vertical leaves affect the results less.

## 11.3.2 Spectra and Indices of Reflectance

There are many different types of fungal infections that can occur and these may affect the reflectance in different ways. Even within the same crop, different fungi can cause different **reflectance spectra**. This can be seen when the spectra of three different fungal infections of sugar beet leaves – cercospora leaf spot, powdery mildew and rust – are viewed (Fig. 11.3). The reflectances of the artificially infected leaves were recorded in a laboratory with increasing severities of the diseases (Mahlein et al. 2010).

The courses of the spectra for the non-infected leaves correspond to the known pattern within the visible and near-infrared wavelengths (Fig. 6.3). However, the deviations from these courses that result from the infections are different for each fungus. The powdery mildew infections cause rather uniform increases in reflectance in the visible as well as in the near-infrared range. The deviations for the infections of the cercospora and rust fungi do not comply with this in the range from about 720 to 900 nm wavelength. Within this range, the infections by these fungi result in decreases of the reflectance. And an unusually steep rise in the red reflectance evolves from cercospora fungi for highly infected leaves. But it has to be pointed out that these are results from one trial and its course of the reflectance could also be interpreted as the fact that *e.g.* 20 % of cercospora severity damages the crop the same as 100 % rust.

For the sensing of many crop properties, the use of **reflectance indices** has led to success. Generally indices allow to limit the sensing to just a few, narrow

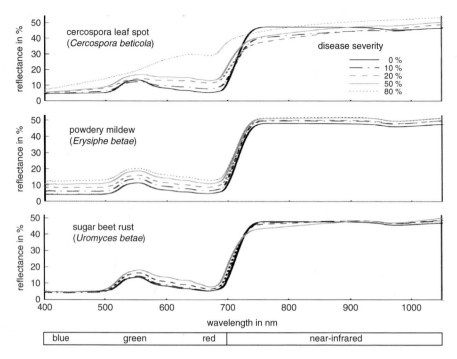

**Fig. 11.3** Reflectance spectra of sugar beet leaves infected with various fungi (From Mahlein et al. 2010, altered)

wavelengths and thus to simplify the technology and – if properly selected – still can provide a high precision.

Attempts with **standard reflectance indices** have been made for sensing fungal infections in wheat. The leaves were artificially infected. Among many standard indices, the best **differentiations** between infected and healthy leaves for yellow rust (*Puccinia striiformis*) were provided by either

- the Photochemical Reflectance Index = $\dfrac{R531 - R570}{R531 + R570}$ (Huang et al. 2007) or

- the Anthocyanin Reflectance Index = $\dfrac{1}{R550} - \dfrac{1}{R700}$ (Devadas et al. 2009).

It should be noted that all standard reflectance indices originally were developed for defined purposes. The photochemical reflectance index was created for assessing the efficiency of radiation-use in photosynthesis (Penuelas et al. 1995). And the anthocyanin reflectance index was developed for estimating the anthocyanin accumulation in plants (Gitelson et al. 2001). The fact that these indices depend on the development of rust fungi does not imply that they are the best possible choice for any reflectance indices that are conceivable. Because none of the standard reflectance indices was developed with the aim of sensing fungi.

It might be helpful to remember that in-season nitrogen sensing also started with standard reflectance indices that too originally also were created for other purposes (Heege and Reusch 1996). It turned out that the prediction of these standard indices could be improved by special indices, which solely were determined for nitrogen sensing (Sect. 9.4.3.1, Table 9.5). These special nitrogen reflectance indices were developed by systematically checking all theoretically possible ratios of narrow reflectance bands from the visible range plus the adjacent near-infrared range (Reusch 2003, 2005; Müller et al. 2008).

Hence for sensing of fungi it probably is reasonable as well to systematically check mathematical combinations of discrete narrow wavelengths along a sensible full spectrum for their sensitivities in this respect. The efforts needed for this **systematic searching** will be immense, since many different fungi and various crops should be considered. A start in this direction has been made by Mahlein et al. (2013) for fungal diseases of sugar beets.

Instead of sensing by indices from narrow wavelengths, sometimes **image sensing** for the detection of fungi is proposed. These images rely on broad wavelength bands, *e.g.* the red, green and blue (RGB) bands of the spectrum. An obvious advantage of imaging is that the locations of the infected loci or discrete spots directly can be seen on the records. A disadvantage of images is that the broad wavelength bands tend to hide early effects of fungal infections. The detection is possible not until when symptoms appear as obvious lesions. Yet with suitable spectral indices from narrow wavelengths, the sensing can start already slightly before the first sporulation (Moshou et al. 2011). And a still earlier detection might be possible when fluorescence provides the signals. This will be dealt with in the next sections.

## 11.4 The Discrete-Spot Sensing Concept Based on Fluorescence

### 11.4.1 Indirect Measuring with In-Situ Sensor System

Fluorescence is the emission of radiation that follows an absorption of light energy. The emitted radiation has longer wavelengths than the absorbed light. Fluorescence requires an adequate substance called **fluorophore**. Normally, the sensing of fungi by means of fluorescence is an indirect method. This means that the fungi are non fluorescent or only show a very weak fluorescence according to their low concentration in the leaf when compared with fluorescent plant pigments. Only some biotrophic fungi like powdery mildew and yellow rust show a blue fluorescence if excited with UV light (Zhang and Dickensen 2001).

The normally called "plant fluorescence" originates from substances insides the leaves that are natural fluorophores. This is different from the fluorescence detection tools used in serological or molecular methods where the samples must be prepared to show fluorescence.

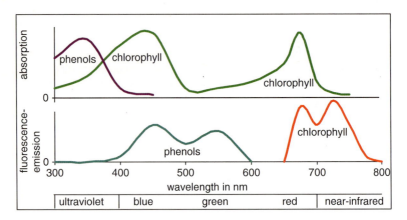

**Fig. 11.4** Schematic diagram of plant fluorescence: The light absorption in the ultraviolet region by phenols (*violet line at top*) induces fluorescence in the blue-green (*blue line at bottom*), the light absorption in the blue and red region by chlorophyll (*green line at top*) induces fluorescence in the far-red region (*red line at bottom*)

In higher plants, the natural fluorophores are mainly chlorophyll and phenols. The latter are organic compounds that develop in plants during decomposition. The chlorophyll absorbes primarily in the blue as well as in the red and it emits fluorescence in the far red, whereas the phenols absorb dominantly in the ultraviolet and emit in the blue and green. The typical spectral characteristics of these optical properties are shown in Fig. 11.4.

A non-invasive measurement is only possible from the whole leaf or plant. This is of course not only a fluorescent dilution, but it is a very complex optical system with many other compounds in separated compartments and with a typical geometrical structure. For a first approach, one can summarise up to three types of fluorescence with their related measurement techniques:

- With excitation in the ultraviolet, the whole fluorescence **emission spectrum** is possible to measure. The excitation wavelength, which is also reflected, can be separated from the emission with wavelength-selective filters. But with blue or red excitation, only the chlorophyll fluorescence is measureable. Because these wavelengths are not short enough to induce phenolic fluorescence. So the fluorescence emission provides information about chlorophyll and/or phenols.
- Another method is measuring the **excitation spectra,** which induce the fluorescence. The excitation wavelength is changed consecutively to discrete bands or scanned continuously and the fluorescence is measured at a fixed emission wavelength. So a kind of absorption spectrum for substances inside the leaf is obtained. Normally the emission is detected in the far red (approximately 650–750 nm). So chlorophyll serves as sensor inside the leaf. Mainly two excitations are used: one within the ultraviolet and one within the visible region.
- As mentioned before in Sect. 6.4 "Fluorescence Sensing", the fluorescence is also temporally variable. This so called "**Kautsky effect**" is typically measured

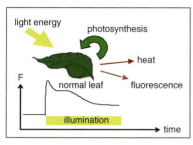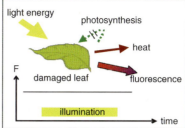

**Fig. 11.5** Schematic of variable fluorescence (F) in a normal leaf and in a damaged leaf where photosynthesis is totally blocked. In the former, the fluorescence shows typical kinetics during illumination. In contrast to this, the damaged leaf has a higher but constant fluorescence

when a dark adapted plant is illuminated with intense light. The effect in normal plants is that the fluorescence rises quickly up to five times higher than an initial low level and after a few seconds it decreases to an intermediate stationary level. This is due to the processing of the light in photosynthesis, and it is obvious that a damaged leaf which is "dead" or "nearly dead" shows no or limited variable fluorescence (Fig. 11.5).

So in summary: there exist not only the spectral emission and excitation spectra of fluorescence, but also the temporal kinetic behaviour of the plant's fluorescence.

## 11.4.2 Fungi-Plant-Interaction and Physiology of Infected Plants

There exist many fungi (about 1.5 Mio) and plant species (250,000) and these have had much time during their evolution to develop strategies against each other. Fungi cause a **defence mechanism** in the infected plant. Elicitors (*e.g.* reactive oxygen species – ROS – that are produced by fungi) trigger a hypersensitive response of the plant. This could be the production of special substances – callose as a barrier for the fungi or phenols with antifungal properties – or the dying of hypersensitive cells.

With common fungi in cereals, the defense reactions are either **necrotrophic** or **biotrophic**. For the first case: septoria leaf spot and fusarium head blight are typical examples in wheat. These fungi are destroying the plant cell walls with toxic substances and lead to necrosis of the plant leaf. Hereby compounds in the plant cells transform into phenols (Osbourn 1996).

The obligate biotrophic fungi need living plant material as they assimilate the plant's nutrients by haustoria (extraction by roots). Typical examples are powdery mildew (*Blumeria*) and rust (*Puccinia*) fungi. The effects on the plant parameters are normally reduced photosynthesis but increased respiration. The biotrophic fungi often form "green islands" on the leaf with a nutrient sink and higher nonphotochemical quenching (energy dissipation of the light excitation) than the surrounding area (Scholes and Rolfe 1996).

**Table 11.1** Types of plant fluorescence and their applications

| Fluorescence | Subtype and index | Properties | Trend when infection rises |
|---|---|---|---|
| Emission | Red: F680/F730 | Chlorophyll | Long term falling |
| | Blue green : F450/F730 | Phenols | Short term rising |
| Excitation | $F_{UV}/F_{VIS}$ | Phenols | Short term rising |
| Kinetics | $(F_M - F_0)/F_M$ | Photosynthesis | Decreased activity |

Hypersensitive death of invaded cells is known as the most typical feature of rust resistance (Heath 1982), and this increases the concentration of phenols.

## 11.4.3 Fluorescence Indices Related to Infection

Indices are normally ratios of simultaneously measured radiation intensities. They are rather insensitive to changes in measurement geometry, to sensor drifting or to environmental conditions. Each index can be assigned to the corresponding measurement technique:

- The maxima in the fluorescence **emission spectra** correspond to chlorophyll and phenols (Fig. 11.4, bottom). The ratio of the intensities of the maximum chlorophyll fluorescence supplies a simple but powerful index. This relation of **F680/F730** is a measure for the chlorophyll content (Gitelson et al. 1999). As described before in Sect. 6.4.1, this ratio is negatively correlated to the chlorophyll content due to self absorption of the radiation below 700 nm. The blue fluorescence around 450 nm related to the far red fluorescence 680 or 730 nm – *e.g.* **F450/F680** – is related to the content of phenols (Lichtenthaler and Schweiger 1998).
- The dual **excitation** of chlorophyll with light in the ultraviolet and visible region leads to the $F_{UV}/F_{VIS}$ quotient. For this the fluorescence intensity in the far red excited with UV light is related to the one that is excited with VIS – normally blue or red light. This is a measure for the UV absorbing pigments in the leaf epidermis (Cartelat et al. 2005), namely the phenols.
- The **kinetics** of the Kautsky effect are very complex and can be influenced by the light scheme during measurement. The main parameter is the variable fluorescence, *i.e.* the difference between maximum ($F_M$) and minimum fluorescence ($F_0$) related to the maximum fluorescence, hence $(F_M - F_0)/F_M$. This variation occurs in the time of some seconds. The initial rise in the millisecond range can be described with a special method (Strasser et al. 2004).

In general the changes in plant properties – *e.g.* chlorophyll and phenol content or photosynthesis – are detectable with the corresponding fluorescence measurement techniques. This is summarised in Table 11.1. In detail there is a lot of literature dealing with stressed plants and fluorescence. In the following, only a few selected publications that deal with fungal infections as the stressor are listed:

- Lüdeker et al. (1996) analysed wheat and barley infected with rust (*Septoria*) and mildew (*Blumeria*) with ultraviolet excitation and found the blue **fluorescence**

**emission** ratio F440/F730 higher in infected leaf areas. They related the effect directly to the fungi, especially for the symptoms of mildew. But also for necrotrophic fungi like septoria leafspot it was found that this fluorescence ratio F440/F730 increases compared to uninfected wheat leaves even before symptoms are visible (Thiessen 2002). For the classification in healthy and infected leafs the whole fluorescence emission spectra could be used with multivariate analysis. Römer et al. (2011) demonstrated that there are mainly differences in the **blue-green region** of fluorescence spectra comparing healthy and rust (*Puccinia*) infected wheat leafs. This is due to the change in phenols in the early first days after infection. Differences in the far-red region due to chlorophyll degradation show up significantly later. This is in agreement with many other results in literature (*e.g.* Kuckenberg et al. 2009) where the fluorescence emission F680/F730 was used to quantify the disease. With this index, changes show up within a time period in which also visual symptoms and chlorosis appear. This could also be measured by reflectance.

- Scholes and Rolfe (1996) studied the **variable fluorescence** on oat leaves infected with crown rust (*Puccinia coronata*). They discriminated infected and healthy regions by differences in the fluorescence kinetics. The **fast kinetic method** shows differences between drought stress and infection with a fungal infection of wine – called esca disease – compared to the control (Christen et al. 2007).

## 11.4.4  Problems and Discussion

The **environmental light** also induces fluorescence that is reflected into the sensor. Normally this fluorescence is eliminated by modulating the active excitation light (*e.g.* by using **pulsed lasers**) and relating the detected fluorescence only to these light pulses. This works even in field conditions. But for the variable fluorescence, the uncontrolled **sun light** generates an additional effect to the photosynthetic state. And for this method, of course the measuring duration of some seconds is a problem when sensing in the field from a driving vehicle.

Fluorescence indices are also influenced by other factors like abiotic stress, environmental condition or even age of the leaves (Gorbe and Calatayud 2012). So it is difficult to state that *e.g.* "F440/F730>0.8" means "infected" and "F440/F730<0.8" means "healthy". As with other indirect measurements of plant parameters like reflectance or mechanical resistance, the fluorescence is a **relative measure** and needs to be compared with "reference" plants. So a calibration for the special situation is essential. One possible solution to differentiate between stressors is a multivariate calibration. For this, more then one index is measured, *i.e.* some spectral region or even a complete spectrum is obtained and analysed according to its whole profile.

Another fact is that an infection shows a spatial pattern on the leaf. If symptoms appear pronounced in small spots (like sporulation or insertion points of the fungi), the use of sensors with a high spatial resolution – in the range of some mm – might be needed. Consequently, the variation of the detected property along the leaf is an

**Fig. 11.6** Experimental robot in maize (Photo: Ruckelshausen)

identifier for an infection. This can be detected with a spot sensor by successive scanning or with images from cameras (Kuckenberg et al. 2009; Moshou et al. 2005). The image analysis of fluorescence pictures is a very promising technique and could be automated with increasing computational capacity.

For an effective sensing in a field, the **sample size** is an important aspect: Fungicides are applied typically in an area within the working width of a sprayer. However, optical sensors for fluorescence detect an area in the range of only some $mm^2$ up to a few $cm^2$ when acting with artificial light. Small plots of infections might start with a diameter of maybe some meters. Hence if the sensor is operating in an on-the-go manner, there might be problems:

- The measured area is only a very small spot sample of the whole field (*e.g.* a line of some mm of the 24 m working width).
- The concerned leaf area (which is affected by fluorescence changes due to infection) is too small and the sensed surface of some $cm^2$ shows up in an average signal, which is not significantly different from healthy plants.
- The infected crop area is not measured, because the driving direction of the sensor is not in line with the infected plot.

The solution could be smaller sensed sample areas that are obtained with **scanning laser techniques** *or a* **series of small sensors** perpendicular to the driving direction. A large size of the sensed area would be a good choice if the infection and its fluorescence signals extend over whole leaves or plants. Though this probably is not the rule, it still might be necessary to have a signal which describes the average plant property of an area where the application should be done. Depending on the spraying technique, this could be within the range of the working width, within sections of the boom length or even only within the range of individually controlled nozzles (Vondricka and Schulze Lammers 2009).

The best way out of this dimensional problem between current spraying techniques and the infected loci probably is to leave the detection and perhaps also the treatment to small **scouting robots** (Fig. 11.6). In the future, these small robots might loiter through the fields and carefully inspect as well as treat individual plants.

## 11.4.5 Sensors for Practice and Research

For an on-line use, the **MiniVegN** (Fritzmeier, Germany) or **Laser-N-Detector** (Planto, Germany) are commercially available. These systems are able to measure the fluorescence emission ration F680/F730 of a small spot which is scanned beyond the tramline. This principle is sensible to the chlorophyll content and therefore mainly used for nitrogen application.

When using handheld commercial sensors, it is possible to measure almost everything. Spectralfluorimeters (*e.g.* RF5001PC, Shimadzu, Japan) with ultraviolet excitation can be applied in a laboratory or a field to monitor the fluorescence emission. For full **excitation spectra of plant fluorescence,** the **Dualex** and **Multiplex** (Force-A, France) or the **UV-PAM** (Walz, Germany) are available. The **kinetics** can be measured *e.g.* with the **PAM** (Walz, Germany) or the fast kinetics with the **PEA** (Hansatech, England).

## 11.5 Differentiation Between N Deficiency and Effects of Fungi

For any site-specific application of farm chemicals there is the problem that the control signals of sensors may be ambiguous. Nitrogen deficiencies in crops cause increases of reflectances in the visible range. The same effect can result from fungal infections (Fig. 11.3). And there can be further examples of abiotic stress factors – *e.g.* lack of water – that affect the signals of a sensor in a similar way as biotic stress factors do. Because all signals of optical sensors are substitutes for the respective site-specific crop properties. These substitutes can be influenced by other factors than the one that the control should rely on.

The interaction which thus might arise between the effects of **nitrogen and fungi** is especially disturbing. Because it is general experience that high nitrogen rates promote the development of fungi. Hence if the control of site-specific nitrogen application results in increasing rates because of high reflectances in the visible range that are caused by fungal infections, the result on the development of the crop might be disastrous.

To some extent, the present reflectance indices for nitrogen sensing do prevent this misinterpreting of the situation. Because these indices do not only include details of the visible spectrum, but of the near-infrared radiation as well. This even holds if the indices are extracted from the red edge range. In the near-infrared and red edge range, the effects of fungi can deviate from the result of nitrogen (see Figs. 9.19 and 11.3).

Reusch (1997) designed a field experiment with winter barley and strips that were either sprayed against fungi or not. In addition, different nitrogen application rates were included on all strips. The strips where no fungicides had been applied and where higher nitrogen rates had been given were more severely

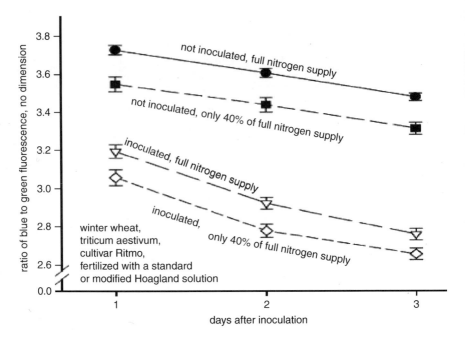

**Fig. 11.7** Effects of nitrogen supply and inoculations by powdery mildew – *Blumeria graminis* – on the ratio of blue to green fluorescence (From Bürling et al. 2011, altered)

infected by leaf blotch (*Rhynchosporium secalis*), leaf rust (*Puccinia hordei*) and powdery mildew (*Erysiphe graminis*). However, the effects on the **red edge inflection point indices** were much higher for the nitrogen rates than for the mode of fungicide application. The red edge inflection point indices thus represented mixed effects from nitrogen as major contributor and fungi as minor participants.

**Mixed effects** of nitrogen and fungi may also be hidden in fluorescence signals. However, the weighting of the contributors nitrogen and fungi for the sensed index might be reversed (Bürling et al. 2011). In Fig. 11.7, the sensing index is the ratio of the blue to green fluorescence emission amplitudes (Fig. 11.4 bottom, blue curve). The results were obtained in a laboratory and refer to wheat leaves from plants that were artificially infected or not by powdery mildew and either fully or only partially supplied with nitrogen. There is an effect of the nitrogen supply on the fluorescence index, yet it is small. The influence of the fungal infections on the index is much more pronounced.

So regarding reflectance sensing by a red edge inflection index, the signals represent predominantly the nitrogen supply, whereas instead with the blue to green fluorescence index mainly the fungal infections matter. In the experiments by Bürling et al. (2011) this held not only for powdery mildew (*Blumeria graminis*) but for leaf rust (*Puccinia triticina*) as well. The discrimination among the infected and not infected groups was possible as early as 1 or 2 days after inoculation for powdery mildew and leaf rust respectively.

In short, the problems with mixed effects from nitrogen supply and from fungal infections might be dealt with by **fusing** reflectance and fluorescence sensors. The latter, however, are not yet capable to sense fungi in an on-the-go mode in fields. And the multitude of different fungi and various crops still calls for much research.

## 11.6 Summary and Prospects

A pinpointed, site-specific application of fungicides is one of the most ambitious and challenging aims in precision farming. Because extraordinary high spatial- as well as temporal resolutions are needed. If these resolutions can be realized in an on-the-go mode, very substantial savings in fungicides and consequently environmental reliefs might become feasible.

Yet as long as such extraordinary high spatial resolutions cannot be obtained in practice, site-specific application based on the varying biomass or leaf-area-indices of crops must be regarded as the sole present alternative. This method fits well to a precautionary, prophylactic approach of spraying against fungi. Though prophylactic methods hardly provide options for drastic reductions in the use of fungicides, they might even be needed in combination with pinpointed, site-specific applications. The timeliness of both methods might be greatly improved by local epidemic forecasts of extension services.

The pinpointed approach might rely either on reflectance or on fluorescence. Both methods still are not state of the art, and its prospects for applications in practice are difficult to predict. Solely reflectance sensing for biomass and for nitrogen is well established. This technique depends largely on leaf-area-indices and chlorophyll. Fluorescence can be based on either emission spectra that sense chlorophyll, on excitation spectra that are influenced by phenols or on its kinetics that depends on the photosynthesis.

The earliest detections of fungal infections theoretically can be provided when the indications occur via phenols or via the photosynthetic process. Hence in this respect, sensing by the blue-green fluorescence or by the kinetics of the variable fluorescence is an advantage. Whenever the indications rely on biomasses, leaf-area-indices and chlorophyll, the earliest detection is postponed by a few days until visual symptoms like chlorophyll degradation or other canopy alterations occur. This holds regardless of the sensing method that is used for these indicators.

Whether in practice any earliest possible detection as outlined above will be feasible and reasonable, depends not only also on the ability of sensing but also on control techniques that deal with the tiny initially infected loci. Since managing the spreading of fungicides with a temporal precision of 1 or 2 days might be difficult in practice anyway, the use of reflectance sensing methods should not be excluded. For any pinpointing approach, sophisticated scanning, processing and spraying techniques will be essential.

# References

Bürling K, Hunsche M, Noga G (2011) Use of blue-green and chlorophyll fluorescence measurements for differentiation between nitrogen deficiency and pathogen infection on winter wheat. J Plant Physiol 168(14):1641–1648

Cartelat A, Cerovic ZG, Goulas Y, Meyer S, Lelarge C, Prioul JL, Barbottin A, Jeuffroy MH, Gate P, Agati G, Moya I (2005) Optically assessed contents of leaf polyphenolics and chlorophyll as indicators of nitrogen deficiency in wheat (*Triticum aestivum* L.). Field Crops Res 91(1):35–49

Christen D, Schönmann S, Jermini M, Strasser RJ, Défago G (2007) Characterization and early detection of grapevine (*Vitis vinifera*) stress responses to esca disease by in situ chlorophyll fluorescence and comparison with drought stress. Environ Exp Bot 60(3):504–514

Devadas R, Lamb DW, Simpfendorfer S, Backhouse D (2009) Evaluating ten spectral vegetation indices for identifying rust infection in individual wheat leaves. Precis Agric 10:459–470

Ehlert D, Dammer KH (2006) Widescale testing of the crop-meter for site-specific farming. Precis Agric 7:101–115

Ehlert D, Dammer KH, Völker U (2004) Application according to plant mass (in German). Landtechnik 59(2):76–77

Gitelson A, Buschmann C, Lichtenthaler HK (1999) The chlorophyll fluorescence ratio F735/F700 as an accurate measure of the chlorophyll content in plants. Remote Sens Environ 69(3):296–302

Gitelson A, Merzlyak MN, Chivkunova OB (2001) Optical properties and nondestructive estimation of anthocyanin content in plant leaves. Photochem Photobiol 74(1):38–45

Gorbe E, Calatayud A (2012) Applications of chlorophyll fluorescence imaging technique in horticultural research: review article. Sci Hortic 138:24–35

Heath MC (1982) Host defense mechanisms against infection by rust fungi. In: Scott KJ, Chakravorty AK (eds) The rust fungi. Academic, London, pp 223–245

Heege HJ, Reusch S (1996) Sensor for on-the-go control of site-specific nitrogen top dressing. International Meeting in Phoenix, AZ. American Society of Agriculture Engineering, St. Joseph, MI, Paper No. 961018

Huang W, Lamb DW, Niu Z, Zhang Y, Liu L, Wang J (2007) Identification of yellow rust in wheat using in-situ spectral reflectance measurements and airborne hyperspectral imaging. Precis Agric 8:187–197

Kuckenberg J, Tartachnyk I, Noga G (2009) Detection and differentiation of nitrogen-deficiency, powdery mildew and leaf rust at wheat leaf and canopy level by laser-induced chlorophyll fluorescence. Biosyst Eng 103(2):121–128

Lichtenthaler HK, Schweiger J (1998) Cell wall bound ferulic acid, the major substance of the blue-green fluorescence emission of plants. J Plant Physiol 152:272–282

Lüdeker W, Dahn HG, Günther KP (1996) Detection of fungal infection of plants by laser-induced fluorescence: an attempt to use remote sensing. J Plant Physiol 148(5):579–585

Mahlein A-K, Steiner U, Dehne H-W, Oerke E-C (2010) Spectral signatures of sugar beet leaves for the detection and differentiation of diseases. Precis Agric 11:413–431

Mahlein A-K, Rumpf T, Welke P, Dehne H-W, Plümer L, Steiner U, Oerke E-C (2013) Development of spectral indices for detecting and identifying plant diseases. Remote Sens Environ 128:21–30

Moshou D, Bravo C, Oberti R, West J, Bodria L, McCartney A, Ramon H (2005) Plant disease detection based on data fusion of hyper-spectral and multi-spectral fluorescence imaging using Kohonen maps. Real-Time Imag 11(2):75–83

Moshou D, Bravo C, Oberti R, West J, Ramon H, Vougioukas S, Bochtis D (2011) Intelligent multi-sensor system for the detection and treatment of fungal diseases in arable crops. Biosyst Eng 108:311–321

Müller K, Böttcher U, Meyer-Schatz F, Kage H (2008) Analysis of vegetation indices derived from hyperspectral reflection measurements for estimating crop canopy parameters of oilseed rape (Brassica napus L.). Biosyst Eng 101:172–182

Osbourn AE (1996) Preformed antimicrobial compounds and plant defense against fungal attack. The Plant Cell 8:1821–1831

Penuelas J, Filella I, Gamon JA (1995) Assessment of photosynthetic radiation-use efficiency with spectral reflectance. New Phytol 131:291–296

Reusch S (1997) Development of an optical reflectance sensor for recording the nitrogen supply of agricultural crops (in German). Doctoral dissertation, University of Kiel, Forschungsbericht Agrartechnik der Max-Eyth-Gesellschaft Agrartechnik im VDI 303

Reusch S (2003) Optimization of oblique-view remote measurement of crop N-uptake under changing irradiance conditions. In: Stafford J, Werner A (eds) Precision agriculture. Papers from the 4th European Conference on Precision Agriculture, 16–18 June 2003, Berlin. Wageningen Academic Publishers, Wageningen

Reusch S (2005) Optimum waveband selection for determining the nitrogen uptake in winter wheat by active remote sensing. In: Stafford J (ed) Precision agriculture '05. Wageningen Academic Publishers, Wageningen, extended paper of Precision Agriculture '05 conference, pp 261–266

Reusch S (2009) Use of ultrasonic transducers for online biomass estimation in winter-wheat. In: van Henten EJ, Goense D, Lockhorst C (eds) Precision agriculture '09. Wageningen Academic Publishers, Wageningen, pp 169–175

Römer C, Bürling K, Hunsche M, Rumpf T, Noga G, Plümer L (2011) Robust fitting of fluorescence spectra for pre-symptomatic wheat leaf rust detection with support vector machines. Comput Electron Agric 79(2):180–188

Scholes JD, Rolfe SA (1996) Photosynthesis in localized regions of oat leaves infected with crown rust (*Puccina coronata*): quantitative imaging of chlorophyll fluorescence. Planta 199:573–582

Strasser RJ, Srivastava A, Tsimilli-Michael M (2004) Analysis of the chlorophyll a fluorescence transient. In: Papageorgiou G, Govindjee C (eds) Chlorophyll a fluorescence: a signature of photosynthesis, vol 19, Advances in photosynthesis and respiration. Springer, Dordrecht, pp 321–362

Thiessen E (2002) Optical sensing techniques for site-specific application of farm chemicals (in German). Doctoral dissertation, University of Kiel, Kiel, Germany. Forschungsbericht Agrartechnik, VDI-MEG 399

Vondricka J, Schulze Lammers P (2009) Real-time controlled direct injection system for precision farming. Precis Agric 10:421–430

West J, Bravo C, Oberti R, Lemaire D, Moshou D, McCartney HA (2003) The potential of optical canopy measurement for targeted control of field diseases. Annu Rev Phytopathol 41:593–614

Zhang L, Dickinson M (2001) Fluorescence from rust fungi: a simple and effective method to monitor the dynamics of fungal growth in planta. Physiol Mol Plant P 59(3):137–141

# Chapter 12
# Site-Specific Recording of Yields

Markus Demmel

**Abstract** Site-specific recording of yields, also known as geo-referenced yield detection or yield monitoring, contributes to site-specific precision farming concepts by delivering fundamental information about the diversity of the yield potential. It is feeding back the effects of site-specific management on yields and allows to calculate the exports of nutrients.

In the early 1990th it started with combine harvesters and continuously working systems for recording yields in combination with the position of the machine. It followed site-specific yield measurement systems for forage harvesters, cotton pickers or strippers, potato and sugar beet harvesters, peanut and grape harvesters as well as sugar cane harvesters. Attempts were made to determine site-specific yield of manually harvested cultures like oranges, apples or coffee. Testing procedures to determine the accuracy of material flow sensors and yield measurement systems in the laboratory and in the field have been standardized.

Systems for site-specific recording of yields in combine-harvesters, forage choppers and cotton pickers are available at the market. The adoption in professional farming is concentrating on combinable crops. Efforts are needed to make systems available for all important crops, to integrate the sensing of essential crop ingredients and to standardize data formats as well as algorithms for data filtering and data analysis.

**Keywords** Site specific • Geo-referenced • Yield detection • Yield measurement • Yield monitoring • Yield recording

---

M. Demmel (✉)
Institute for Agricultural Engineering and Animal Husbandry,
Bavarian State Research Center for Agriculture, Voettinger Street 36,
85354 Freising, Germany
e-mail: markus.demmel@lfl.bayern.de

## 12.1 Introduction

Site-specific recording of yields, also known as **geo-referenced yield detection** or yield measurement, was the first technology in precision farming that was available as an accessory for combine harvesters in the early 1990th. The objective is to automatically get information about the heterogeneity of crop yields within fields.

In the context of site-specific precision farming, geo-referenced detection of yields has three major objectives:

- It delivers information to determine the **site-specific yield potential** of fields that were managed uniformly before the transition to site-specific management.
- It gives a **feed-back of the effects on yield** of site-specific management.
- It makes it possible to calculate the **export of nutrients** for balancing and management of plant nutrition.

To fulfill these tasks, geo-referenced yield data have to be collected continuously covering the whole area of a field with an acceptable accuracy.

## 12.2 Principle of Site-Specific Yield Recording

Recording site-specific yield data from crops can be provided by some extra technology on the harvesting equipment. The principle and the components for most local yield measurement systems include:

- Product output sensor (*e.g.* grain output)
- Area sensing (speed and working width)
- Position detection system (usually differential GNSS)
- Data processing, -monitoring and -storing unit
- Data transfer to office computer.

The arrangement and interaction of the subsystems can be seen in Fig. 12.1, which shows an example for site-specific yield recording in a combine harvester.

The systems to detect the **product output** (t/h) have to be adapted to the specific crop and to the harvesting process and equipment, while sensing the **area capacity** (ha/h) is realized by measuring the speed of the harvester and multiplying this by the working width. Dividing the product output by the area capacity provides the actual local yield data. After a first statistical smoothing (*e.g.* moving average) this site-specific yield in mass per unit area (as well as the product output and the area capacity) is presented to the machine driver on the information display and stored typically every second together with the positional data of the harvester in an electronic device.

Within some time, at least at the end of the harvesting season, the data are transferred to an office computer for further processing. The media for data transfer have often changed during the last two decades due to fast technological progress.

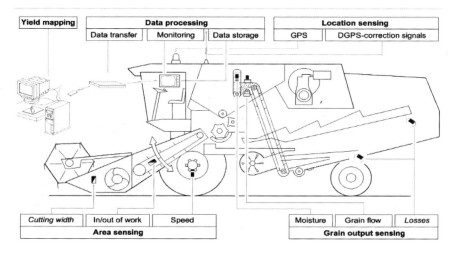

**Fig. 12.1** Subsystems for site-specific yield recording in a combine harvester

Filtering, analyzing and mapping of the geo-referenced yield data is normally realized on the office computer with specific software.

For the first time Bae et al. (1989) automatically recorded grain yields and the position of the combine harvester in a field using a yield measurement system and a radio beacon positioning system.

A comprehensive overview on systems for yield sensing of combinable crops, forage crops and root crops was presented by Demmel in Schmidhalter et al. (2005).

## 12.3 Yield Measurement for Combinable Crops

For yield detection within combine harvesters, several grain output sensors have been developed, evaluated and introduced into practical use. They work on the continuous flow principle and are installed in the upper part or in the head of the clean grain elevator (Auernhammer et al. 1993). The measuring systems on the market are either based on the volume flow principle or on the mass flow principle (Fig. 12.2).

With the **volume flow measurement principle**, the mass of the grain output is obtained by converting volume flow via specific gravity or specific density into mass flow. The volume is recorded by determining the grain pile volume on the elevator paddles.

The first continuously working yield measurement system on the market, which is not in production any more, used a paddle/bucket wheel with a magnetic clutch that was fed from a widened elevator head serving as a hopper (Claydon Yield-O-Meter, Fig. 12.2, system 1). As soon as enough grain was in the hopper, the paddle wheel started turning and conveyed the material into the grain auger. The frequency of the buckets was detected and provided the volume flow. This was converted into

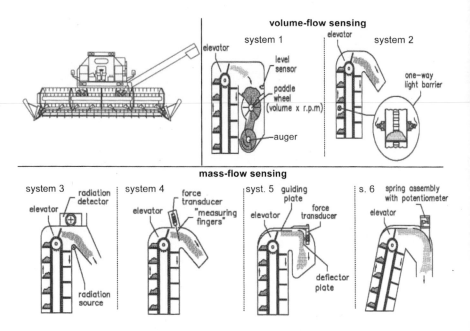

**Fig. 12.2** Grain output and grain flow measurement systems for combine harvesters

mass flow by accounting for the grain density (hl-weight) that was determined with a measuring cylinder and a spring scale.

Two systems on the market (Claas Quantimeter and RDS Ceres) operate with a light barrier in the upper part of the feed-flow side of the clean grain elevator (Fig. 12.2, system 2). The grain piles on the elevator paddles interrupt the light beam. From the length of the dark phase and from calibration functions, the height and hence the volume of the grain load on the paddles is calculated. The zero tare value is obtained from the darkening rate when the elevator is running empty. A tilt sensor is used to compensate for the influence of non-uniform loading of the elevator paddles on a side slope. The hl-weight for transforming into **mass-flow** is determined with a beam-balance. In all recording systems, this is converted into the area yield by referencing to the harvested area, which is calculated from the cutting width and the harvesting speed (wheel sensor). Finally, with all combine yield sensing systems, the grain flow and the yield can be adjusted to standard moisture by using a continuously operating moisture sensor in the clean grain elevator.

For directly **determining the mass** of the grain flow, either a force/impact measurement principle or on the absorption of gamma rays by mass in a radiometric measuring system is used.

One early measurement system, which was taken from the market a few years ago, operated according to the radiometric principle (Massey Ferguson Flowcontrol). The grain discharged from the elevator paddles passed through a region between a weakly radioactive source (Americium 241, activity 35 MBq) and a radiation sensor. As it did so, radiation was absorbed (Fig. 12.2, system 3). The degree of absorption corresponded to the mass per unit area of the grain in the region of the measuring

## 12 Site-Specific Recording of Yields

**Table 12.1** Errors of yield sensing systems for combine harvesters in practical use with field experiments

| Sensing system (in bold), Meter (in italics), Manuf. (in brackets) | Period of study, total area, number of grain tank loads | Combine harvester models, grain spec. | **Relative calibrat. error** in % (av.) | **Stand. dev. of rel. error** in % |
|---|---|---|---|---|
| **Light-barrier, system 2,** CERES 2, (RDS) | 3 years, 140 ha, 179 tank loads | 3 models, 4 grain species | −0.14 | 3.43 |
| **Radiation, system 3,** FLOWCONTROL, (MASSEY FERG.) | 2 years, 140 ha, 132 tank loads | 2 models, 2 grain species | −1.01 | 4.07 |
| **Force, system 5,** YM 2000, (AGLEADER) or LH565 (LH AGRO) | 3 years, 130 ha, 182 tank loads | 3 models, 4 grain species | −1.83 | 4.06 |

For system numbers see Fig. 12.2 (Demmel 2001)

window. The material velocity, which was obtained from the elevator speed, was used to calculate mass flow. Similar systems are used in food processing.

A number of **yield measurement systems** developed in the USA use the force/impact measurement and is likewise fitted into the elevator head in the discharge path of the corn. The sensor consists either of a baffle plate that is fitted to a force-measuring cell (AgLeader, Case) or of a curved plate fitted to a spring element measuring the displacement way (John Deere) or of a curved plate mounted in patented geometry to a force measurement cell to compensate varying friction force (Strubbe et al. 1996; Strubbe 1997). Corn hitting the baffle plate or the curved plate causes a force effect to the bending bar, the spring element or the load cell, which is electrically sensed with strain gauges or the displacement sensor (Fig. 12.2, systems 4–6). Since this impact is the product of mass and velocity, it is possible to calculate the mass flow. The material velocity is obtained, in turn, from the elevator velocity.

All systems consist of the sensor element, processing, monitoring and data storage units in the cab and have the possibility to integrate information from a moisture sensor. Today most of the systems are factory installed accessories; some products can be retrofitted to some combines (Demmel 2001).

Some authors report the evaluation of a single yield sensor under specific situations in the field (e.g. Perez-Munoz and Colvin 1994, Senaci and Yule 1996).

Extensive studies on the **measuring accuracy** of several sensing systems were carried out in the years 1991–1995 (Demmel 2001). The field experiments (Table 12.1) were supplemented by joint test bench trials in 2000 and 2001 (Kormann et al. 1998; Demmel 2001).

The level of accuracy in the field experiments was determined by counter weighing the grain tank loads on calibrated and certified platform scales. The measuring systems were examined, in part, on different combine harvester models with different grain types in lightly to medium rolling fields (Table 12.1).

The mean relative error represents a measure of the calibration quality. It should be zero ideally, or at least close to zero. This requirement was successfully achieved

**Table 12.2** Errors of yield sensing systems for combine harvesters at different throughputs in the 2000/2001 test bench studies

| Sensing system (in bold), *Meter* (*in italics*), Manufacturer (in brackets) | Relative calibration error in % (average) | Standard deviation of relative error in % |
|---|---|---|
| **Light-barrier, system 2,** *CERES 2*, (RDS) | −0.57 | 5.50 |
| **Light-barrier, system 2,** *QUANTIMETER*, (CLAAS) | −2.71 | 1.72 |
| **Light-barrier, system 2,** *PRO SERIES 2000*, (RDS) | −3.89 | 5.54 |
| **Radiation, system 3,** *FLOWCONTROL*, (MASSEY FERG.) | −1.64 | 3.02 |
| **Force, system 4,** *FIELDSTAR*, (DRONNINGB./AGCO) | −0.22 | 1.52 |
| **Force, system 5,** *YM 2000* (AGLEADER), *LH 565* (LH AGRO) | −1.71 | 3.65 |
| **Force, system 6,** *GREENSTAR*, (JOHN DEERE) | −2.89 | 2.81 |

Flat standing position; 10, 15, 20, 25 and 30 t h$^{-1}$ throughput; 5 repetitions/variant; n=25/m; reference mass/variant=1 t; winter wheat. For system numbers see Fig. 12.2 (Demmel 2001)

by all meters. The standard deviation (s) is the measure of the sensing accuracy. It indicates the range of error, within which around 2/3 of all measurements lie. Despite the different measuring principles, all sensing systems are characterised by approximately equal ranges of error between ±3.5 and ±4 %.

In the **test bench studies**, the accuracy of the systems was determined under identical, clearly defined conditions. Special consideration was given to the effect of different throughput levels and of transverse and longitudinal tilt (Table 12.2).

When the measuring accuracy of the various yield measuring systems was checked in the test bench under flat conditions at different throughputs, mean calibration errors <3 % are obtained with two exceptions. Larger deviations (3–10 %) only occur at lower throughputs (10 t h$^{-1}$). This indicates that the calibration curves plotted in the instruments are not optimally matched to low throughputs.

Across all throughputs, the standard deviations varied between 2 and 6 % (Table 12.2). At the individual throughput levels, the standard variations were only between 0.5 and 3 % (not in Table 12.2). Distinct differences between the sensing systems do not exist.

However, **lateral and longitudinal tilts** of the combine harvesters at constant throughputs (20 t h$^{-1}$) exert a much greater influence upon the accuracy of the meters (Table 12.3).

The least deteriorating influence of tilt on the errors results when radiation is used for the sensing (system 3). In order to compensate for the influence of tilt, some volumetric measuring systems are equipped with one or two axle tilt sensors. Nevertheless, the errors caused by lateral and longitudinal tilt cannot successfully be compensated under all conditions. In this regard, the force measuring systems occupy an intermediate position between radiometric and volumetric meters.

Taylor et al. (2011) collected data from 29 test plots for corn on farms in Kansas, Alabama and Iowa, which were harvested using different yield measurement

**Table 12.3** Errors of yield sensing systems for combine harvesters at different tilts in the 2000/2001 test bench studies

| Sensing system (in bold), *Meter* (in italics), Manufacturer (in brackets) | Relative calibration error in % (average) | Standard deviation of the relative error in % |
|---|---|---|
| **Light-barrier, system 2**, *CERES 2*, (RDS) | −3.38 | 8.07 |
| **Light-barrier, system 2**, *QUANTIMETER*, (CLAAS) | −0.91 | 3.74 |
| **Light-barrier, system 2**, *PRO SERIES 2000*, (RDS) | −0.90 | 11.73 |
| **Radiation, system 3**, *FLOWCONTROL*, (MASSEY FERG.) | −1.11 | 2.17 |
| **Force, system 4**, *FIELDSTAR*, (DRONNINGB./AGCO) | −0.02 | 2.38 |
| **Force, system 5**, *YM 2000* (AGLEADER), *LH 565* (LH AGRO) | −0.24 | 4.31 |
| **Force, system 6**, *GREENSTAR*, (JOHN DEERE) | −1.36 | 3.37 |

Constant throughput of 20 t h$^{-1}$; lateral tilt of 5, 10 and 13° to the left and to the right or longitudinal tilt forward and back as well as combinations thereof; 5 repetitions/variant; n = 60/m; reference mass/variant = 1 t; winter wheat. For system numbers see Fig. 12.2 (Demmel 2001)

systems on combine harvesters and simultaneously counter-weighed using a weigh wagon or certified scale. A mean calibration error of −1.1 % was detected with an average error standard deviation of 3.6 %.

Each individual trial delivered between 4 and 39 control weights. The calibration errors for these varied between +15.2 and −27.1 %. This shows that the precision of calibration is a key issue, although most of the mean errors remained between ±10 %. At two locations the standard deviation of the error was above 10 %. The results show an accuracy level comparable to the results reported by Demmel 2001 and emphasise the need of a repeated and accurate calibration.

To make the results of an evaluation of yield measurement systems for combines comparable, ASABE Standard S578 (2007) provides the basic requirements for a uniform procedure to measure and report yield monitor accuracy. It describes a series of repeatable tests that may be used to evaluate and compare grain flow sensors in laboratory testing. In addition ASABE Standard S579 (2012) defines a "Yield Monitor Field Test Engineering Procedure" and provides the basic requirements for field evaluating the accuracy of yield monitors (ASABE 2007, 2012).

## 12.4 Yield Measurement for Forage Crops

Typical crop rotations in many regions of the world, especially in Western Europe, do not only consist of combinable crops. To get information on the variability of yields of different kind of crops, yield measurement systems for other harvesting equipment than combine harvesters is necessary.

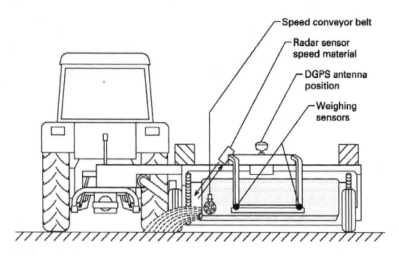

**Fig. 12.3** Material flow measurement based on belt weighing technology for a grass mower (Demmel et al. 2002)

Some research groups have published work on the development of material flow measurement technology for tractor mounted **grass mowers**. The systems were either based on a belt weighing technique (Fig. 12.3), on force and torque measurement or on the application of a dielectric sensor for mass flow detection. Until now, none of these developments is available commercially (Kumhala et al. 2001, 2007; Demmel et al. 2002; Wild et al. 2004).

Since 1990 research on material flow sensors for **forage choppers** is reported. First investigations and results were published by Vansichen and De Baerdemaeker (1993). In 1995 Auernhammer et al. presented results of a comparison of material flow measurement systems based on the clearance between upper and lower feed rolls (volumetric), the power consumption of the cutter drum, the power consumption of the blower and a radiometric measurement system (mass flow) in the spout (Fig. 12.4). Barnett and Shinners (1998) investigated similar sensor configurations on a trailed forage chopper.

As with yield measuring systems for combines, the material flow is converted into the area yield (t ha$^{-1}$) by referencing to the harvested area that is obtained from the cutting width and harvested distance (wheel sensor). In addition, the harvested area is used to determine the area output (ha h$^{-1}$). Continuous yield sensor readings (often 1 Hz) are combined with information from a satellite positioning system for geo-referencing.

As an alternative for material flow detection in a forage chopper, Missotten (1998) developed and evaluated a system with a friction compensated curved plate in the spout.

Some manufacturers of self propelled forage choppers (CLAAS, JOHN DEERE and KRONE) are offering material flow and yield recording systems based on displacement sensing of the feeder rolls – thus on volume flow – as options.

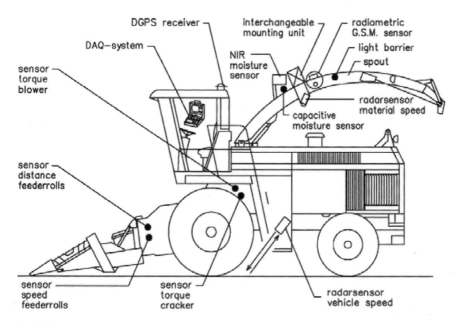

**Fig. 12.4** Systems for material flow measurement in forage harvesters

Auernhammer et al. (1997) reported on the accuracy of a sensing system that relies on gamma ray absorption by the flow of maize forage in the spout of a forage harvester. The evaluation was based on control weighing a very high number of trailer loads. After optimising of the meter, the standard deviation of the relative error was 3.3 %. For a system that relies on the distance of the feeder rolls and with whole crop barley, Ehlert (1999, 2002) obtained a standard deviation of the relative error of 5.9 %. Finally when used with maize, the standard deviation of the relative error with the friction compensated curved plate in the spout as developed by Missotten et al. (1997) was 2.7 %.

Research on systems for local yield detection in round **balers**, square balers and **self-loading trailers** (Auernhammer and Rottmeier 1990; Behme et al. 1997; Sauter et al. 2001; Shinners et al. 2003) initiated a development which resulted in the commercial availability of weighing systems for large square balers. These systems are able to deliver the weight of each single large square bale by integrating load cells in the bale chute and to record this information together with the position (and moisture) of the bale in the field. One manufacturer offers the individual identification of the bales using RFID tags – that are capable to locate via **r**adio-**f**requency – together with the recording of the bale weight, material moisture and bale location. However, because the systems are not able to correctly measure the increase of the weight of a bale until it is tied and passes the chute, it is not possible to obtain the site-specific yield in the field. The time- and distance lags that are involved prevent this.

## 12.5 Yield Sensing for Root Crops

For georeferenced yield sensing of root crops like potatoes and sugar beets, various measurement devices have been integrated into the harvesting equipment, tested and evaluated (Fig. 12.5).

A number of authors successfully developed and evaluated conveyor belt weighing systems combined with GPS positioning systems and data processing units (Campbell et al. 1994; Walter et al. 1996; Demmel and Auernhammer 1999).

Godwin and Wheeler (1997) used a trailer equipped with load cells to obtain yield data based on the mass accumulation rate.

Kromer and Degen (1998) obtained the sugar beet yield by estimating the volume of the beets on a conveyor belt using a laser scanner.

Hennes et al. (2002) adapted the friction compensated curved plate principle to the conditions at a rotating cleaning turbine of a sugar beet harvester.

A compilation of results on the **accuracy of measurement systems for root crops** shows that a majority of authors has used the conveyor weighing system Harvestmaster version HM 500 (Table 12.4). This implement can be integrated in very different harvesting machines and hence be used for a variety of root crops. The level of errors as listed is rather similar. Actually this holds for all systems that are included in Table 12.4.

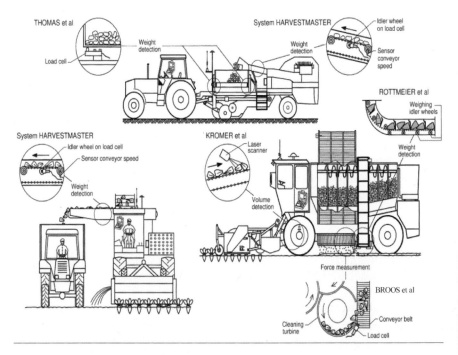

**Fig. 12.5** Systems for material flow measurement in potato and sugar beet harvesters. For details see the literature cited in the text

Table 12.4 Accuracy of systems for yield measurement used with root crop harvesters

| Measurement system | Harvester type, crop | Reference | Relative calibration error in % (average) | Standard deviation of relative error in % |
|---|---|---|---|---|
| Mass accumulation system "Silsoe" | Trailer, sugar beets, potatoes | Godwin and Wheeler (1997) | 1.1 | 4.0 |
| Basket weighing system "Tifton" | Trailed two row basket, peanuts | Durance et al. (1998) | 0.2 | 3.1 |
| Conveyor weighing "Harvestmaster" | Trailed two row side loading, potatoes | Rawlins et al. (1995) | Not recorded | 4.9 |
| Conveyor weighing "Harvestmaster" | Trailed six row side loading, sugar beets | Hall et al. (1997) | −1.0 | 2.2 |
| Conveyor weighing "Harvestmaster" | Trailed one row bunker, potatoes | Demmel and Auernhammer (1998) | −1.3 | 4.1 |
| Conveyor weighing "Harvestmaster" | Self propelled six row loading, sugar beets | Demmel et al. (1998) | 1.0 | 3.7 |
| Conveyor weighing "Rottmeier" | Self propelled six row tanker, sugar beets | Demmel et al. (1998) | 2.1 | 5.6 |
| Friction compensated, curved plate | Self propelled tanker loader, sugar beets | Broos et al. (1998) | 0.4 | 1.6 |
| Laseroptical volume system "Bonn" | Self propelled cleaner loader, sugar beets | Kromer and Degen (1998) | Not recorded | 4.0 |

Although Harvestmaster has stopped production of the HM 500 yield measurement system (conveyor weighing principle), a number of potato custom harvesters in the USA still is using the system to control the load of the transport trucks to avoid fines for overload.

In Germany, an increasing portion of self propelled sugar beet cleaning loaders are equipped with conveyor weighing systems also in order to avoid overloading the trucks that take the sugar beets to the factory.

So presently there are no commercially available site-specific yield recording systems offered for root crops. The fact that root crops very often grow in fields that are rather uniform and flat may have contributed to this.

## 12.6 Yield Measurement for Other Crops

In the USA and Australia, yield measuring systems for self propelled **cotton pickers and cotton strippers** have been intensively developed and evaluated since the early 1990th. Cotton flow is determined either by measuring the increase of

Table 12.5 Evaluated systems for yield measurement in sugar cane harvesters

| Measurement system | Placement of system | Authors | Relative calibration error in % (average) | Stand. dev. of relative error in % |
|---|---|---|---|---|
| Pressure-measurement | Chopper and elevator | Cox et al. (1998) | 10.0 | No record |
| Weigh scale | Elevator floor | Benjamin et al. (2001) | 11.0 | No record |
| Weigh scale | Elevator floor | Molin and Menegatti (2004) | −3.5 | 6.1 |
| Weigh scale | Elevator floor | Cerri and Magalhaes (2005) | 1.0 | 4.7 |
| Fiber optic sensor | Elevator floor | Price et al. (2011) | 7.5 | 6.3 |

cotton weight as it accumulates in the basket (Searcy et al. 1997) or with optical or microwave sensors in the cotton conveyor pipes. The sensor readings are converted into mass flow by calibration algorithms (Wilkerson et al. 1994, 2001; Durrance et al. 1998; Khalilian et al. 1999; Vellidis et al. 2003; Sui et al. 2004). The evaluation results ranged from excellent to less than ideal. The most often reported problems involved dust and debris build-up on sensor faces. Presently systems for site-specific sensing of yields are available commercially for cotton pickers and cotton strippers from AGLEADER, AGRIPLAN, CASE IH and JOHN DEERE.

A high activity in sensor research, development and evaluation was oriented at making systems available to continuously detect mass flow and yield of **sugar cane** in self propelled chopper harvesters (Table 12.5). Cox et al. (1998) developed and evaluated a hydraulic pressure measurement system placed in the chopper and elevator system to determine the material flow rate. A number of other concepts was based on weighing systems placed in the elevator floor of the harvesters (Benjamin et al. 2001; Molin and Menegatti 2004; Cerri and Magalhaes 2005; Mailander et al. 2010). Problems observed with the weighing systems were the compensation of tilt and vibration as well as varying tare caused by soil, sediment and debris sticking. A sugar cane yield sensing system that is based on fiber optic flow sensing was described and evaluated by Price et al. (2007, 2011).

Other crops for which continuously sensing material flow and yield measurement systems have been developed and evaluated are peanuts (Perry et al. 1998; Vellidis et al. 2001; Thomasson and Sui 2004; Porter et al. 2012), peas (Glancey et al. 1997), tomatoes (Pelletier and Upadhyaya 1998), grapes (Tisseyre et al. 2001), and pistachios (Rosa et al. 2011).

## 12.7 Quality Sensing of Harvested Material

In the future besides geo-referenced yield measurement, online detection of the quality of harvested products will gain in importance. Only the combination of quantity and quality allows site-specific evaluation and targeted control of plant production. To measure the site-specific moisture content of grain for up to 30 % on a wet basis, continuously operating capacitance sensing within an elevator bypass is state of the art.

For higher moisture levels as well as to detect the content of protein, starch, oil and energy, measurement systems based on near infrared spectroscopy have been calibrated as well as evaluated (Kormann and Auernhammer 2001; Reyns 2002; Welle et al. 2003) and are also used commercially.

## 12.8 Processing and Mapping of Yield Data

During the harvesting process, data such as material flow, yield, moisture and travel speed from several sensors are continuously recorded by the yield monitor together with positional information provided by the GPS receiver. The information in the yield data files varies with products and manufacturers. Data stored in the files are pre-processed to a different extent. Information about the methods and the algorithms of pre-processing is not available from the manufacturers. Investigations (Steinmayr 2002; Noack 2003) indicated that the systems are filtering or/and averaging in different ways.

Yield data collected with yield monitors can contain erroneous measurements since the sensors are operating under difficult conditions. Distance and time lags during sensing must accounted for as with other site-specific operations (Sect. 9.4.8).

Errors occurring during the process of yield data recording have been well described and classified by Blackmore and Marshall (1996). The removal of potentially erroneous yield data from yield recordings is a prerequisite for the creation of accurate yield maps. A number of authors have presented filters for yield data records (Rands 1995; Blackmore and Marshall 1996; Taylor et al. 2000; Thylen et al. 2000; Beck et al. 2001; Noack et al. 2001; Steinmayr 2002). In 2012 Sudduth et al. presented a software tool for automatic removal of yield map errors.

The yield map should represent the yield variation within a field. Different kinds of yield maps can be found. A very simple one is the **point map**. At every position, the yield data are ranged and printed as colored points. Because no correlation is created to neighboring points, it is critical to deduce yield zones from such a map. In **block yield maps,** a yield value is assigned to each grid (rectangle) of a specific size. These are typically created with methods that interpolate by kriging. Some details to these methods are in Sect. 2.4.

## References

ASABE (2007 and 2012) American Society of Agricultural and Biological Engineers. Standards, S 578 and S 579. http://elibrary.asabe.org/standards.asp

Auernhammer H, Rottmeier J (1990) Weight determination in transport vehicles – exemplary shown on selfloading trailers. In: Technical Papers and Posters Abstracts of AgEng'90, 24–26 October 1990, Berlin, pp 100–101

Auernhammer H, Demmel M, Muhr T, Rottmeier J, Wild K (1993) Yield measurement on combine harvesters. Paper No. 931506, ASAE, St. Joseph, MI

Auernhammer H, Demmel M, Pirro PJM (1995) Yield measurement on self propelled forage harvesters. Paper No. 951757, ASAE, St. Joseph, MI

Auernhammer H, Demmel M, Pirro PJM (1997) Throughput and yield sensing in self-propelled forage harvesters (in German) VDI Bericht 1356:135–138

Bae YH, Borgelt SC, Searcy SW, Schueller JK, Stout BA (1989) Mapping of spatially variable yield during grain combining. Trans ASAE 32(3):826–829

Barnet NG, Shinners KJ (1998) Analysis of systems to measure mass-flow-rate and moisture on forage harvesters. Paper No. 981118, ASAE, St. Joseph, MI

Beck AD, Searcy SW, Roades JP (2001) Yield data filtering techniques for improved map accuracy. Appl Eng Agric 17(4):423–431

Behme JA, Schinstock JL, Bashford LL, Leviticus LI (1997) Site specific yield for forages. Paper No. 971054, ASAE, St. Joseph, MI

Benjamin CE, Price RR, Mailander MP (2001) Sugar cane monitoring system. Paper No. 011189, ASAE, St. Joseph, MI

Blackmore S, Marshall C (1996) Yield mapping; errors and algorithms. In: Robert PC, Rust AH, Larson WE (eds) Proceedings of the 3rd international conference on precision agriculture, 23–26 June 1996, Minneapolis. ASA; CSSA; SSSA. Madison, pp 403–415

Broos B, Missotten B, Reybrouck W, De Baerdemaker J (1998) Mapping and interpretation of sugar beet yield differences. In: Robert PC, Rust RH, Larson WE (eds) Proceedings of the 4th international conference on precision agriculture. ASA, Madison

Campbell RH, Rawlins SL, Han S (1994) Monitoring methods for potato yield mapping. Paper No. 943184, ASAE, St. Joseph, MI

Cerri GP, Magalhaes PG (2005) Sugarcane yield monitor. Paper No. 051154, ASAE, St. Joseph, MI

Cox GJ, Harris HD, Cox DR (1998) Application of precision agriculture to sugar cane. In: Robert PC, Rust RH, Larson WE (eds) Proceedings of the Fourth International Conference on Precision Agriculture. ASA; CSSA; SSSA, St. Paul, MN

Demmel M (2001) Yield recording in combines – yield determination for site-specific yield sensing (in German). DLG Merkblatt 303. Hrsg: Deutsche Landwirtschafts-Gesellschaft, Fachbereich Landtechnik, Ausschuss für Arbeitswirtschaft und Prozesstechnik, Deutsche Landwirtschafts-Gesellschaft, 20 p

Demmel M, Auernhammer H (1998) Local yield recording with potatoes and sugar beets (in German) VDI-Berichte 1449:263–268

Demmel M, Auernhammer H (1999) Local yield measurement in a potato harvester and overall yield pattern in a cereal – potato crop rotation. Paper No. 991149, ASAE, St. Joseph, MI

Demmel M, Auernhammer H, Rottmeier J (1998) Georeferenced data collection and yield measurement on a self propelled six row sugar beet harvester. Paper No. 983103, ASAE, St. Joseph, MI

Demmel M, Schwenke T, Böck J, Heuwinkel H, Locher F, Rottmeier J (2002) Development and field test of a yield measurement system in a mower conditioner. EurAgEng Paper Number 02-PA-032, AgEng Budapest

Durrance JS, Perry CD, Vellidis G, Thomas DL, Kvien CK (1998) Evaluation of commercially available cotton yield monitors in Georgia field conditions. Paper No. 983106, ASAE, St. Joseph, MI

Ehlert D (1999) Throughput measurements for yield mapping in forage harvesters. Agric Eng Res 5(1):1–7

Ehlert D (2002) Advanced throughput measurement in forage harvesters. Biosyst Eng 83(1): 47–53

Glancey JL, Kee WE, Lynch M (1997) A preliminary evaluation of yield monitoring techniques for mechanically harvested processed vegetables. Paper No. 971060, ASAE, ST. Joseph, MI

Godwin RJ, Wheeler PN (1997) Yield mapping by mass accumulation. Paper No. 971061, ASAE, St. Joseph, MI

Hall TL, Backer LL, Hofmann VL, Smith LJ (1997) Monitoring sugar beet yield on a harvester. Paper No. 973139, ASAE, St. Joseph, MI

Hennes D, Baert J, De Baerdemaeker J, Ramon H (2002) Yield mapping of sugar beets with a momentum type flow rate sensor. In: Proceedings of the conference agricultural engineering, Halle 2002. VDI, Düsseldorf, pp 247–252

Khalilian A, Wolak FJ, Dodd RB, Han YJ (1999) Improved sensor mounting technology for cotton yield monitors. Paper No. 991052, ASAE, St. Joseph, MI

Kormann G, Auernhammer H (2001) Continuous determination of ingredients in self-propelled forage harvesters (in German) VDI-Berichte 1636:279–284

Kormann G, Demmel M, Auernhammer H (1998) Testing stand for yield measurement systems in combine harvesters. International AgEng Conference 98, Oslo, Paper No. 98-A-054

Kromer KH, Degen P (1998) Volume and scale based measuring machine capacity and yield and soil tare of sugar beet. Paper No. 983107, ASAE, St. Joseph, MI

Kumhala F, Kroulik M, Hermanek P, Prosek V (2001) Yield mapping of forage harvested by mowing machines. VDI Berichte 1636, Düsseldorf:267–272

Kumhala F, Kroulik M, Prosek V (2007) Development and evaluation of forage yield measure sensors in a moving-conditioning machine. Comput Electron Agric 58(2):154–163

Mailander M, Benjamin C, Price R, Hall S (2010) Sugar cane yield monitoring system. Appl Eng Agric 25(6):965–969

Missotten B (1998) Measurement systems for the mapping and evaluation of crop production performance. Doctoral dissertation, Katholieke Universiteit Leuven, Faculteit Landbouwkundige en Toegepaste Biologische Wetenschappen

Missotten B, Broos B, Strubbe G, De Baerdemaeker J (1997) A yield sensor for forage harvesters. In: Stafford JV (ed) Precision agriculture '97. Bios Scientific Publishers, Oxford, pp 529–536

Molin JP, Menegatti LAA (2004) Field-testing of a sugar can yield monitor in Brazil. Paper No. 041099, ASAE, St. Joseph, MI

Noack PO, Muhr T, Demmel M (2001) Long term studies on determination and elimination of errors occurring during the process of georeferenced yield data collection on combine harvesters. In: Grenier G, Blackmore S (eds) Proceedings of the third European conference on precision agriculture. Agro Montpellier 2001, vol 2, pp 833–837

Noack PO, Muhr T, Demmel M (2003) Relative accuracy of different yield mapping systems installed on a single combine harvester. In: Stafford J, Werner A (eds) Precision agriculture – Proceedings of the European conference on precision agriculture 2003. Wageningen Academic Publishers, Wageningen, pp 451–457

Pelletier MG, Upadhyaya SK (1998) Development of a tomato yield monitor. In: Robert PC et al (eds) Proceedings of 4th international conference on precision agriculture. American Society of Agronomy, Madison, pp 1119–1129

Perez-Munoz F, Colvin TS (1994) Continuous grain yield monitoring. Paper No. 941053, ASAE, St. Joseph, MI

Perry CD, Durrence JS, Vellidis G, Thomas DL, Hill RW, Kvien CS (1998) Experiences with a prototype peanut yield monitor. Paper No. 983095, ASAE, St. Joseph, MI

Porter WM, Taylor RK, Godsey CB (2012) Application of an Ag Leader cotton yield monitor for measuring peanut yield. Paper No. 12–1338357, ASABE, St. Joseph, MI

Price RR, Larsen J, Peters A (2007) Development of an optical yield monitor for sugar cane harvesting. Paper No. 071049, ASAE, St. Joseph, MI

Price RR, Johnson RM, Viator RP, Larsen J, Peters A (2011) Fiber optic yield monitor for a sugar cane harvester. Trans ASABE 54(1):31–39

Rands M (1995) The development of an expert filter to improve the quality of yield data. Unpublished Masters thesis. Department of Agricultural and Environmental Engineering, Silsoe College, UK

Rawlins SL, Campbell GS, Campbell RH, Hess JR (1995) Yield mapping of potatoes. In: Robert PC, Rust RH, Larson WE (eds) Proceedings of the second international conference site-specific management for agricultural systems, ASA, Madison, WI, pp 59–69

Reyns P (2002) Continuous measurement of grain and forage quality during harvest. Dissertation, Katholieke Universiteit Leuven, Belgium, Faculteit Landbouwkundige en Toegepaste Biologische Wetenschappen

Rosa UA, Rosenstock TS, Choi H, Pursell D, Gliever CJ, Brown PH, Upadhyaya SK (2011) Design and evaluation of a yield monitoring system for pistachios. Trans ASABE 54(5):1555–1567

Sanaei A, Yule IJ (1996) Yield measurement reliability on combine harvesters. Paper No. 961020, ASAE, St. Joseph, MI

Sauter GJ, Kirchmeier H, Neuhauser H (2001) Yield recording in a cubic big baler (in German) Landtechnik 56(1):24–25

Schmidhalter U, Maidl FX, Heuwinkel H, Demmel M, Auernhammer H, Noack PO, Rothmund M (2005) Precision farming – adaptation of land use management to small scale heterogeneity. In: Schröder P, Pfadenhauer J, Munch JC (eds) Perspectives for agroecosystem management. Elsevier, Amsterdam, pp 121–186

Searcy SW, Motz DS, Inayattullah A (1997) Evaluation of a cotton yield mapping system. Paper No. 971058, ASAE, St. Joseph, MI

Shinners KJ, Huenink BM, Behringer CB (2003) Precision agriculture as applied to North American hay and forage production. In: Electronic proceedings of the international conference on crop harvesting and processing, ASAE, St. Joseph, MI

Steinmayr T (2002) Error analysis and error correction with site-specific yield sensing in combines for the development of a standardized algorithm for yield recording (in German). Doctoral dissertation, TU München, p 227. http://mediatum.ub.tum.de

Strubbe GJ (1997) Mechanics of friction compensation in mass flow measurement of bulk solids. Dissertation, Katholieke Universiteit Leuven, Belgium, Faculteit Landbouwkundige en Toegepaste Biologische Wetenschappen

Strubbe GJ, Missotten B, De Baerdemaker J (1996) Mass flow measurement with a curved plate at the exit of an elevator. In: Robert PC, Rust RH, Larson WE (eds) Proceedings of the third international conference on precision agriculture, Madison, WI, pp 703–712

Sudduth KA, Drummond ST, Myers DB (2012) Yield editor 2.0: software for automated removal of yield map errors. Paper No. 12-1338243, ASABE, St. Joseph, MI

Sui R, Thomasson JA, Mehrle R, Dale M, Perry C, Rains G (2004) Mississippi cotton yield monitor: beta test for commercialization. Comput Electron Agric 42:149–160

Taylor RK, Kastens DL, Kastens TL (2000) Creating yield maps from yield monitor data using multi-purpose grid mapping (MPGM). In: Robert PC (ed) Proceedings of the fifth international conference on precision agriculture and other precision resources management, 16–20 July 2000, Bloomington/Minneapolis

Taylor R, Fulton J, Mullenix D, Darr M, McNaull R, Haag L, Stauggenborg S (2011) Using yield monitors to assess on-farm test plots. Paper No. 1110690, ASABE, St. Joseph, MI

Thomasson JA, Sui R (2004) Optical peanut yield monitor. Development and testing. Paper No. 041095, ASABE, St. Joseph, MI

Thylen L, Algerbo PA, Giebel A (2000) An expert filter removing erroneous yield data. In: Robert PC (ed) Proceedings of the fifth international conference on precision agriculture and other precision resources management, 16–20 July 2000, Bloomington/Minneapolis

Tisseyre B, Mazzoni C, Ardoin N, Clipet C (2001) Yield and harvest quality measurement in precision viticulture – application for selective vintage. In: Grenier G, Blackmore S (eds) Third European conference on precision agriculture, Montpellier, France, pp 133–138

Vansichen R, Baerdemaeker D (1993) A measurement technique for yield mapping of corn silage. J Agric Eng Res 55(1):1–10

Vellidis G, Perry CD, Durrence JS, Thomas DL, Hill RW, Kevin CK, Hamrita TK, Rains G (2001) The peanut yield monitoring system. Trans ASAE 44(4):775–785

Vellidis G, Perry CD, Rains GC, Thomas DL, Wells N, Kvien CK (2003) Simultaneous assessment of cotton yield monitors. Appl Eng Agric 19(3):259–272

Walter JD, Hofmann VL, Backer LF (1996) Site-specific sugar beet yield monitoring. In: Robert PC, Rust RH, Larson WE (eds) Proceedings of the third international conference on precision agriculture, Madison, WI

Welle R, Greten W, Rietmann B, Alley S, Sinnaeve G, Dardenne P (2003) Near-infrared spectroscopy on chopper to measure maize forage quality parameters online. Crop Sci 43:1407–1413

Wild K, Ruhland S, Haedicke S (2004) A conveyor belt based system for local yield measurement in a mower conditioner. In: Proceedings of AgEng '04, Leuven, Belgium

Wilkerson JB, Kirby JS, Hart WE, Womac AR (1994) Real-time cotton flow sensor. Paper No. 941054, ASAE, St. Joseph, MI

Wilkerson JB, Moody FH, Hart WE, Funk PA (2001) Design and evaluation of a cotton flow rate sensor. Trans ASABE 44(5):1415–1420

# Chapter 13
# Fusions, Overlays and Management Zones

**Hermann J. Heege**

**Abstract** Site-specific farming can be based on various signals about soil- and crop properties. And some farming operations might need a control that relies simultaneously on several properties. However, the control of an individual farming operation requires singular and unambiguous site-specific signals. Hence, information about soil- and crop properties must be merged. Means to achieve this are sensor-fusion, map-overlay and management zones.

Sensor-fusion can be a reasonable approach for signals independent of their temporal stability, preferably for online control, yet also for delayed control via mapping. Map-overlay as well as management zones should be based on properties that are temporally stable. All these technologies for merging the information depend on a solid logic for the fusion of site-specific signals. This logic should be oriented at high yields in an environmentally sustainable manner.

**Keywords** Cluster analysis • Data cleaning • Data fusion • Intra-crop transfer • Inter-crop transfer • Perennial maps • Precision agriculture cycles • Predictive approach • Reactive approach • Sensor fusion

## 13.1 Crop Growth Factors, Sensing and Information Use

The dominating aim in site-specific farming is to provide for high crop yields in an environmentally sustainable manner. Accordingly, the site-specific control of the growth factors for a locally optimal crop development must be provided, *e.g.* the supply with water, nutrients and the protection against weeds plus pests. The signals

---

H.J. Heege (✉)
Department of Agricultural Systems Engineering, University of Kiel,
24098 Kiel, Germany
e-mail: hheege@ilv.uni-kiel.de

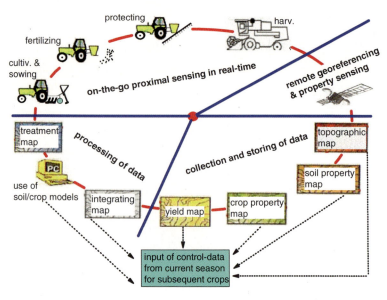

**Fig. 13.1** Information transfer within a precision agriculture cycle and between subsequent crops (Compiled and complemented with details from Stafford 2006)

for this site-specific control can be sensed, recorded and applied within **precision agriculture cycles** (Stafford 2006) that surround every growing season.

In addition, from the yield of the present crop, again information can be derived for site-specific operations that hold for subsequent crops. Examples for this are long-term mapped records that can be used for many years and especially fertilizing according to yields or nutrient removals of previous crops. In an ideal sense and for a holistic control, the information transfer should take place within precision agriculture cycles as well as between successive cycles. The information transfer might even extend over several crop rotations. Hence **intra-crop** as well as **inter-crop information transfer** might be needed. The processing of signals between farming operations or crop growing seasons too must be considered.

However, it always should be kept in mind that information transfer has no value in itself. Whenever any information transfer does not improve the control of precision farming, it should be left out. Cemeteries of unused data as well do not help, they might be a great nuisance. Condensing the information transfer into the actual need, *e.g.* by adequate averaging of signals and sorting out of useless data, might ease the situation.

Furthermore, the variety and abundance of steps as listed in Fig. 13.1 should be regarded as a list of possibilities – not of essentials – in order to exploit the sources of information. Where the site-specific conditions vary greatly, it might be reasonable to use all the information transfer routes. However, there are also many cases where only some of the routes listed are needed. And the start into site-specific control of precision farming will always be a gradual procedure.

Regarding the actual **crop growth factors**, the conditions too can differ greatly. The simplest situation is that just one growth factor is important, *e.g.* the water supply. And if the site-specific information about this factor can be sensed and processed on-the-go while the control of the farming operation takes place, the respective transfer route is very short. Mapping the situation for this transient growth factor alone seldom will be reasonable. But mapping and in addition averaging as well as further processing of signals makes sense whenever it comes to factors or properties that are temporally stable such as topography, soil texture, soil organic matter and immobile nutrients like phosphorus. And the transfer of information to subsequent crops by maps anyway is reasonable only for temporally stable properties (Fig. 13.1, bottom).

In many cases, the sensed signals do not precisely correspond to growth factors. Instead, they hold for defined physical units. The sensed signals might not even comply with single soil- or crop properties, which each in turn again might stand for several growth factors. The texture of soils *e.g.* affects the water, oxygen and nutrient supply of crops. And single crop properties also can be correlated to several growth factors. The leaf-area-index of a crop affects its photosynthesis, but also its energy loss via respiration and its infections. Finally, the widely sensed signals like electrical conductivities or radiation reflectances each are related to several soil or crop properties (see Chaps. 5, 6 and 9). So the rationale or the theoretical connection between sensed signals and actual growth factors might not be simple.

However, what is essential for site-specific farming? The important point is that the sensed signals are suitable **empirical indicators** for either growth factors or for properties of soils and plants that influence the development of crops. The signals must be able to serve as **surrogates** for direct indications of growth factors or of properties of soils or crops. And if the sensed signals are correlated to several growth factors simultaneously and counteractions between these do not occur, successful control can be possible.

The complexity of the sensing and the transfer of information calls for solutions that allow **simplifications**. Steps in this direction are concepts of sensor fusion, map overlay and management zones in site-specific farming. Sensor fusion aims at getting more information with proximal sensing within one pass through the field by combining hitherto separated sensing operations. This procedure can be reasonable for signals irrespective of their temporal stability, since the control of the farming operation might occur within real-time when this is necessary, but also in a time-delayed mode after mapping. Methods of map overlay and of farming by management zones inherently rely on temporal stability of the properties or their surrogates.

## 13.2 Sensor Fusion – Solutions and Approaches

The simplest example of sensor fusion has become quite common: it is used self-evidently when any real-time soil- or crop property sensor operates in combination with georeferencing. This mode of sensor fusion is a base of site-specific farming. An additional basis is the simultaneous use of topographic information that is available either from real-time GNSS signals or from maps. These basics will be routine matters in the future.

**Table 13.1** Conceivable applications for sensor fusion, grouped for site-specific farming objectives

| Primary sensing techn. | Second. sensing techn. | Tertiary sens. techn. |
|---|---|---|
| **Depth of primary cultivation** | | |
| Texture by electr. cond. | Soil org. matter by refl. | Soil hydromorphology |
| **Intensity of secondary cultivation** | | |
| Clod sizes by tine forces | Water by capacitance | |
| **Depth of stubble cultivation in regions with humid climate** | | |
| Residues by reflectance | Residues via grain yield | Water by capacitance |
| **Seed density** | | |
| Texture by electr. cond. | Water by capacitance | |
| **Seeding depth** | | |
| Water by capacitance | Drying zone sensing | Depth by ultrasonics |
| **Irrigation** | | |
| Crop temp. by therm. IR | Crop water by NIR | Soil water by capacit. |
| **Liming** | | |
| Ions by select. electrodes | Buffer pH by el. cond. | |
| **Fertilizing based on removal of phosphate, potash and some micronutr.** | | |
| Yield of previous crops | Texture by electr. cond. | |
| **Fertilizing of nitrogen and phosphates based on soil sensing** | | |
| Ions by select. electrodes | Ions by IR reflectance | |
| **Nitrogen fertilizing primarily based on crop-canopy sensing** | | |
| N by reflectance index | Soil texture by electric. conductivity | Crop water by infrared reflectance |
| N by fluorescence index | | |
| LAI by crop bending or by crop height (ultrason.) | | |
| **Spraying against weeds** | | |
| Leaf shape by reflect. | Soil org. matter by refl. | |
| **Spraying against fungi** | | |
| LAI by crop bending or by crop height (ultrason.) | N sensing by reflect. (adapting application of fungicides to N supply) | |
| Infected loci by fluor. | | |

Potential candidates for sensor fusion are sorted in an order of priorities and therefore as primary-, secondary- and tertiary sensing techniques. Please note that neither georeferencing nor topographic sensing are listed, though very often these techniques will be included in a fusion as well. The fusions in green are state of the art

Abbreviations: *LAI* leaf-area-index, *IR* infrared, *NIR* near-infrared

However, the variety of sensing techniques that has been developed for getting information about soil- and crop properties provides many prospects for sensor fusion. Some feasible and conceivable applications for sensor fusion are listed in Table 13.1.

The list is oriented on control needs and sensing concepts for farming operations that were dealt with largely in previous chapters. A concept of **sensor fusion** can be, but must not be congruent with **data fusion** or **data merging**. The latter terms stand for the processing of the data into singular signals for the control. The reasonable uses of the data will be different for transient properties like water content of soils on the one hand or for temporally stable properties like the topography or organic matter content of soils on the other hand. All properties can be sensed simultaneously and might also be useful for fused on-the-go control. However, only the temporally stable properties lend themselves well for delayed use, for averaging with previous records and hence for mapping. So sensor fusion can involve data fusion, but it might also be reasonable to complement this technique with **data splitting**. And a fused control of a farming operation can be reasonably based partly on sensing in an on-the-go mode that is supplemented with signals obtained from maps.

The actual need for sensor fusion must depend on the information variety that assists in the site-specific control of farming operations. The **primary sensing** for a site-specific objective in Table 13.1 stands for the most important control signals, which in turn act as substitutes for particular soil and crop properties. Consequently, the **secondary** and **tertiary sensing** are for signals or respective properties that might be needed to correct or supplement this result. But in some cases, these secondary and tertiary signals might even be alternatives or surrogates for those from the primary sensing

The approaches listed go far beyond the presently in practice used technologies, which are marked in green. Up to now, tertiary sensing is not state of the art. This can be due to an absence of suitable sensing techniques. As an example, for controlling the depth of primary cultivation adequately, sensing the hydromorphic soil properties can be very reasonable (Sect. 7.2.2.1). But the presently used methods of sensing electrical conductivities in soils do not allow for this

Several approaches listed for secondary sensing hold for signals that indicate the **water supply** in soils or crops. This probably characterizes the special possibilities and the particular potential of sensor fusion: it makes it possible to include information about a **transient factor** into the control process. This information hardly can be accounted for within due time if delays due to maps and its processing occur. An urgent case for including the transient factor water via sensor fusion and secondary sensing is in-season nitrogen fertilizing. For details to this see Sect. 9.4.10.

Sensing the electrical conductivity is currently the most frequently used method of recording soil properties. When this method is used in humid areas, it is mainly the clay and the water content of the soils that affect the signals. Hence a temporally stable and a transient property act together. For long-term use, the information that is derived from electrical conductivity sensing would be more valuable if the effect of the temporally varying water content could be normalized or taken care of by an adequate logical mathematical processing. A prerequisite for this is simultaneous yet separated sensing of electrical conductivity as well as of water.

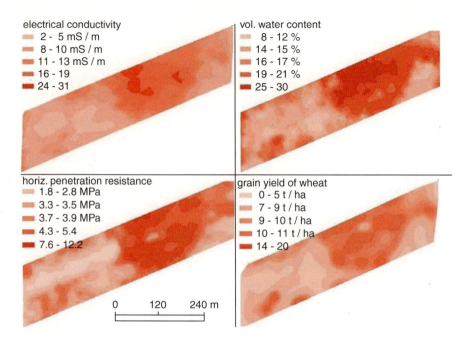

**Fig. 13.2** Maps of soil electrical conductivity, soil volumetric water content, soil horizontal penetration resistance and wheat grain yield of a rainfed field in Warleberg near Kiel, Germany. The soil is a carbon depleted glacial loam. The electrical conductivity was recorded by an EM38 sensor operating in a deep mode, while the water content and the penetration resistance were measured via dual sensing of capacitance and force by a tine that operated at a depth of 20 cm from the surface (see Sect. 5.2.3.4). Note that for all soil- or crop properties the coloured scales are contiguous, though the legends do not indicate this (From Sun et al. 2011, rearranged and altered)

The maps in Fig. 13.2 are based on simultaneous sensing of electrical conductivity, water content and horizontal penetration resistance (Sun et al. 2011). The recording took place immediately after the harvest of a winter-wheat crop so that a correlation of its site-specific yield to the sensed properties still makes sense.

From the maps, it can be seen that similarities between the sensed soil properties and the grain yield exist, however, the visual impression provides only very rough estimates. The respective **coefficients of determination** ($r^2$) in relation to the site-specific wheat yield were

- 0.811 for the soil electrical conductivity
- 0.713 for the volumetric soil water content
- 0.501 for the horizontal soil resistance.

So from the signals recorded, the smallest influence on the yield comes from the soil resistance. This soil property itself is highly dependent on the water content (Sect. 7.2.1.5). When taking into account the positive experiences that have been made on a worldwide basis with reduced- or even no-tillage methods, the question is, whether soil resistance is a property that needs to be sensed within the topsoil.

On the other hand, it is general experience that with well managed farming the **water supply** of crops has a dominant influence on yields. The success of electrical conductivity sensing for site-specific characterizing and delineating of soils in humid areas probably is due to the fact that this method at least for sandy and loamy conditions indicates the ability of providing water for crops. Yet a flaw of electrical conductivity sensing is that the results to some extent vary like those of a snap-shot. This is since in humid areas the electrical conductivity senses texture as well as water of soils. And because of the water, incidental rain affects the results. So the electrical conductivity sensing aims at signals that indicate the ability of soils to provide long-term plant available water capacity. But this objective is influenced by short-term weather variations.

Yet if the site-specific water content is sensed in addition and separately, it might be possible to remove this transient feature from the signals of electrical conductivity. Hence basically this method of sensor fusion makes very well sense. But **sensor fusion** alone is not sufficient. It must include **data fusion**. Because the control of farming operations needs singular signals. But how should the data from electrical conductivity sensing on the one hand and from the separate water sensing on the other hand be fused to a common signal? The best mathematical combination of the data from both sensors is not known yet and deserves efforts.

The fact that sensor fusion alone does not help much but that it should end up in data fusion applies to many of the approaches that are listed in Table 13.1. And here too **logical mathematical concepts** for the respective data fusion are needed.

## 13.3 From Properties to Treatment – Map Overlay

Any control via mapping allows the farmer to inspect the sensed results, to compare them with those from other records and to decide about the application of diverse processing methods. Whereas with on-the-go control via sensor fusion the processing of the signals is determined *a priori*, with mapping this decision can be made *ex post*. The expertise of the farmer or of his adviser can still be included in the control after the sensing of the data has occurred. Hence a more flexible control is possible.

With temporally stable soil properties, another feature is that maps can spare sensing operations. Over the years, it might be advisable to create "maps of the mean" in order to eliminate stochastic or incidental errors (Sects. 5.1 and 5.2.2.4). But once reliable "maps of the mean" have been obtained, sensing for this soil property can be omitted. Instead, "**perennial maps**" can take over.

Any **data fusion in mapping** or any creation of maps of the mean needs a precise procedure in processing. The objective is to combine the information from several maps and thus to improve the site-specific control of farming operations. Visually, this objective is defined as "**map overlay**". The aim of it is a single integrating or treatment map for the control of farming operations (Fig. 13.1).

The term "map overlay" makes it easy to grasp the objective. However, this should not hide the fact that several logical steps must be considered in order to get reliable results:

- Before any further processing of signals takes place, **data cleaning** might be needed. Artefacts such as extremely low or exceedingly high values, which

obviously result from technical noise and not from the variability of the respective soil- or crop property, should be removed. Such artefacts can be generated when *e.g.* the tractor stops or starts.
- All maps must be accurately geo-encoded so that the geographic coordinates precisely match. Distance- or time lags that are inevitable with sensing and control operations must be compensated for (see Sect. 9.4.8).
- Most software programs start with the creation of a **point-based map**, in which each point represents a signal. But map overlay and the corresponding data fusion process need a **grid-** or **block based map**, for which the data around each point are spread out in a square. When the grids or blocks are contiguous, a surface map is created.
- Such a block based surface map requires to determine adjacent points by interpolation. Computers can easily do this automatically, however, this interpolating should never occur beyond the range of the variogram that holds for the respective signals (see Sect. 2.3). **Kriging** must be included in the processing, and for this a special computer program (VESPER) from the Australian Centre for Precision Agriculture of the University of Sydney can be helpful (Whelan et al. 2002; Whelan and Taylor 2010).

These steps are state of the art in geostatistics, but nevertheless indispensable prerequisites for any data fusion by map overlay. The most important prerequisite however is that the combination of properties or of signals must make sense. This might depend on the mathematical logic that is used to merge the respective information as with any data fusion.

There exist numerous applications for this geostatistical route from property maps to a treatment map. An example for site-specific soil cultivation is outlined in Sect. 7.2.2.1. And a rather simple application that is oriented at environmental control in the application of a pre-emergence herbicide named Isoproturon is presented below.

The objective was the prevention of **herbicide leaching** into the groundwater (Mertens 2008). For this, the absorbing capacity of the soil is important. As for the soil in Fig. 13.3, the silt-loam on top of sandy gravel deposits has a largely varying vertical thickness, so the leaching of the herbicide Isoproturon as a result of this thickness was determined. Hence the site-specific information about the risk of leaching could be derived from the map about the vertical thickness of the silt-loam. This map was obtained via sensing of electrical conductivity by the deep induction method. Though generally this method does not provide information about the resolution in a vertical direction, in this case it was possible. Because there were roughly just two vertical layers, the loess-soil on top and the sandy gravel below.

A logical procedure would be to put in addition a site-specific map about the local **weed infestation** underneath the stack in Fig. 13.3. In most cases, the weed infestation has a patchy pattern. It would then be possible to see how the weed patches correspond to the areas of different risks for leaching. In case the weed patches would not fall into areas of high or very high leaching risk, the application of Isoproturon might be acceptable if it is done in a site-specific mode. But if weed patches are within

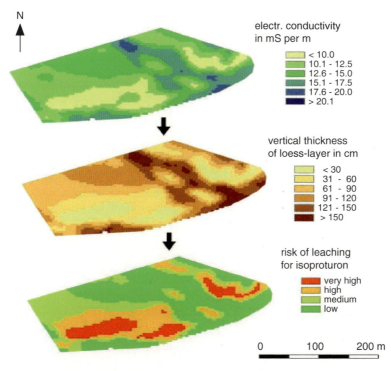

**Fig. 13.3** Map overlay for the control of herbicide leaching by site-specific application. The soil is a silt-loam of aeolian origin that is resting on sandy gravel deposits from the Rhine river and is located near Cologne in Germany. The herbicide Isoproturon is a frequently used chemical for pre-emergence as well as for post-emergence weed control (From Mertens 2008, altered)

areas of high leaching risk, choosing a herbicide that is better absorbed would be necessary. A similar adaptation to the absorption of herbicides would be feasible based on **soil organic matter**, which can easily be sensed by visible and infrared reflectance (Sect. 5.3). However, in humid regions it can be assumed that a positive correlation between the clay- and the organic matter content of a soil exists. So with soils that contain some clay, either sensing by electrical conductivity or sensing by reflectance for organic matter may be suitable, whereas for sandy soils the sensing based on organic matter seems more reasonable.

The most frequently mapped crop property is the yield. For modern combines, **grain yield sensors** have become a standard accessory. In fact, maps about a spatially resolved yield are indispensable for the determination of the site-specific need of phosphorus and – with the exception of sandy soils – also of potassium. Because for these nutrients no other sensing methods – that are well developed – are available.

And there may be additional sensible uses for grain yield maps. From records about the local grain yield mass, maps about the site-specific residue mass easily can be obtained. The mass of crop residues is positively correlated to the mass of grain. In humid regions, site-specific **residue maps** can be useful for controlling the depth of stubble cultivation (see Sect. 7.4).

The relation between the crop residue mass and the recorded grain mass can be derived from the harvesting index, which is the fraction of the total aboveground biomass that is allocated to the grain. For most crops and its varieties, this harvesting index is well known or can easily be determined (Kemanian et al. 2007). The residue index – the biomass-fraction of the aboveground residues – is simply 1 (one) minus the harvesting index. Furthermore it holds:

$$\frac{yield\ of\ residue\ mass}{yield\ of\ grain\ mass} \approx \frac{residue\ index}{harvesting\ index} \approx \frac{1 - harvesting\ index}{harvesting\ index}$$

So if the mass of grain is mapped and the harvesting index is known, it is rather easy to get from this to a residue map. And using this for precisely controlling the site-specific depth of stubble cultivation can avoid waste of energy. However, an accurate map about crop residues requires considering the time that the straw as well as the grain spend within the combine, hence distance lags must be compensated for in the georeferencing processes.

It has been expected that **yield maps** could be useful for predicting the site-specific performance of crops constantly. This expectation has been based on the assumption that permanent soil properties such as texture, organic matter and topography would always behave in the same way each year. If the site-specific yields were temporally stable, hence the yield maps would be a useful management tool. The input of agrochemicals could be oriented at such yield maps. But so far, this expectation of a **constant site-specific yield pattern** per field has not been met.

Blackmore et al. (2003) tried to predict the actual site-specific yield map pattern from yields in previous harvests in four fields and over 6 years for wheat, barley and rape (colza) in rainfed areas of England. The site-specific cell size was 20 m × 20 m. The mean yields of the whole fields differed from year to year due to good or bad weather or because of varying disease or pest attacks. However, the inter-year offsets that resulted from these different average yields were neutralized by the data processing.

The coefficients of determinations between the site-specific yields of the previous harvests and the respective actual harvest of the fields went from a very low level of $r^2 = 0.02$ up to a maximum of $r^2 = 0.43$ with an average for 11 comparisons of only $r^2 = 0.20$. It was concluded that yield maps cannot predict the corresponding site-specific yield in the following year. However, spatial and temporal maps could help to identify larger **homogeneous management zones**. The authors recommend based on these findings that the growing crop should be managed according to its current needs. And these current needs would have to be derived from the actual crop itself.

Experiments from Rothamsted Research in England (Milne et al. 2012) confirm these findings. These experiments were limited to winter-wheat that was grown in 3 years – 2002, 2004 and 2006 – in a rotation with rape and peas. Only the results with wheat are shown (Fig. 13.4) for site-specific cell sizes of 10 m × 10 m. Some

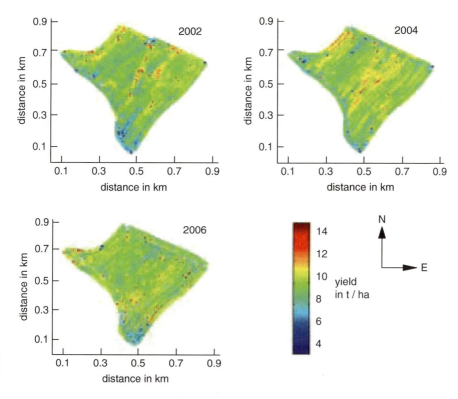

**Fig. 13.4** Site-specific yield maps for winter-wheat on a field with clay soils in Flawborough, Nottinghamshire, England (From Milne et al. 2012)

similarities between the years are evident. However, a close visual observation reveals also many differences in the site-specific yields between the years. The coefficients of determination ($r^2$) of the harvesting results for the site-specific cells between the years were extremely low (Milne et al. 2012):

- between the year 2002 and the year 2004        $r^2 = 0.07$
- between the year 2002 and the year 2006        $r^2 = 0.03$
- between the year 2004 and the year 2006        $r^2 = 0.03$.

It may be surprising that despite the influence of the temporally constant soil properties such as texture, organic matter and topography, any correlations in the site-specific yields between the years practically do not exist. The explanation for this is that these soil properties exert its influence on the growth of crops always together with the weather. It is the **interaction** between soil properties and the weather that creates the conditions for the growth of crops in rainfed areas. In a wet year, the best growth is obtained in a soil that drains freely. *Vice versa* in a dry year, the soil with the best water-holding capacity excels. So the weather steadily alters the soil qualities for crop growth. And the influence of the weather on additional growth factors such as the visible radiation and the temperature should also be mentioned.

It can be expected that the prediction of site-specific yields by means of maps from previous harvests is better in irrigated areas, since here the soil water supply varies less.

## 13.4 From Properties to Treatment – Management Zones

Map overlay can be used to create **management zones**. These can be defined as areas within a field to which a particular common treatment may be applied. The basis of management zones are generally **temporally stable soil properties** such as electrical conductivity, elevation, slope and organic matter content. The objectives can be *e.g.* the control of cultivating, sowing and fertilizing operations. Contrary to classical on-the-go control in real-time that might be based on just one soil- or crop property, management zones always rely on several properties simultaneously with the emphasis on soil attributes. Crop properties are less suited as components for management zones as a result of their more transient character. However, site-specific information about past crop yields might be included.

The preparatory steps for generating maps of individual properties have been dealt with in Sect. 13.3. In a **treatment map**, several properties that have been individually mapped must be merged to singular site-specific signals. With the properties mentioned, this is generally not possible by applying simple mathematical procedures. Because these properties differ substantially and hence do not allow for this. Instead, the generating of a treatment map or a management zone for a field requires applying an algorithm. Such an algorithm consists of a set of steps in similar way as outlined in Sect. 7.2.2.1 for the control of the depth in primary cultivation. Essential is that within such an algorithm the georeferenced properties from the individual maps are sorted into groups that are called clusters.

The processing of the grouped property signals includes a **cluster analysis**, which is a classification in such a way that association between the signals is maximal if these belong to the same group and minimal otherwise. The clustering aim is to obtain a high internal homogeneity within the groups combined with large external heterogeneity between the means of the groups. The cluster analysis thus simply discovers structures in the data without explaining why these exist.

For further details to the procedures in the delineation of management zones see Fridgen et al. (2004), Schepers et al. (2004), Taylor et al. (2007) and Whelan and Taylor (2010).

The number of management zones that are created per field can be preset in a computer program, however, always is less than 8. The example in Fig. 13.5 shows just two management zones. A similarly high resolution as with on-the-go operations that are based on single properties cannot be obtained. The loss in resolution that occurs may be due to the collating process and to limitations in the algorithm. However, present state of the art is that the same management zones are used for

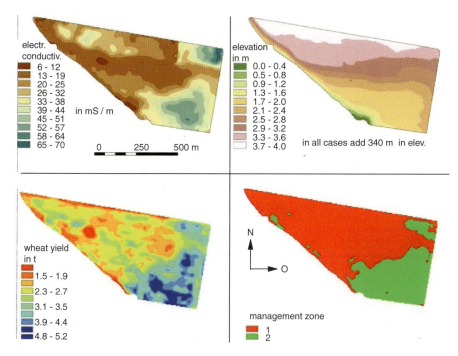

**Fig. 13.5** Electrical conductivities, elevations, crop yields and management zones for a 40 ha field in Australia (From Whelan and Taylor 2010, altered)

several successive farming operations such as soil cultivating, sowing and fertilizing. This more general application of management zones makes the control with high spatial resolutions difficult. The prerequisites for a control with small cell sizes are better, if this is oriented just at one farming operation, *e.g.* fertilizing nitrogen.

It deserves attention that site-specific control of farming operations can be realized either by a predictive- or by a reactive approach. The basis of a **predictive approach** are temporally stable soil properties such as texture, organic matter content, cation-exchange-capacity, elevation and slope orientation. These properties lend themselves for map-overlay and management zones.

In contrast to this, a **reactive approach** relies in addition on temporally varying phenomena, such as the water situation in soils or crops and canopy properties that can be detected via reflectance or otherwise. Hence, a reactive approach is based on the "reaction" to weather and crop development. Since it depends largely on transient properties, mapping the sensed results can be omitted. Instead, site-specific control in an on-the-go manner and in real-time is an important objective. Whenever sensible, sensor- and data fusion should be included.

Within this break-down into a predictive- and a reactive approach, **yield sensing** has an ambiguous position. It is part of a predictive approach when past site-specific yields are included in the generating of management zones (Fig. 13.5). But if yields are used to detect the removal of nutrients and hence to control fertilizing rates

(Sect. 9.1) or to determine the site-specific straw mass left in the field and thus the depth of stubble cultivation, they are part of a reactive approach.

A disadvantage of predictive approaches is that – contrary to reactive approaches – its accuracy often is impaired by the unforeseeable influence of the weather. Mainly because of this, predictive approaches are more difficult to realize with high resolutions than reactive approaches are. The challenge for the future will be to find intelligent combinations of predictive- and reactive approaches, *e.g.* by simultaneous use of maps plus on-the-go sensing in real-time. In this way, erroneous controls that result from the influence of the unpredictable weather probably can be reduced.

## References

Blackmore S, Godwin RJ, Fountas S (2003) The analysis of spatial and temporal trends in yield map data over six years. Biosyst Eng 84(4):455–466
Fridgen JJ, Kitchen NR, Sudduth KA, Drummond ST, Wiebold WJ, Fraisse CW (2004) Management zone analyst (MZA): software for subfield management zone delineation. Agron J 96:100–108
Kemanian AR, Stöckle CO, Huggins DR, Viega LM (2007) A simple method to estimate harvest index in grain crops. Field Crop Res 103:208–216
Mertens FM (2008) Spatial variability of soil properties and its significance for the performance of herbicides – an optimizing approach for precision in plant protection (in German). Doctoral dissertation, University of Bonn, Bonn
Milne AE, Webster R, Ginsburg D, Kindred D (2012) Spatial multivariate classification of an arable field into compact management zones based on past crop yields. Comput Electron Agric 80:17–30
Schepers AR, Shanahan JF, Liebig MA, Schepers JS, Johnson SH, Luchiari A Jr (2004) Appropriateness of management zones for characterizing spatial variability of soil properties and irrigated corn yields across years. Agron J 96:195–203
Stafford JD (2006) The role of technology in the emergence and current status of precision agriculture. In: Srinivasan A (ed) Handbook of precision agriculture. Principles and applications. Food Products Press, New York, pp 19–56
Sun Y, Druecker H, Hartung E, Hueging H, Cheng Q, Zeng Q, Sheng W, Lin J, Roller O, Paetzold S, Schulze Lammers P (2011) Map-based investigation of soil physical conditions and crop yield using diverse sensor techniques. Soil Till Res 112:149–158
Taylor JA, McBratney AB, Whelan BM (2007) Establishing management classes for broad acre agricultural production. Agron J 99:1366–1376
Whelan BM, Taylor J (2010) Precision agriculture education and training modules for the grains industry. Australian Centre for Precision Agriculture, University of Sydney for the Grains Research and Development Corporation. Sydney, Australia
Whelan BM, McBratney AB, Minasny B (2002) VESPER 1.5 – Spatial prediction software for precision agriculture. In: Robert PC, Rust RH, Larson WE (eds) Proceedings of the 6th international conference on precision agriculture, Madison, WI

# Chapter 14
# Summary and Perspectives

**Hermann J. Heege**

Future farming has to provide high yields in order to feed the growing world population, but should do this with an environmental impact that is sustainable. With the limited resources of the planet earth, this dual challenge can only be met by precise adaptation to the locally varying conditions. And since the field boundaries by no means are precise environmental borders, this necessitates site-specific farming within single fields.

The prospects in site-specific farming depend largely on the possibilities to sense the properties of soils and crops. An impressive variety of site-specific sensing techniques has been developed and is ready for its use. The spatial resolutions that can be obtained by these techniques allow up to 100 signals per ha and sometimes much more, which is a substantial improvement to the present situation of treating the whole field uniformly. However, in addition to the spatial resolution, the temporal situation is important.

Some soil properties can be regarded as being temporally stable, such as texture, organic matter content, cation-exchange-capacity, elevation, slope and its orientation. Hence these properties or their sensing surrogates lend themselves for mapping on a rather permanent basis. It may be reasonable to improve the mapping precision for these properties by repeating the recording and thus creating **maps of the mean** or by correcting the maps for the influence of *e.g.* the transient factor water. But once these maps of the means or corrected maps have been created, they can be used as **perennial maps**. The challenge is the adequate processing of the mapped information for the control of farming operations.

But besides perennial maps, seasonal maps are a reasonable approach for soil properties that can be regarded as being constant only on a short-term basis.

---

H.J. Heege (✉)
Department of Agricultural Systems Engineering, University of Kiel,
24098 Kiel, Germany
e-mail: hheege@ilv.uni-kiel.de

These **seasonal maps** can serve as control instruments for just a crop season or a rotation of crops. Properties that lend themselves for these maps are *e.g.* the lime-, phosphorus- and potash requirement of soils as well as the weed infestation. Though it may be reasonable to keep these seasonal maps for long-term records, their use as control devices is temporally limited.

Finally there are the **transient soil- and crop properties**. To these belong the soil water content, the nitrogen status and fungal diseases of crops. The short-term validity of these soil- or crop properties calls for immediate action, hence for proximal on-the-go control in real-time. So the control mode should fit to the temporal stability of the respective properties or their sensed surrogates. It is possible to map the site-specific situation of these soil- and crop properties simultaneously as well. However, for these transient properties, such a mapping procedure seems more reasonable when the sensing occurs in a remote mode from satellites instead of a proximal mode from tractors or vehicles.

The needs as to the frequency of sensing for these transient properties may be quite different. The moisture content of soils and the water supply of crops that depends on this are rather critical factors for high yields. Hence ideally, both factors should be monitored throughout the growing season with sensing intervals of not more than a few days or weeks. Such a procedure does not lend itself to proximal sensing from vehicles or farm machines as long as these need drivers. Instead of this, **moisture monitoring** should be aimed at via remote sensing from satellites in a similar way as georeferencing is being done. Principally, radar waves can do this independent of cloud covers. However, the processing of the signals still must be improved and the number of radar satellites does not suffice yet. There are alternatives to remote water detection via radar waves, such as remote or proximal sensing by infrared radiation and finally proximal sensing by electrical capacitance. None of the methods mentioned presently are state of the art.

It can be expected that many farming operations in the future will include a site-specific control system.

In **primary cultivation**, the depth of operation is the main objective of control. Soil properties that can supply signals for a site-specific control are texture, organic matter content, slope and eventually the hydromorphic situation. For **secondary cultivation**, the tilth or the break-up of the seedbed is an important criterion since it affects the emergence of crops. This seedbed property can be sensed in a site-specific mode rather effectively by means of the standard deviation of forces that act on a tine of the cultivating implement. In site-specific **sowing**, the seed-density per unit area of the field and the seeding depth are objectives of control. For the seed-density, the control should primarily be oriented at the soil texture. Especially in regions with continental climate, the seeding depth should be such that the seeds are deposited underneath the drying front of the soil. Hence the vertical distance of this drying front from the soil surface must be sensed. Approaches to do this rely either on near-infrared radiation or on electrical resistance.

Site-specific **fertilizing** relies either on the removal of nutrients by preceding crops or on the supply in the soil as well as in the growing plants. The removal of nutrients by preceding crops can be determined via site-specific yield sensing.

This sensing technique is state of the art with many crops. Basing the fertilizer application on the site-specific yield can be recommended provided the respective nutrients are not lost otherwise in an unregulated manner, *e.g.* by leaching. Hence this method is feasible for site-specific fertilizing of phosphorus and with the exception of sandy soils also of potash. With nitrogen, the situation is completely different because of its high disposition to unpredictable leaching, at least in areas with humid climate. In these areas therefore, **site-specific nitrogen fertilization** is a prime candidate for proximal sensing. This technique can be based on either sensing the nitrogen supply in the soil or in the crop. Determining the supply in the soil might rely on the use of ion-selective electrodes or on the reflectance of mid-infrared radiation. For sensing the supply of the crop, reflectance indices that are based on narrow wavelengths from the red-edge range between the visible and near-infrared radiation are reliable indicators. This technique of in-season site-specific nitrogen application based on a control in real-time is used commercially and allows small yield increases or better nitrogen use efficiencies and hence a reduction of nitrate leaching into ground-waters. The latter result is very important since less nitrate in drinking waters is a very urgent environmental objective in humid regions.

Site-specific control in the **application of fungicides** can be based on spatial differences of the crop-biomass or on infected loci. The first approach is state of the art. The second concept – treating only the infected loci – is still in an experimental stage, but should be strived for.

The variety of interdependencies between soil-and crop properties as well as between various farming operations that rely on these calls for **holistic solutions**. These should, however, avoid information overloads. Concepts that point in this direction are sensor fusion, map overlay and management zones. **Sensor fusion** aims at recording two or several soil- and crop properties simultaneously in order to improve the control of farming operations. And urgent approach along this line is simultaneous sensing and processing information about the nitrogen and water supply of crops. Because without the uptake of water, any nitrogen fertilization is useless. **Map overlay** holds for the merging of site-specific information from several maps of the same field. And **management zones** within single fields are created from maps about temporally stable soil properties. All these techniques condense the site-specific information, yet depend on an adequate processing of the data.

Finally, it always should be kept in mind, that precision farming techniques must comply to spatial, temporal and rate resolutions. The latter term refers to the rate at which the various farming operations fulfill their function. In short, the challenge is to do the **right thing** at the **right place** and in the **right time**. Meeting these challenges will provide the means to feed the world population and to protect the environment.

# Index

**A**
Absorbance, 17
Absorbed radiation, 17
Accuracy, 96
Accuracy per signal, 97
Active guidance, 47
Active sensors, 23, 120
Adapting crops to herbicides, 290
Adapting the mass, 288
Adjusting the formulation, 288
Aerial platforms, 20–23
Aggregated patches, 273
Agronomic benefits of clay, 72
Analysis of the spectrum, 18
Anchored crop residues, 163
Area basis, 56
Arid regions, 66, 76
Arithmetic averages, 53
Artificial, additional time
    delay, 216
Artificial illumination, 229
Artificial light, 232
Atmospheric windows, 19
Attitude, 46
Autocorrelation, 67
Automatic guidance system, 38
Availability, 29
Averages, 10, 53
Azimuth angle, 230

**B**
Backscatter, 134
Balers, 321
Bare soil, 239
Bending resistance, 297

Benefits, 152, 260
Biomasses, 106, 136, 233, 245, 296
Biotrophic, 303
Bi-spectral camera, 278, 279
Black boxes, 211
Block-kriging, 11
Blue-green region
    of fluorescence, 305
Bornim Institute, 221
Bound water, 79
Broadcasting, 177
Buffering properties, 212
Bulk drilled crops, 176, 177
Byproduct, 53, 75
By-product of photosynthesis, 118

**C**
Calibration, 91, 211
    procedure, 218
    quality, 317
Cane, 324
Canopy chlorophyll, 114
Capacitance, 76–87
Capacity of the combine, 265
Carbon, 97
Carrier phase signals, 32
Cation-exchange-capacity, 201
Cell-shapes, 248
Cell sizes, 6, 8–10
    for site-specific
        distributing, 246
    for site-specific sensing, 246
Centrifugal spreaders, 248, 250
Change detection, 65, 83
Chemical weed control, 145

Chlorophyll, 106
  concentration, 241
  content, 111, 221
  fluorescence, 118
  index, 114
  per unit of ground area, 223
Circular closing operator, 279
Classification algorithm, 281
Classifiers, 283
Clay content, 150
Climate, 145
Clod break-up, 152
Closed crop canopies, 124, 240
Cloud-cover, 228
Clouds, 20
Cluster analysis, 342
Coefficients of determination, 209, 215
Color of soils, 89
Combination of control techniques, 292
Conditions for harvesting, 263
Conservation of soil moisture, 162
Constant site-specific yield pattern, 340
Contact sensing, 184
Continental climate, 183
Contour farming, 38
Control-algorithm, 251
Controlled traffic farming, 166
Control-line, 252
Control of nitrogen application, 251
Correction of the azimuth-effect, 230
Correlations, 56
Costs, 260, 264
Cotton pickers, 323
Cotton strippers, 323
Crop classification, 137
Crop growth factors, 332
Crop plasticity, 174
Crop productivity potential, 114
Crop properties, 103–137, 220
Crop residues, 145
Crop water stress index, 130
Cross-polarized, 25, 135
Cross-track-error, 39
Cultivating intensity, 154
Cultivation depth, 146, 150
Cultivation induced change, 88
Cultivations, 76

**D**
Data cleaning, 337
Data fusion, 335
Data fusion in mapping, 337
Data merging, 335

Data splitting, 335
Decomposition of the organic matter, 148
Deep response curves, 68
Default slope, 253
Defence mechanism, 303
Defined point estimation, 97
Depth of incorporation, 163
Depth of light penetration, 114
Depth of sensing, 61–65, 81
Depth of the primary cultivation, 145
Detectives of patterns, 296
Dielectric constant, 77, 78
Dielectric properties, 23
Difference image, 280
Difference of the inversions, 112
Differences of differences, 130
Differential positioning, 31
Differentiating between soil and plants, 107
Digital elevation models, 53
Digital image analysis, 278
Direct injection, 296
Direct injection system, 288
Discrete narrow waveband, 219
Discrete-spot sensing, 298, 301
Discrete spot spraying, 296
Discrete waveband approach, 18, 90,
  91, 108, 212
Dissimilarity, 8
Dissolvable salts, 66
Distance-and time lag, 249
DLR system, 221
Dried soil samples, 214
Driving pattern, 41
Dryland regions, 76
Dual sensing strategy, 259
Dual soil property sensing, 98
Dynamic properties, 65
Dynamics of weed populations, 284

**E**
Economically optimal nitrogen rate, 257
Economics, 260
Efficacy of weed control, 274
Efficiency of nitrogen use, 255, 261
Electrical capacitance, 57, 77
Electrical frequency, 64, 86
Electrical permittivity, 77
Electric current, 56
Electric/electrical conductivity, 57, 65–76
Electrochemical cell, 203
Electrochemical series of potential, 197
Electromagnetic induction, 57, 59–61
Electromagnetic radiation, 15–32, 56

Elimination of soil errors, 117
Emergence, 177, 183
Emission spectrum, 302, 304
Emitted radiation, 17
Emitted thermal radiation, 126, 130–133
Energy per photon, 16
Erosion, 52
Erosion control, 162
Excitation spectra, 302, 304
Exports of nutrients, 313, 314
Extractant, 209

**F**
Factory-fitting, 41
Farm-specific georeferencing method, 53
Feature set, 281
Feed-back of the effects on yield, 314
Feelers, 36
Fertilizing nitrogen, 220
Field capacity, 66
Field machines, 20–23
Field of view, 239, 298, 299
Field-trafficability-sensor, 166
Fixed line guidance, 37
Flat surface sensing, 215
Flattened soil surface, 95
Fluorescence, 301
　absorption, 241
　emission, 304–305
　indices related to infection, 304
　sensors, 118–125, 277
Fluorescence/reflectance, 124–125
Fluorescence peaks, 241
Fluorophore, 301
Forage choppers, 320
Fractionated seedbed, 161
Fraunhofer lines, 121
Free water, 79
Frequencies, 78
Fresh, wet soils, 214
Full-area preventive concept, 296
Full reflectance spectra, 212
Full spectrum approach, 18, 90, 108
Fungal infections, 237
Fungicide spraying, 245
Fungi-plant-interaction, 303
Fusing reflectance and fluorescence
　　sensors, 309

**G**
Galvanic contact, 58
Geometrical position, 29

Geometric means, 229
Geophilus electricus, 63
Georeferenced maps, 151
Geo-referenced yield detection, 313
Georeferencing, 28–32
Geosynchroneous path, 21
Germination, 183
Global navigation satellite systems
　　(GNSS), 28, 36
GPS, 28
Gradients with depth, 63
Grain yield sensors, 339
Grass mowers, 320
Green chlorophyll index, 113
Green NDVI, 234
Green seeking, 108, 109
Grid-/block based map, 338
Gross primary productivity, 114
Group the operations, 95
Growing degree days, 118
Growth stage, 245
Guidance, 35

**H**
Half the range, 9
Handheld meters, 252
Hardpans, 149
Heading, 46
Heat dissipation, 120
Herbicide leaching, 338
Herbicide resistant crops, 290
Heterogeneity, 3–13
Heterogeneity of crop yields, 314
Hofmeister series, 206
Homogeneous management zones, 340
Humid areas, 66
Humid regions, 74
Hydromorphic properties, 150
Hydromorphic soils, 147
Hyperspectral precision, 128

**I**
Image database, 281
Image sensing, 301
Imaginary part of the permittivity, 77
Impact forces, 154
Impact sensing, 156
Improvements in the seed distribution, 177
Incidence angle, 83, 135
Indicator of a vegetation cover, 107
Indices, 25
Indices of reflectance, 299

Indirect measuring, 301
Inertial georeferencing, 53
Inertial sensors, 46
Infrared radiation, 126–133
Infrared to red indices, 108
Inorganic carbon, 99
In-season crop properties, 220
In-situ sensor system, 301
Integrals, 62
Interaction, 151, 220, 341
Interaction between soil and water, 75
Inter-crop information transfer, 332
Interferences, 210
Interfering factors, 227
Intermediates, 56
Intermittent sensing operation, 198
Interpolation, 11, 199
Inter-row sowing, 186
Inverse square law, 18
Inversion, 64
Inversion of electrical conductivity, 63
Inverted reflectance, 112
Ion absorbing capacity, 200
Ion-selective electrodes, 197, 202
Irradiance, 122
Irrigation, 76

**K**
Kautsky effect, 123, 302
Key narrow bands, 90
Kiel system, 221
Kinetics of the Kautsky effect, 304
Kriging, 11, 338

**L**
Laser light, 53
    flashes, 125
Late dressing, 255
Lateral and longitudinal tilts, 318
Layered soils, 62
Leaching of nitrogen, 265
Leaf-area-indices, 106, 111, 221, 243, 245, 276, 296
Lightbar guidance system, 38
Like-polarized, 25, 135
Lime requirement, 200, 212
Limiting rates, 252
Limits for drinking water, 262
Liquid water, 128
Lodging, 263
Longitudinal cell side lengths, 248
Long stubbles, 187

Loose residue sizes, 188
Losses by leaching, 195
Lower baseline, 130

**M**
Malting barley, 255
Management zones, 342
Managing the variations, 10–13
Maps, 26–28
    of the means, 74, 88, 345
    overlay, 337
Maritime climate, 182
Markers, 36
Mask angle, 30
Mass-flow, 316
Mean weight diameters, 157
Measuring accuracy, 317
Microwaves, 23, 29, 133–137
Mid-infrared range, 89
Minimum growth stage, 240, 297
Mixed effects of nitrogen and fungi, 308
Mixtures of soil and water, 198
Moisture seeking control algorithm, 182
Moisture sensing, 24
Motion memory, 46
Multiple field sprayer, 288
Multiple soil property sensing, 89, 90
Multivariate calibration, 305

**N**
Narrow bands, 212, 225
Narrowing the view, 240
Natural illumination, 229
Natural light, 225
Naturally moist soil, 198, 201
Natural soil properties, 51–99
NDVI. *See* Normalized difference vegetation index (NDVI)
Near-infrared domain, 89
Near-infrared range, 104
Near-infrared region, 127
Near-infrared to green, 235
Near-infrared to red, 239
Necrotrophic, 303
Nested pattern, 4
Network base receiver stations, 32
Network of RTK-GPS, 32
New reflectance indices, 234
Nitrate ions, 218
Nitrogen
    fertilizing, 223, 245
    and fungi, 307

# Index

indices, 224
  sensing by reflectance, 223
  and water, 257
Nitrogen-rich strips, 254
Non contact sensing, 184
Non-steady state mode, 123–124
Non-vegetated soil, 241
Normalized difference vegetation
    index (NDVI), 109, 225
Normalized reflectance indices, 256
No-tillage, 165–168, 185
Nugget variance, 7
Number of fluorescence signals, 242
Numerical features, 288

## O

Oblique view, 240
Offset distance, 249
One point calibration concept, 253
On-the-go control, 26–28
On-the-go dual sensing, 259
Optical fingerprint, 18
Organic matter content, 88, 148, 150
Orienting herbicides at weeds, 290
Overlapping, 42
Overview, 26

## P

Passive guidance, 47
Passive sensing, 121
Patches, 296
Patch spraying systems, 274
Penetration depth, 134
Perennial maps, 337, 345
Perennial weeds, 284
Permittivity, 24, 76–87
Phenols, 305
Phosphorus requirement, 214
Photosynthesis, 104
Photosynthetically active radiation, 115
Physiology of infected plants, 303
Pinpointing approach, 298
Pitch, 45
Plant available phosphorus, 214
Plant density, 242, 244
Plant density per unit area, 172
Plant growth function, 256
Plough, 144
Pneumatic spreaders, 248, 250
Point-kriging, 12
Polarization, 135
Polarizing the radiation, 24

Polar orbit, 21
Positional lag error, 250
Positioning systems, 28–32
Post-calibration, 201
Potassium ions, 218
Potato, 322, 323
Precision agriculture cycles, 332
Precision of positioning, 31
Precision sown crops, 176
Predictive approach, 343
Pre-emergence site-specific
    applications, 284
Preferential water flow, 64
Preparation, 144
Pressure, 179
Primary cultivation, 144
Primary ion, 207
Primary magnetic field, 60
Primary sensing, 335
Prior pass guidance, 36
Product output, 314
Prolongation of wavelengths, 118
Protein content, 263
Proximal sensing, 117
Proximal site-specific nitrogen
    sensing, 225
Pulsed lasers, 305
Pulsing nozzles, 132
Punctual basis, 11

## Q

Quality-dressing, 255

## R

Radar, 23, 260
Radar interferometry, 52
Radar satellites, 80
Radar-waves, 133
Radiance, 123
Radiation of photons, 16
Radii of curvature, 37
Rake angle, 156, 161
Random pattern, 4
Ranges, 7, 10
Rapid detection, 125
Re-absorption of photons, 120
Reaction times, 288
Reactive approach, 343
Real-time, 209, 212
Real-time kinematic differential GPS, 32
Real-time kinematic georeferencing, 53
Real-time sensing, 27

Recording of yields, 313
Recording the heterogeneity, 246
Red edge, 106
    chlorophyll index, 113
    inflection point indices, 308
    inflection points, 109, 225, 226, 229, 235, 236, 259
    position, 276
    range, 237
    ratio index, 237
    ratios, 259
Reflectance, 17, 87–99, 104, 114–118, 214, 297, 298
    indices, 299
    of soils, 209
    spectra, 299
Reflectance ratio minus one, 112
Reflected radiation, 17, 80, 126
Regression analyses, 211
REIP, 226
Relatedness, 8
Relative overlapping, 42
Relative permittivity, 24, 77
Remote productivity sensing, 117
Remote sensing, 94, 224
Removal of nutrients, 194
Removal rates, 196
Removing noise, 210
Repeatability, 32
Residue management, 162
Residue maps, 340
Resistance of the canopy against bending, 245
Resistance to penetration, 149–150
Resolution, 3
Response curves, 61
Retrofit, 40
Robots, 167, 291
Rod-weeder, 164
Roll, 46
Roller-crimpers, 187
Root development, 220
Row cleaners, 185
RTK-GPS georeferencing, 32, 54
Run-off, 52

**S**
Saline soils, 66
Sampling, 9
Satellites, 20–23, 80–84
Saturation effect, 111, 233, 240
SAVI, 225
Saving of nitrogen, 264
Savings for nitrogen fertilizers, 260
Savings in fungicides, 298
Scanning laser techniques, 306
Scanning technique, 242
Scouting purposes, 168
Scouting robots, 306
Seasonal maps, 346
Secondary and tertiary sensing, 335
Secondary cultivation, 144, 152–162
Secondary magnetic field, 61
Sectional rate control, 248
Section control, 49
Seed clusters, 172
Seed costs, 175
Seed counting, 173
Seed distribution over the area, 172, 176
Seeding depth, 178
Seeding direction, 188
Seeding into cover crops, 187
Seeding underneath undercutters, 190
Seed-rate, 159
Seed-rate/seed-density, 172
Segregation by sifting, 161
Self-loading trailers, 321
Semivariance, 6–7
Semivariogram, 6–7
Sensing-depth, 157
Sensing for a water line, 56
Sensing for fungicide spraying, 295
Sensing from farm machines, 95
Sensing from satellites, 128
Sensing in fields, 94
Sensing in laboratories, 94
Sensing nitrate in slurries, 202
Sensing nitrogen by fluorescence, 240
Sensing of carbon, 97
Sensing of several properties, 212
Sensing of the drying front, 183
Sensing the water supply, 126–133
Sensing underneath a cultivator sweep, 95
Sensing water, 88, 130–133
Sensitivity, 19
Sensor fusion, 333
Separating the effects, 70
Sequence freezing-thawing, 147
Series of small sensors, 306
Setting a moisture line, 164
Several frequencies, 86
Shape features, 278
Short-wave infrared region, 104, 127
Signal noise, 210
Signal resolution, 5
Sill variance, 7

Index 355

Simple ratios, 108, 225
Single property sensing, 212
Single reflection, 135
Single soil property, 89
Site-specific correction, 86
Site-specific dry matter, 196
Site-specific fertilizing, 193
Site-specific fungicide applications, 296
Site-specific irrigation, 132
Site-specific maintenance
    application, 194
Site-specific soil to crop relation, 220
Site-specific weed control, 273–292
Site-specific yield potential, 314
Sizes of loose residues, 189
Skeleton, 281
Slope induced erosion, 150
Slope of the control-line, 252
Slopes, 44–49, 148
Slurries, 198
Sodium, 70
Soil
    break-up, 159
    buffer pH, 200
    conductivity, 255
    layers, 61–65, 75
    line, 108
    moisture, 69, 182
    nutrients, 209
    organic matter, 339
    penetrating radar, 57
    properties, 65–76
    properties by volume, 55–57
    salinity, 69
    slurry, 203
    solution, 70
    surface, 77
    texture, 133, 147, 173
    tilth, 153–160
    volumes, 77
    water pH, 200
Soil adjusted vegetation index, 225
Soil-brightness, 228
Solutions, 198
Sowing, 171
    depth, 172
    and planting, 76
Spacers, 179
Spatial pattern, 305
Spatial resolutions, 6, 22, 55, 97, 199
Special nitrogen indices, 224
Species-specific spraying, 288
Spectra, 299
Spectral reflectance, 276
Spectrometers, 276
Spot spraying, 108, 239
Standard deviation of the forces, 157
Standard indices, 224
Standardized semivariograms, 9
Standard NDVI, 234
Standard reflectance indices, 225, 300
Standard track width, 167
Static properties, 65
Steady state mode, 120–123
Strain gage, 154
Strip spraying, 109, 239
Stubble-/fallow cultivation, 144
Substitute of texture, 174
Sufficiency index, 256
Sunsynchroneous path, 22
Surface-fluorescence, 243
Surface-reflectance, 243
Surface roughness, 24, 85
Surface scattering, 135
Surface sensing, 89–91
Survival strategies, 284

**T**

Tactical inspections, 27
Tailing out, 48
Targeted decomposing, 259
Temperature difference, 130
Temporal frequency, 22
Temporally stable soil properties, 342
Temporal precision, 22, 225
Temporal resolution, 4, 22
Temporal variations, 55, 69, 104
Test bench studies, 318
Texture based slope, 255
Texture of soils, 255
Time-bridges, 13
Time domain reflectometry, 57, 80
Time-lag distance, 216, 249, 250
Top leaf, 299
Topography, 52–55, 133
Topp's equation, 80
Top-surfaces, 87, 93
Tramlines, 35, 43
Transient factor, 335
Transient property, 205
Transient soil-and crop properties, 346
Transmittance, 17
Transmittance of radar, 134
Transmitted radiation, 17
Transpiration, 130
Treatment map, 342
Two point calibration concept, 253

## U

Ultrasonics, 297
  sound waves, 180
  transducer, 245
Under-cutter, 165
Unmanned farm machinery, 167
Upper baseline, 130
Upper limit of cell side length, 246
Upper limit of the cell-size, 9
Use of absorbed light, 119

## V

Validation, 211
Variable fluorescence, 305
Variable orifice, 132
Variation, 3
Variation in the protein content, 255
Varying seeding depth, 181
Varying the coil orientation, 64
Vegetation line, 108
Vertical cross-section, 56, 87–88
Vertical discs, 185
Vertical soil profile, 93
View directing, 117, 230
Virtual tramlines, 36
Visible and infrared reflectance, 104–108
Visible region, 89
Voltage output, 198
Volume basis, 56
Volume flow measurement principle, 315
Volume-reflectance, 243
Volume scattering, 135

## W

Water content, 146
Water in the seedbed zone, 56
Water sensing, 76–87
Water supply, 335
Water tension, 79
Water vapor, 128
Wavelength, 16, 23, 134
  of 7148 nm, 219
  ratios, 128

Weather, 146
  forecasts, 146
  satellites, 159
  sequences, 147
Weed control, 162
Weed density reduction, 285
Weed distribution maps, 285
Weed growth, 177
Weed identification, 276
  shape, 276
Weed infestation, 338
Weed mapping, 274–284
Weed patches, 287
Weed population dynamics, 285
Weed populations models, 286
Weed seedling distributions,
  285, 287
Weed seedling populations, 275
Weighted averages, 53
Wenner array, 58

## X

Xeromorphic characteristics, 258

## Y

Yaw, 46
Yield maps, 340
Yield measurement for combinable
  crop, 315
Yield measurement for forage
  crops, 319
Yield measurement systems, 317
Yield monitoring, 313
Yield potentials, 114–118, 223, 313
Yield-predicting-maps, 75
Yields, 65–76, 152, 175, 177,
  260, 264
Yield sensing for root crops, 322

## Z

Zenith angle, 230
Zenith-angle of the sun, 228

Printed by Printforce, the Netherlands